THE
PLANT MILDEWS

THE PLANT MILDEWS
A Colour Handbook

S.G. Borkar
Retd. Dean and Principal
College of Agriculture, Ambajogai-431517, Maharashtra
Vasantrao Naik Marathwada Agricultural University
Parbhani, Maharashtra

Ajayasree T.S.
Department of Plant Pathology
Kerala Agricultural University
P.O, N.H. 47, Vellanikkara, Thrissur -680656
Kerala

CRC Press is an imprint of the
Taylor & Francis Group, an *informa* business

NEW INDIA PUBLISHING AGENCY
New Delhi – 110 034

First published 2021
by CRC Press
2 Park Square, Milton Park, Abingdon, Oxon, OX14 4RN
and by CRC Press
6000 Broken Sound Parkway NW, Suite 300, Boca Raton, FL 33487-2742

© 2021, New India Publishing Agency

CRC Press is an imprint of Informa UK Limited

The rights of S.G. Borkar and Ajayasree T.S. to be identified as authors of this work has been asserted by them in accordance in accordance with sections 77 and 78 of the Copyright, Designs and Patents Act 1988.

Reasonable efforts have been made to publish reliable data and information, but the author and publisher cannot assume responsibility for the validity of all materials or the consequences of their use. The authors and publishers have attempted to trace the copyright holders of all material reproduced in this publication and apologize to copyright holders if permission to publish in this form has not been obtained. If any copyright material has not been acknowledged please write and let us know so we may rectify in any future reprint.

All rights reserved. No part of this book may be reprinted or reproduced or utilised in any form or by any electronic, mechanical, or other means, now known or hereafter invented, including photocopying and recording, or in any information storage or retrieval system, without permission in writing from the publishers.

For permission to photocopy or use material electronically from this work, access www.copyright.com or contact the Copyright Clearance Center, Inc. (CCC), 222 Rosewood Drive, Danvers, MA 01923, 978-750-8400. For works that are not available on CCC please contact mpkbookspermissions@tandf.co.uk

Trademark notice: Product or corporate names may be trademarks or registered trademarks, and are used only for identification and explanation without intent to infringe.

Print edition not for sale in South Asia (India, Sri Lanka, Nepal, Bangladesh, Pakistan or Bhutan).

British Library Cataloguing-in-Publication Data
A catalogue record for this book is available from the British Library

Library of Congress Cataloging-in-Publication Data
A catalog record has been requested

ISBN: 978-1-032-00746-5 (hbk)

Acknowledgement

The authors duly acknowledge the help rendered by his doctorate students at department of Plant Pathology, Mahatma Phule Agriculture University, Rahuri, Dist-Ahmadnagar during the preparation of this manuscript on plant Mildews. We thanks all our fellow researchers all over the world whose references and scientific texts are used in the preparation of this manuscript. We also acknowledge the contribution of various anonymous scholars who shared their mildews photos on social media and different website for the public domain use, some of which are included at appropriate place as photo of the disease. We thanks all these contributors including the websites.

The help rendered by Miss. Ayushi Bhagwat and Mrs. Kavita Dhawade who was associated with me during the preparation of this manuscript in its final form is acknowledged with appreciation. Those of my colleagues who directly or indirectly help us during the work on this book, their efforts is being acknowledge herewith.

Authors

Preface

Mildews on plants include downy mildews and powdery mildews and are one of the important diseases of crop plants. Mildews occurs on cereal crops, pulse crops, oilseed crops, fiber crops, cash crops , vegetable crops, fruit crops, spices crop, ornamental and flowering plants, medicinal and aromatic plants, forage crops and forest trees and causes enormous losses in the absence of suitable control measures. These losses includes the loss of crop foliage (in leafy vegetables), grains (in cereals), deformation of fruits (in fruit crops), death of plants (in cereals and vegetables) and quality of yield (in fruit crops). The losses reported due to mildews are as high as 60 percent in grapevine in term of fruit quality and crop monetary returns. Total crop loss can occur in case of susceptible variety of the crop to mildews. The mildew disease occurs in all the continents of the world on all kind of crop plants. Application of appropriate fungicides are important to control these diseases as many mildew pathogens have developed the fungicidal resistant.

The book "The Plant Mildews: A Colour Handbook" includes 24 chapters. 11 chapters are on downy mildews of various crop plants with 55 downy mildew diseases while 13 chapters are on powdery mildew diseases of various crops with 92 powdery mildew diseases. The mildew diseases around the world included in the book are 147 with detailed information on each mildew disease including pathogen responsible, its host range, geographical distribution, disease symptoms, epidemic and losses, disease cycle, weather parameters and disease forecasting, and disease managements.

This is the only book on mildews of crop plants which include both downy mildew and powdery mildew diseases and will be useful for researcher, students, extension workers and farming communities.

Authors

Contents

Acknowledgement .... *v*

Preface .... *vii*

SECTION I: DOWNY MILDEW

1. An Introduction to Downy Mildew **1**

  1.1. History of Downy mildew 2

  1.2. Epidemics of Downy mildew 2

  1.3. Losses due to Downy mildew 2

  1.4. Symptomatology 3

  1.5. The fungal pathogen 3

  1.6. Important characteristics of the Downy mildew generas 7

  1.7. Disease cycle and epidemiology 8

  1.8. Disease management 10

2. Downy Mildew of Cereals **11**

  2.1. Downy mildew of sorghum 11

  2.2. Downy mildew or green ear disease of pearlmillet 16

  2.3. Downy mildew of maize (Corn) 21

  2.4. Brown stripe Downy mildew of maize 24

  2.5. Downy mildew of wheat and minor millets 28

3. Downy Mildew of Pulses **33**

  3.1. Downy mildew of peas 33

  3.2. Downy mildew of lima bean 36

4. Downy Mildew of Oilseed Crops **41**

  4.1. Downy mildew of sunflower 41

  4.2. Downy mildew of soybean 46

  4.3. Downy mildew of mustard 49

x The Plant Mildews

5. Downy Mildew of Fiber Crop **53**
    5.1. Downy mildew of hemp 53

6. Downy Mildew of Cash Crops **55**
    6.1. Downy mildew of sugarcane 55
    6.2. Downy mildew or blue mold of tobacco 59

7. Downy Mildew of Fruit Crop **63**
    7.1. Downy mildew of grape 63
    7.2. Downy mildew of melons (watermelon, muskmel and sweetmelon) 71

8. Downy Mildew of Vegetables **77**
    8.1. Downy mildew of cruciferous vegetables 77
    8.2. Downy midew of cucurbits 85
    8.3. Downy mildew of Suger beet 95
    8.4. Downy mildew of Carrot 99
    8.5. Downy mildew of spinach 99
    8.6. Downy mildew of lettuce 102
    8.7. Downy mildew of fenugreek 106
    8.8. Downy mildew of onion and garlic 107
    8.9. Downy mildew of capsicum 112

9. Downy Mildew of Spices **115**
    9.1. Downy mildew of opium poppy 115

10. Downy Mildew of Medicinal and Aromatic Plants **119**
    10.1. Downy mildew of *Plantago psyllium* 119
    10.2. Downy mildew of *Atropa belladonna* 121
    10.3. Downy mildew of *Hyoscyamus* sp. 122
    10.4. Downy mildew of basil 123

11. Downy Mildew of Ornamental and Flowering Plants **129**
    11.1. Downy mildews of anemone (Windflower) 129
    11.2. Downy mildew of antirrhinum (Snap-dragon) 131
    11.3. Downy mildew of rose 133
    11.4. Downy mildew of alyssum saxatile 134
    11.5. Downy mildew of chrysanthemum 135
    11.6. Downy mildew of cineraria 135
    11.7. Downy mildew of clarkia (*Clarkia elegans*) 136

11.8. Downy mildew of viola (violet, pansy) 137

11.9. Downy mildew of helleborus (Christmas rose) 137

11.10. Downy mildew of matthiolae 138

11.11. Downy mildew of Mecanopsis 139

11.12. Downy mildew of papaver (poppy) 139

11.13. Downy mildew of veronica (Speed well) 140

11.14. Downy mildew of mimulus 141

11.15. Downy mildew of arabis (Rock cress) 141

11.16. Downy mildew of argemone (Prickly-poppy) 141

11.17. Downy mildew of centaurea (Cornflower) 142

11.18. Downy mildew of gramineae(Grasses) 142

11.19. Downy mildew of houstonia 143

11.20. Downy mildew of lagneria 143

11.21. Downy mildew of myosotis (Forget-me-not) 143

11.22. Downy mildew of paeonia (Peony) 143

11.23. Downy mildew of parthenocissus (Boston ivy & Virginia creeper) 144

11.24. Downy mildew of rudbeckia (Goldenglow) 144

11.25. Downy mildew of stelbaria holostea (Easter bells) 145

11.26. Downy mildew of impatiens 145

Literature Cited for Downy Mildew 151

SECTION II: POWDERY MILDEW

12. An Introduction to Powdery Mildews **171**

12.1. History of powdery mildews 171

12.2. Losses due to powdery mildew 172

12.3. Symptoms 172

12.4. The pathogen 173

12.5. Disease cycle 182

12.6. Disease control 183

13. Powdery Mildew of Cereals **185**

13.1. Powdery mildew of wheat (*Triticum aestivum* L.) 186

13.2. Powdery mildew of barley (*Hordeum vulgare* L.) 191

13.3. Powdery mildew of rye (*Secale cereale* L.) 194

13.4. Powdery mildew of oat (*Avena sativa* L.) 196

xii The Plant Mildews

14. Powdery Mildew of Pulse Crops 197
14.1. Powdery mildew of pea 197
14.2. Powdery mildew of pigeon pea 199
14.3. Powdery mildew of green gram (Mungbean) 202
14.4. Powdery mildew of black gram (Urdbean) 203
14.5. Powdery mildew of lathyrus *(Lathyrus sativum* L.) 205
14.6. Powdery mildew of lentil 205
14.7. Powdery mildew of cowpea *(Vigna unguiculata)*, Mothbean (*Vigna aconitififolin*) and Horsegram *(Macrotyloma uniflorum)* 206
14.8. Powdery mildew of minor legumes 207

15. Powdery Mildew of Oil Seed Crops 211
15.1. Powdery mildew of rapeseed and mustard 211
15.2. Powdery mildew of sesame 213
15.3. Powdery mildew of sunflower 214
15.4. Powdery Mildew of castor 216
15.5. Powdery mildew of olive 217
15.6. Powderymildew of safflower 218
15.7. Powdery mildew of ;inseed (Flax) 219

16. Powdery Mildews of Fibre Crops 223
16.1. Powdery mildew of cotton 223
16.2. Powdery mildew of jute 225
16.3. Powdery mildew of sunhemp 227

17. Powdery Mildew of Cash Crops 231
17.1. Powdery mildew of tobacco 231
17.2. Powdery mildew of rubber 234
17.3. Powdery mildew of betelvine 235

18. Powdery Mildew of Vegetables 237
18.1. Powdery mildew of cucurbits 240
18.2. Powdery mildew of lettuce 247
18.3. Powdery mildew of carrot 248
18.4. Powdery mildew of beans 251
18.5. Powdery mildew of okra (Leadyfinger) 252
18.6. Powdery mildew of cruciferous vegetables 253
18.7. Powdery mildew of tomato 255

18.8. Powdery mildew of onion (*Allium cepa*) 257

18.9. Powdery mildew of potato (*Solanum tuberosum* L.) 258

18.10. Powdery mildew of capsicum and Green house pepper 260

18.11. Powdery mildew of cluster bean 263

18.12. Powdery mildew of eggplant 265

18.13. Powdery mildew of turnip and rutabaga (*Brassica* sp.) 266

19. Powdery Mildew of Fruit Crops 269

19.1. Powdery mildew of grapevine 269

19.2. Powdery mildew of citrus 281

19.3. Powdery mildew of papaya 282

19.4. Powdery mildew of mango 285

19.5. Powdery mildew of jujube (Ber) 291

19.6. Powdery mildew of apple 294

19.7. Powdery mildew of strawberry 303

19.8. Powdery mildew of gooseberry and currant 305

19.9. Powdery mildew of peach and apricot 308

19.10. Powdery mildew of cherry 310

19.11. Powdery mildew of plum 312

19.12. Powdery mildew on blueberries 314

19.13. Powdery mildew of stone fruit (Almond and others) 315

19.14. Powdery mildew of mulberry (Morus species) 316

19.15. Powdery mildew of passion fruit 318

19.16. Powdery mildew of pomegranate 319

19.17. Powdery mildew on watermelon 320

20. Powdery Mildew of Spices Crop 323

20.1. Powdery mildew of black pepper 323

20.2. Powdery mildew of coriander 325

20.3. Powdery mildew of fenugreek 326

20.4. Powdery mildew of fennel 327

20.5. Powdery mildew of cumin 329

20.6. Powdery mildew of dill 330

20.7. Powdery mildew of chilli 331

21. Powdery Mildews of Flowering and Ornamental Plants 333

21.1. Powdery mildew of rose 334

21.2. Powdery mildew of chrysanthemum 338

xiv   The Plant Mildews

21.3.   Powdery mildew of gerbera 342
21.4.   Powdery mildew of marigold 345
21.5.   Powdery mildew of dahlia 347
21.6.   Powdery mildew of aster 350
21.7.   Powdery mildew of anthurium 353
21.8.   Powdery mildew of begonia 354
21.9.   Powdery mildew of delphinium 356
21.10.   Powdery mildew of ornamental plants 358

22. Powdery Mildew of Medicinal and Aromatic Plants 371
22.1.   Powdery mildew of plantago 371
22.2.   Powdery mildew of rauvolfia 372
22.3.   Powdery mildew of opium poppy 373
22.4.   Powdery mildew of ginseng 374

23. Powdery Mildews of Forest Trees 377
23.1.   Powdery mildew of eucalyptus 377
23.2.   Powdery mildew of teak 379
23.3.   Powdery mildew of oak 381
23.4.   Powdery mildew of arjuna 382
23.5.   Powdery mildew of khair (*Acacia catechu*) 383
23.6.   Powdery mildew of sisso 383
23.7.   Powdery mildew of tamarind (*Tamarindus indica*) 384
23.8.   Powdery mildew of sandal wood 385

24. Powdery Mildew of Forage Crop 387
24.1.   Powdery mildew of clover 387

Literature Cited for Powdery Mildew 391

Section I: Downy Mildew

1

An Introduction to Downy Mildew

Downy mildew is refered to a condition of plant disease where the fungal growth and infection appears on the dorsal side of the leaf surface while the corresponding ventral side of the leaf shows faint yellowish color. It is also refered to the fungus which causes this disease.The word mildew in Latin language means mil + dew i.e honey like dews.The fungal growth on plant surface resembles to the milky dews in the beginging which turn to golden brownish honey like dews in later stage. Subsequently the affected areas turn grayish brown and dry. Besides the infection of leaves, the fungus also infects the fruits. The downy mildew infection on fruits causes the deformities on the affected fruits. The infection of the downy mildew pathogen on some of the cereals like pearlmillet causes the symptoms like green ear disease .

Downy mildew pathogens can cause major damage in the nursery, in greenhouse, in field crops and in landscape in the form of leaf spots, blights, and distortions with mildew growth. Downy mildews are primarily foliage blights with mildew growth of the fungus on affected portion. The fungus attack and spread rapidly in young, tender green leaf, twig, and fruit tissues. The fungal development are severe when a film of water is present on the plant tissues and the relative humidity in the air is high during cool or warm, but not hot, periods. Downy mildews can cause severe losses in short periods of time. Downy mildews often cause rapid and severe losses of young crop plants still in the seedbed or in the field. They often destroy from 40 to 90% of the young plants or young shoots in the field, causing heavy or total losses of crop yields. The severity of loss depends on the prolonged presence of wet, cool weather during which the downy mildews fungus sporulate profusely, cause numerous new infections, and spread into and rapidly kill young succulent tissues.

The downy mildew diseases are characterized by the white, gray or brown, 'downy (hairy) growth' of sporangiophores and sporangia on the lower surface of the infected leaves while the corresponding area on the upper side of leaf show faint yellowish coloration which subsequently become brown, grayish and die. The infected leaf/leaves become brown, shrivel and die.The severely affected plant parts distorts and die resulting in heavy losses.

The downy mildew fungi are obligate parasites (biotrophs) and have the absorptive mode of nutrition. The downy mildew diseases are major deterrent to the cultivation of some of the very important crops like maize, grapes, sorghum, soybean, cucurbits, hops, Lucerne, peas, tobacco, rose and many others.

1.1. History of Downy mildew

The occuranec of mildews on cereals and legumes are mentionrd in numerous passages in the old testaments (about 750 BC) of the Bible. Homer (1000 BC) mentioned mildews and its control with blasting of sulphur. It was not until 1780 that the British records mentioned a plant disease stating that the crops were attacked by mildews and that the harvest therefore was less (Stratton, 1969). The mildew, were confused with other diseases and no reliable record of them could be found even though the name mildew was occasionally used in medieval period (Megenberg,ca.1388). Since then the downy mildews are known to occur on one or other crops in different countries.

1.2. Epidemics of Downy mildew

The most of the epidemics of downy mildews are reported on grapevine. The epidemics of grapevine downy mildew occurred in France in as early as 1879 and subsequent epidemics followed in 1900, 1910, 1915 and 1927. In Australia the downy mildew epidemic on grapevine occurred in 1917 in north east Victoria.

In other crops epidemics of downy mildew are not reported though the incidences of heavy losses are reported.

1.3. Losses due to Downy mildew

The downy mildew of grapevine caused heavy losses in U S A and Europe in 1870-1880. A loss of 70 percent of potential grape yield was reported during epidemic of 1915 in France alone. It also caused serious losses in some years in northern Africa, in South Africa, in parts of Asia, in Australia and in South America as well. In Georgia, the losses of 9.5 percent are reported while in India the losees are reported to the tune of 100 percent in heavily infected grapevine gardens in Maharashtra region (Anonymous, 2006).

The losses ranging from 2.3 to 30.1 percent are reported in pearlmillet due to downy mildew (Thakur, 1986). Severe losses are reported due to maize downy mildew in phillipines.

1.4. Symptomatology

The initial symptoms appear as yellow spots on the upper surface of the leaves that some times look oily and are called "oily spots". The spots grow in size and cover most of the leaf surface. White to grayish 'downy growth', of the fungus comprising of sporangiophores and sporangia, appear on the corresponding lower surface of the yellow spots (Fig.1.1.). The downy growth can best be seen with a hand lens. The lesions on the leaves become brown and necrotic with age. Severely infected leaves fall prematurely.

Fig.1.1: Symptoms of downy mildew on upper side and lower side of leaf.

In some cases, like crazy top of maize and the green ear disease of pearl millet, the effects of the infection are drastic. The flower parts become leafy and defunct. Tillering, stunting of the plants, phyllody (in ears) are the other drastic symptoms

1.5. The fungal pathogen

The fungal pathogen grows intercelluarly and become systemic. They derive nutrition from the adjoining living cells through haustoria. The hyphae fill the intercellular spaces and assume irregular shape. In humid and cool weather, which occurs during dewy nights, the sporangiophores emerge, singly or in groups, through the stomata on the lower surface (occasionally on both surfaces) of the infected leaves. The sporangiophores are of determinate (definite) growth, (except in Sclerophthora), and branch typically to bear sporangia (Fig.1.2), which is a diagnostic character for the genus of the downy mildew fungus.

Skalicky (1966) has given a key for the identification of the genera by characteristics of oogonia and oospores, but the genera of downy mildew fungi continue to be identified by their sporangiophores and sporangia.

4 The Plant Mildews

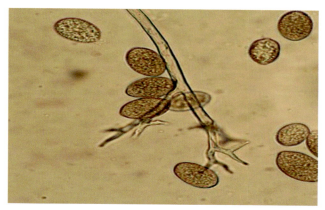

Fig.1.2: Sporangiophore and sporangia of downy mildew fungi.

The Downy mildew fungus belongs to

Kingdom: Chromista (Stramenopiles)

The general characteristic of the organisms in this kingdom are unicellular or multicellular, filamentous or colonial, primarily phototrophic, some with tubular flagellar appendages or with chloroplasts inside the rough endoplasmic reticulum or both. The kingdom includes brown algae, diatoms, oomycetes, and some other similar organism.

Phylum: Oomycota

The organisms of this phylum have biflagellate zoospores, with longer tinsel flagellum directed forward and a shorter whiplash flagellum directed backward. It has diploid thallus, with meiosis occurring in the developing gametangia. Gametangial contact produces thick-walled sexual oospore. Cell walls of the fungus are composed of glucans with a small amounts of hydroxyproline and cellulose.

Class: Oomycetes

This class of fungus includes water molds, white rusts, and downy mildews. The fungus has nonseptate elongated mycelium. Produce zoospores in zoosporangia. Zoospores have two flagella. Sexual resting spores (oospores) produced by the union of morphologically different gametangia called antheridia (male) and oogonia (female)

Order: Peronosporales

The fungal mycelium is well-developed, nonseptate, branching, inter or intracellular, often with haustoria. Zoosporangia are oval or lemon shaped, borne on ordinary mycelium or on sporangiophores. Sporangia in most species germinate by producing zoospores, but in some they germinate directly and produce a germ tube.

Sexual reproduction is by characteristic oogonia and antheridia that fuse and produce an oospore. Oospores germinate by giving rise to a sporangium containing zoospores or to a germ tube, which soon produces a sporagium, depending on the species.

The downy mildew fungi (Peronosporaceae) include number of genera differentiated chiefly by the branching of their sporangiophores.The sporangia in all the downy mildew genera are born on the sporangiophores. The sporangiophores of the peronosporaceae have a determinate growth i.e the mycelium produces a sporangiophore which reaches maturity, stop growning and then produces a crop of sporangia on sterigmata at the apices of its branches. All the sporangia are therefore of approximately the same age. They are round, oval, or lemon shaped and without exception,deciduous and are wind disseminated.

The oospore of the peronosporaceae as a general rule germinates by germ tube except in Peronospora tabacina where it produces a sporangial vesicle which release zoospores.

The downy mildew fungi are placed in two orders and three families,

Order: Peronosporales

Family: Peronosporaceae

Genera: *Peronospora, Bremia, Bremiella, Basidiophora, Benua, Plasmopara, Pseudoperonospora* and *paraperonospora.*

Order: Sclerosporales

Family: Sclerosporaceae

Genera: *Sclerospora, Peronosclerospora*

Family: Sclerophthoraceae

Genera: *Sclerophthora*

These downy mildew genera are mainly identified by the characteristic growth and branching of their sporangiophores, and the nature of the sporangia.

Keys for Identification of genera of Downy mildew Fungi

1. Sporangiophores of indeterminate growth(bear sporangia):

Genus: *Sclerophthora*

2. Sporangiophores of determinate growth:

2.1. Sporangiophores unbranched (bear sporangia)

Genus: *Basidiophora;* sporangiophore emerge from stomato and are thick walled.

Genus: *Benua:* sporangiophore half immersed in plant tissue and are thin walled. The ultimate branchlet is highly reduced.

2.2. Sporangiophores branched:

2.2.1. Branches dichotomous, tapering to fine points:

2.2.1.1. Bearing conidia

Genus: *Peronospora*

Genus: *Paraperonospora*

2.2.1.2. Bearing sporangia

Genus: **Pseudoperonospora**

2.2.2. Branches dichotomous disc-shaped at end: (bearing sporangia)

Genus: *Bremia*

Genus: *Bremiella*

2.2.2.1. Branches at right angles, not equidistant, bearing sporangia

Genus: *Plasmopara*

2.2.2.2. Branches upright at apex of swollen sporangiophore

2.2.2.2.1. Bearing sporangia

Genus: **Sclerospora**

2.2.2.2.2. Bearing conidia

Genus: **Peronosclerospora**

1.6. Important Characteristics of the Downy mildew generas:

1.6.1. Peronospora

The tips of sporangiophores branches are acute. Sporangia germinate by germtube.

1.6.2. Pseudoperonospora

The tips of sporangiophores branches are acute. Sporangia germinate by zoospores.

1.6.3. Paraperonospora

Constantinescu (1989) decribed the genus paraperonospora. Paraperonospora differ from peronospora in having uncolored dissemination units and globose to pyriform haustoria and from plasmopara by the production of sporangia that germinate with a germ tube. Paraperonospora differ from both of these genera in having sporangiophores that are widening towards the ramification. Sporangiophores are hyaline, straight, tree like and mostly dichotomously to monopodially branched in 4 to 5 branchlets.Ultimate branchlets are straight to slightly curved and had truncate or obtuse tips. Sporangia are almost subhyaline, broadly ellipsoidal.

1.6.4 Bremia

Sporangiophore slender, more or less dichotomously branched, mostly at acute angle. Sporangia usually germinate by single germtube or by zoospores. Tips of sporangiophores branches enlarged into disks with sterigmata at the margin. Zoospores are rare. Parasitic on the plants of family Cichoriaceae.

1.6.5. Bremiella

Sporangiophores with determinate growth emerge singly or in group through stomato, are hyaline, branched dichotomously in the upper part, the primary branches are of uniform thickness throughout. The ultimate branchlets slightly inflated at the tip or more rarely truncate. Sporangia borne singly at the branchlet tip, are hyaline, ellipsoidal with a pore and a conspicuous papilla at the distal end, often short pedicellate. Sporangia germinate by zoospores or germinative naked plasma. Oospore aplerotic, sphaerical, endosporium smooth, yellowish; exosporium yellowish brown, variously thickened (G.W.Wilson, 1914).

1.6.6. Basidiophora

Sporangiophore clavate or somewhat cylindrical, somewhat swollen above with numerous short sterigma like branches. Sporangia germinate by formation of zoospore. Parasitic on the plants of family compositae.

1.6.7. Benua

Constantinescu (1998) segregated the monotypic genus Benua from Basidiophora based on the fact that sporangiophores in Basidiophora emerge from stomato and are thick walled, whereas those of Benua are half immersed in the plant tissue and are very thin walled. In contrast to Basidiophora, the ultimate branchlets in Benua are highly reduced.

1.6.8. Plasmopara

Sporangiophore more slender, branching monopodially, usually nearly at right angles. Tip of branches obtuse. Sporangia germinate by formation of a single large naked nonflagellate plasmatic mass which encysts and then germinating by a single germtube.

1.6.9. Sclerospora

Sporangiophores with prominent branches, clustered near the apex, quickly fugacious. Sporangia germinate by germtube or by the formation of zoospores. Parasitic on grasses

1.6.10. Peronosclerospora

Sporangiosphore erect, dichotomously branched or clustered depending upon the species, emerge singly or in group from stomata.Sporangia oval or spherical to subspherical or ovoid to cylindrical or elliptical oblong with round apex depending upon the species. Oospore spherical light yellow or brown in colour or globular yellow depending upon the species.The genera is parasitic on maize, sugarcane and sorghum.

1.6.11. Sclerophthora

The unbranched or sympodial sporangiophores bear citriform or obpyriform sporangia. The mature oospore has a thick epispore which fuses with the oogonial wall to form a thick brown covering over the oospore. The oospore is plerotic which is characteristic of the genus Sclerophthora. The genus is parasitic on cereals and grasses.

1.7. Disease Cycle and Epidemiology

Oospores and sporangia or conidia, formed on collateral hosts, serve as the source of primary inoculum. The infected or infested seeds are the other source of primary inoculum, which give rise to infected plants. In presence of the host plants, the oospores germinate and form a 'germ sporangium' that forms biflagellate zoospores. The zoospores can infect the susceptible host right from the cotyledon stage and cause primary infection (Fig.1.3). The zoospores encyst on the leaf and form a germ tube that penetrates the stomata or the epidermal cells to infect the plant. The fungus grows intercellularly deriving nutrition through hasutoria from the living cells, and becomes systemic. First symptoms appear after 5-7 days of infection.

High relative humidity (90-100%) and cool weather (20-25°C) favours disease development. Dewy nights, followed by cooler days, are ideal for downy growth production. Cooler Nights followed by warmer days fail to induce downy growth. The sporangiophores and the sporangia are formed during night and wither by late morning.

Wind and splashing rains disseminate the sporangia that cause secondary infections. A prolonged wet period enhances the chances and the number of secondary infections. The symptoms accompanying the 'downy growth' develop during this favourable period of disease development. The sporangia are short-lived and their quick drying in 3-4 hours prevents their long distance dispersal.

As the crop season approaches towards the end, the fungus undergoes sexual reproduction leading to production of oospores in necrotic tissues. The remains of the dead plants carry the oospores into the soil, where it remains quiescent till the host is available the next season.

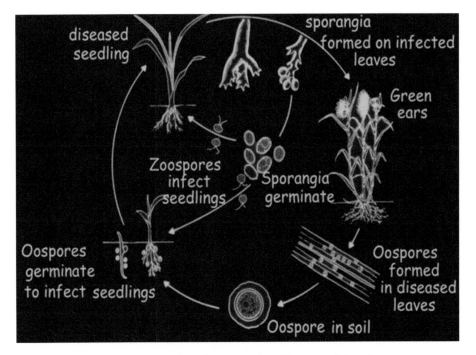

Fig. 1.3: Representative disease cycle of downy mildew disease.

1.8. Disease Management

1.8.1 Cultural practices

These *include*

1. Reduction of primary inoculums, i.e. the oospores by removing and destroying the infected plant remains, or by using 'trap crops' that induce oospore germination, which later meet their death in absence of the host.

2. Elimination of uneconomical collateral hosts to prevent primary infection by sporangia or conidia.

3. Using healthy seeds to prevent infection by seed-borne primary infection.

4. Prevention of humid conditions by:

    i. Choosing sites with good water drainage

    ii. 'Row cropping' to allow air circulation to keep the leaves dry.

    iii. Avoiding overhead irrigation to prevent prolonged wetting of the leaves

5. Rotation with non host crop for 2-3 years that can overcome the viability period of the oospores.

6. Use of resistant, tolerant or less susceptible cultivars.

1.8.2. Chemical control

Both pre-infection (protective) and post- infection (systemic or penetrant) fungicides are widely used. The protective fungicides include the copper-based formulations such as Bordeaux mixture and the dithiocarbamates. There is no report of resistance to these fungicides in the pathogens. The systemic fungicides are used when there is no alternative, and with utmost care, as these generate resistance in the pathogens besides being costly. These fungicides include bis-ethyl aluminium and phenylamides e.g. metalaxyl. Downy mildew forecasts are based on canopy-microclimate data. Disease warning systems have been developed that guide the farmers regarding the use of fungicides.

Novel proteomic and biosensor-based strategies are being devised for detection of downy mildew.

2

Downy Mildew of Cereals

2.1. Downy mildew of Sorghum

Pathogen: *Peronosclerospora sorghi* (W. Weston & Uppal) C.G. Shaw, (1978)

Host plants

Major hosts: *Sorghum bicolor* (sorghum), *Sorghum caffrorum*, *Sorghum Sudanese* (sudan grass), *Zea diploperennis*, *Zea mays.*

Minor hosts: *Andropogon sorghi*, *Panicum trypheron*, *Pennisetum glaucum* (pearl millet), *Sorghum halepense* (Johnson grass), *Zea mexicana* (teosinte)

Geographic Distribution

It is found worldwide in tropical and subtropical regions. *Sorghum downy mildew is* widely distributed in following countries.

Asia: Bangladesh, China, India, Iran, Israel, Japan, Nepal, Pakistan, Philippines, Thailand and Yemen (CMI, 1988; Jeger *et al.,* 1998).In India though the losses are reported from the states of U.P. and Madhya Pradesh, the disease is very common and destructive in Karnataka, Tamil Nadu, Andhra Pradesh, and Maharashtra

Africa: Benin, Botswana, Burundi, Egypt, Ethiopia, Ghana, Kenya, Malawi, Mauritania, Mozambique, Nigeria, Rwanda, Somalia, South Africa, Sudan, Swaziland, Tanzania, Uganda, Zambia and Zimbabwe (CMI, 1988; Jeger *et al.,* 1998)

North America: Mexico and USA (Alabama, Alaska, Arkansas, Georgia, Illinois, Indiana, Kansas, Kentucky, Louisiana, Michigan, Minnesota, Mississippi, Missouri, Nebraska, Nevada, New Mexico, Oklahoma, Tennessee and Texas (Bonde, 1982; CMI, 1988; Jeger *et al.*, 1998)

Central America and Caribbean: El Salvador, Guatemala, Honduras, Nicaragua, Panama and Puerto Rico (CMI, 1988; Jeger *et al,* 1998)

South America: Argentina, Bolivia, Brazil, Colombia, Uruguay and Venezuela (CMI, 1988; Jeger *et al,* 1998).Sorghum downy mildew has not been recorded in Australia.

Disease symptoms

In primary or systemic infection, the disease appears on the seedlings soon after their emergence. The leaves of affected seedlings become pale yellow and narrow and covered with a fine downy growth, which is the sporangial stage of the pathogen. The fur-like downy growth, though may appear on both surface of leaves but is more common on lower surface. Later, when seedlings are 5-6 weeks old, white streaks appear on both surfaces of upper leaves of infected plant. Due to formation of oospores in streaks (veins), these causes shredding of leaves, the most characteristics symptom of the disease. Such leaf tissues separate from the midrib due to shredding. The plants remain stunted and sterile.

The plants that escape systemic infection from the primary source of inoculums (oospores in soil or seed) are attacked when they are two to three months old. These plants are attacked due to secondary infection by the sporangia formed on systemically infected plants. The leaves, especially at the top turn white. The base of lower leaves may also become white, irregular, brown or yellow streaks then appear and the oospores begin to develop abundantly in the tissues. On the lower leaves, pale yellow patches, bearing the sporangial stage of the

Fig. 2.1: Symptoms of downy mildew on sorghum leaves

pathogen, are formed (Fig 2.1). Affected plants may not produce ears. If ears are produced, they are small and bear only few undersized grains.

Epidemics and losses

Jeger *et al.* (1998) demonstrated that the epidemics of P. sorghi in Africa appear to be more sporadic than S. graminicola causing downy mildew in pearl millet. Severe outbreaks of sorghum downy mildew have occured in India, Israel,

Mexico, Nigeria, Texas, Thailand, and Venezuela. In Nigeria, yield loss as high as 90% has been reported. In tropical and subtropical regions it can cause yield loss of 30% or more.

Pathogen's Character

P. sorghi is an obligate parasite systemic in young plant. The mycelium is intercellular, non-septate. Sporangiophores emerge through the stomata in single or in clusters which are stout and dichotomously branched. Spores are single celled, hyaline, globose and thin walled. Oospores are spherical, thick walled and deep brown in colour. The mycelial characters of the species do not much differ from those of *Sclerospora graminicola*. The sterigmata are longer than those of *Sclerospora graminicola*, being up to 16 μm as against 8 μm in the pearl millet downy mildew fungus. Major difference lies with the sporangia which always germinate by germ tube (conidia like), and not by producing zoospores. They are more commonly spherical instead of oval and lack the apical papilla. The size of these conidia is also smaller, being 15-29 μm in diameter. The oospore characters are similar to those of *S. graminicola*.

Conidia of P. *sorghi* are produced in large numbers, they are thin walled, ephemeral, and can cause the rapid build up of an epidemic. Oospores are tough walled, long-lived, and provide a perennating stage for the pathogen, as well as a mechanism for long-distance transport.

Disease cycle

P. sorghi produces oospores in infected host tissue. Oospores are produced less frequently and abundantly on maize than on sorghum, and appear in both hosts only in systemically infected plants (Bigirwa *et al.*, 1998). Oospore populations in soil have been reported as 8–95 oospores per gram of soil following sorghum crops (Pratt & Janke, 1978). Oospores are thought to survive for at least three years under a variety of conditions (Frederiksen, 1980), and can also be dispersed by wind. Oospores germinate in soil by germ tubes that infect underground parts of susceptible seedlings, which then become systemically infected. Infection by oospores does not occur if seedlings emerge in cool soils below 20°C. In arid areas, where soil temperature is higher, oospores initiate infection of seedlings, whereas in areas where a perennial host such as Johnson grass is present, infection can be initiated from conidia (Bigirwa *et al.*, 1998).

Conidia are produced from systemically infected plants at temperatures between 17 and 29°C from midnight to about 8 am (Bock *et al.*, 1998), with an optimum temperature of 24 to 26°C, depending on the geographic isolate (Bonde *et al.*, 1985). Conidial germination requires a saturated atmosphere, and moderate temperatures of 15 to 25°C, depending on the isolate. High levels of systemic infection can occur between 11 and 32°C, with a wet period of 4 hours or longer. Conidia are well adapted to wind dispersal (Bock *et al.*, 1998), but lose their viability after 3 to 4 hours so probably only play a role in short distance spread of the fungus.

The fungus infection in seed is confined in mature seeds to the pericarp and pedicel. Jones *et al.* (1972) found that seed transmission of the disease occurs when infected seeds at the soft dough stage are planted in sterile soil. In contrast, seed transmission of the disease is prevented by reducing the moisture content to 9% and by storage for 40 days prior to plant. Sexually produced oospores generally provide only one cycle of infection per season, whereas asexually produced conidia from an infected plant infect fresh hosts within the same season thus allowing rapid build up of an epidemic of downy mildew of Sorghum.

Weather parameter

Localised infection results from infection of leaves by conidia and may become systemic if hyphae grow into the embryonic leaves (Craig 1986). During cool, humid weather at night, conidiophores bearing conidia are produced in chlorotic areas on the abaxial surface of invaded leaves of systemically-infected plants and on local lesions (Bock and Jeger 1996). Infection by conidia of *P.sorghi* occurs only if there are suitable environmental conditions favourable for sporulation, germination and infection. The conditions required for sporulation are reported as temperatures between 15°C and 23°C for 5-6 h (Bonde *et al.* 1985), relative humidity above 80% (Shetty and Safeeulla 1981), and darkness (Bock and Jeger 1996). The optimum temperature for conidial germination is reported to be 12-20°C for isolates from Texas, India and Brazil, and 12-32°C for a Thailand isolate, for at least 2 h (Bonde *et al.* 1985). A dew period temperature of 10-33°C is required for 4 h for infection to occur (Bonde *et al.* 1978). Bock *et al.* (1999) reported that the optimal temperatures for conidial germination and infection of a Zimbabwean isolate of *P sorghi* was 10-34°C and 14-30°C, respectively.

Disease forecasting

Three basic deliverables of the framework include:

1. Surveillance and monitoring program;

2. Web-based system for information management; and

3. Prediction modeling.

Disease management

Varietal resistance

A number of single-gene sources of resistance have been found, and many resistant varieties are commercially available. Grow moderately resistant varieties like Co25 and Co26. However, new pathotypes of the fungus continue to evolve to overcome different sources of host plant resistance.

Chemical fungicides

Spraying of metalaxyl 25 WP (Ridomil) @ 2 kg a.i/litre water at 10 and 40 or 20 and 50 days after seedling emergence gave complete control. Potassium azide applied @1.12 kg/ha of soil also reduced the disease incidence. Metalaxyl, inhibits protein synthesis in the pathogen, is applied as a foliar spray. Seed treatments with the systemic fungicides metalaxyl and mefenoxam have been widely used to prevent systemic infections of Sorghum Downy Mildew.

Cultural practices

Crop rotation

Include the plants in crop rotation as host and non-host plants which trigger germination of oospores to minimise the infection propugules (Pratt 1978). Bait crops (e.g., *Linum usitatissimum*) grown in infested soil can reduce the incidence of infection in susceptible sorghum crops sown in the same soil (Tuleen *et al.*1980). This may be of value where oospores are the principal source of inoculum and soils are heavily infested.

Deep tillage

Both the incidence of SDM and the oospore content in the upper strata of infested soil are reduced by deep tillage (Tuleen *et al.* 1980, Janke *et al.* 1983). However, it is an expensive operation, and probably not cost-effective as a means of control.

Over-sowing and rouging diseased plants

Sowing at seed rates of up to 50% more than the recommended can lead to an acceptable plant density of healthy plants remaining at harvest after stand losses due to 20-30% disease incidence (Frederiksen *et al.* 1973). It is expedient to sow early in areas where the asexual inoculum rapidly increases as the season progresses. In Israel early-sown crops avoided the disease (Cohen and Sherman 1977). However, Tuleen *et al.* (1980) found a lower incidence of the disease in later-sown crops in the USA, where oospores are the principal source of infection.

Biocontrol practices

Bio-control agents used include *Trichoderma viride, Trichoderma harzianum, Gliocladium virens and Bacillus subtilis*. Under laboratory condition, all bioagents inhibited germination of oospores and conidia of *P. sorghi*. Dual cultures of *T. viride* and *T. harzianum* or *B. subtilus* were the most effective in inhibiting spore germination than the individual one.

A chytrid fungus (*Gaertennomyces* sp.) was found to be effective in reducing the incidence of systemic infection in soils heavily infested with oospores by as much as 58% (Kunene *et al.* 1990).

2.2. Downy mildew or Green ear disease of Pearlmillet

Pathogen: *Sclerospora graminicola*

Host plant

Wild grasses, millet

Geographic Distribution

Downy mildew of pearl millet, sometimes referred to as 'green ear' is the most destructive disease of pearl millet. This disease is widely distributed in temperate and tropical areas of the world, including the U.S.A., Europe, African countries, Iran, Israel, China, Japan, Fiji and Asia.

From India, the disease was first reported in 1907 by Butler in a sporadic form. The disease become serious in 1960s with the cultivation of high yielding varieties. The first epiphytotic occurred in 1971 and, since then, the disease has attracted much attention. In India, the disease is present in all the states where pearl millet is cultivated.

Symptoms

There is considerable variation in the symptoms, which almost always develop as a result of systemic infection. Systemic symptoms generally appear on the second leaf. Once these symptoms appear, all the subsequent leaves and panicles also develop symptoms. The disease can appear on the first leaf also under conditions for severe disease development.

Leaf symptoms begin as chlorosis (yellowing) at the base of the leaf lamina,

Fig. 2.2: Symptoms of downy mildew or green ear disease of pearlmillet

and successively top leaves show greater leaf coverage by symptoms. Severely infected plants are generally stunted and do not produce earheads. When earhead is infected, the floral parts are transformed into leaf like structures, which can be total or partial, hence the name 'green ear' (Fig 2.2).

Epidemics and losses

In India, downy mildew infection due to conditions of high humidity and moderate temperature could be very severe. The disease caused substantial yield losses during 1970s and 1980s. Grain yield losses of 10 to 60% have been reported. The yield reducing potential of downy mildew is very high, and this was evident in HB-3, a popular hybrid in India, where pearl millet grain production was reduced from 8 million tons in 1970-71 to 5.3 million tons in 1971-72 where the yield reduction due to a downy mildew epidemic in some fields were 60 to 70%.

The Pathogen

The disease is caused by *Sclerospora graminicola*, originally named as *Protomyces graminicola,* by P.A. Saccardo. *Sclerospora graminicola* is a biotroph. The mycelium is found in all parts of the systemically affected plants. The mycelium is present in the roots, stems, leaves and inflorescence of infected plants. The hyphae are branched, coenocytic, and intercellular with small bulbous haustoria. In stem, haustoria are not fully formed and are button-shapped. Hyphae are freely present in mesophyll cells of leaf and also may penetrate the bundles. But xylem and phloem are not penetrated. At the time of sporangiophore formation, tufts of hyphae reach the air space beneath the stomata. These develop sporangiophores, which emerge in clusters through the stomata. Each sporangiophore is a broad, short hypha, measuring 100.6-221.2 x 15.3-23.7µm. It is unbranched in the lower part but usually gives out 2-6 thick short branches, di- or trichotomously at the tip (Fig 2.3). Sterigmata are slender, slightly swollen and tapering at the apex bearing single sporangia. The sporangia are hyaline, broadly elliptic, sometimes broadly cylindrical, and slightly pointed or Papillate at the apex, with smooth wall and 19.0-31.6 x 15.8-23.7µm in size.

Oogonia and antheridia develop within the infected tissues, mostly in leaves and malformed floral organs. Oogonium is terminal or intercalary and after fertilization it develops within thick, brown to dark brown irregular walls. Oospores with oogonia measure 47.4-55 µm in diameter, whereas the diameter of the oospores generally ranges between 31.6 and 39.5 µm (average 35 µm). Oval oospores are produced in large numbers and remain scattered in the

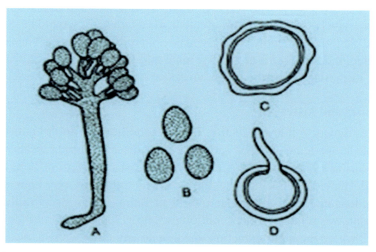

Fig. 2.3 (A): Sporangiophore bearing sporangia (B). Sporangia (C). Oospore and (D). Germinating oospore of *Sclerospora graminicola*.

mesophyll cells. The oogonial wall is persistent which shrinks and surrounds the oospore giving the entire oospore elliptic, angular or irregular shape although majority of oospores usually are oval to spherical. Oospores have three distinct layers in the wall and they do not germinate immediately but have a prolonged resting period. They germinate by one to four germ tubes.

Disease cycle

In North India, temperature and relative humidity prevailing during August-September favour sporangial production. Mature sporangia become detached and soon germinate in water to produce reniform, biflagellate zoospores. Germ tubes from encysted zoospores penetrate the host surface. The oosporic stage is the most common and predominant stage of the fungus in India.

The disease is systemic and primarily soil-borne, though seed-borne inoculum also brings about primary infection. Primary infection and survival of the fungus in nature are affected by oospores. The pathogen is strictly an obligate parasite. Oospores are produced abundantly in infected leaves and floral organs which fall down on the ground with the infected plant debris and perennate during winter. Oospores remain viable in the soil for 3-4 years. However, in the infected host tissue they remain viable up to eight years. The oospores require weathering for good germination and infection of the host. On germination the oospores infect seedlings of around nine day old crop. The pathogen invades the host tissue from underground parts in a systemic manner.

Besides soil-borne oospores, seed-borne inoculum is also important in primary infection. The oospores may also be carried with seed during harvesting and threshing. The oospores carried with seed and present in soil play the main role in the establishment and spread of the disease in the field. Besides oospores, intercellular aseptate mycelium of the fungus has also been detected in seed embryo. Some workers could also trace mycelial fragments inside the seed coat as well as in the embryo of seeds collected from partially infected earheads. It is presumed that the floral infection in bajra may take place like that of loose smut of wheat. Some workers also demonstrated transmission of the pathogen through dormant mycelium carried in the seed. The pathogen thus may be present in seeds either as oospores or as dormant mycelium internally.

Secondary inoculum is believed to be sporangia formed on plants that become infected by primary inoculum from soil or seed. Sporangial formation is favoured by a temperature range of 15°-25°C. Optimum combinations of temperature and relative humidity for sporangial development occur during August-September in northern India and the disease is most commonly seen during these months.

Mature sporangia fall soon and their dispersal takes place by wind, water (including irrigation water flowing in the field) and insects. On host surface, these germinate in the film of water to produce zoospores. The zoospores encyst and form germ tubes to infect the host.

Weather parameter

A temperature of 20-25^{0}c with intermittent rains and humid weather are required for disease development. Downy growths are formed during dewy nights from 1 a.m. to 4 a.m. Oospores are formed late in the season in the necrotic tissues

20 The Plant Mildews

on the leaves. Seeds formed in infected earheads are internally infected.

Disease management

Cultural methods

1. Early planting is recommended to escape the favourable weather for infection during the early crop growth stages.

2. Avoidance of rattoning

3. Roguing (removal of infected plants) and gap filling (transplanting healthy seedlings in the gap created due to roguing)

4. Application of balanced nitrogen to fields

5. Deep ploughing and sun baking of soil

6. Removal of infected crop residue and destruction of infected debris to reduce oospore levels in the soil

7. Cultivation of trap crops and avoiding monoculture. Such practices reduce oospore load in the soil

8. Elimination of collateral hosts to prevent primary infection by sporangia, and

9. Use of healthy disease free certified or disinfected seeds give good control of the disease. Seeds are disinfected by:

    i. Deeping the seeds in 0.1% $HgCl_2$ (clorax 2.6%) for two minutes, followed by several washings with sterile water.

    ii. Treatment with metalaxyl (a phenylamide systemic fungicide) @ 2g/kg of seeds.

Physical methods

Hot water treatment of infected seeds at 55°C for 10 minutes, followed by drying in shade.

Chemical methods

Seed treatment and foliar spray can be used either independently or in combination.Seed treatment with Thiram and Captafol proved very effective. Seed treatment with 0.1% Agrosan GN plus 0.4% Thiram (75% a.i.) gave 50% control of the disease.

For foliar sprays, chemicals of different groups have been tested. Among these, copper oxychloride or ziram did not prove effective, but captan and mancozeb controlled the disease significantly. More recently a new group of chemicals, acylanines proved very effective. Among the group, metalaxyl, with its formulations like Ridomil, Apron, etc. has been found very useful which can be used for seed treatment as well as foliage sprays.

Seed dressing with this chemical followed by foliar sprays gave best results. Ridomil can also be mixed with ziram or mancozeb. Seed dressing with Apron SD-35 (8 g/kg seed) followed by foliar spray with Ridomil MZ-72 WP (metalaxyl + mancozeb) or Ridomil ZM-28- FW (metalaxyl + ziram) was found very effective. Dipping the roots of seedlings in water suspension of Aureofungin (2,000 ppm) and Difolatan for one hr just before transplanting in the gaps reduced infection in adult plants.

Disease-resistant cutivars

Good sources of resistance have been identified in many germplasm accessions (mostly from West Africa) and these have been used for developing disease-resistant varieties of pearlmillet.

Open pollinated cultivars, WC-C 75 and ICTP 8203 have shown durable resistance in Indian fields. ICMH 451 and Pusa 23 also remain free from the disease. AIMP-92901 (Samrudhi), PPC-6 (Sampada) and Hybrids like Shradha, Saburi and Pratibha (AHB-1666) are resistant to downy mildew. Hybrid ICMH 88088 produced by ICRISAT has high level of resistance. Commonly cultivated cultivars HB-5, NHB-10 and NHB-14 are also considered resistant to the downy mildew.

Biocontrol practices

*Pseudomonas fluorescens was effective bioagent in the management of downy mildew of pearl millet. Pseudomonas fluorescens (*formulated in talc powder) treated seeds increased seedling vigour and inhibited sporulation of downy mildew pathogen. The bioagent tested as a foliar spray to pearlmillet downy mildew under greenhouse and field conditions was also effective. *Pseudomonas fluorescens* controlled downy mildew disease both by seed treatment and foliar application, but efficacy was significantly higher when seed treatment was followed by a foliar application. Seed treatment was better than foliar application alone.

2.3. Downy mildew of Maize (Corn)

Pathogen: *Peronosclerospora philippinensis* (W. Weston) C.G. Shaw, 1978

Major hosts: *Zea mays* (corn)

Minor hosts: *Avena sativa* (oats), *Saccharum officinarum* (sugarcane), *Saccharum spontaneum* (wild sugarcane), *Sorghum bicolor* (sorghum), *Sorghum halepense* (Johnson grass), *Zea mexicana* (teosinte)

Geographic Distribution

P. philippinensis is present throughout Asia with limited distribution outside this area. Countries known to have this pathogen are:

Asia: China, India, Indonesia, Japan, Nepal, Pakistan, Philippines, Thailand

Africa: Mauritius, Nigeria

Epidemics and losses

In the southern guinea savanna of Nigeria, loss due to this disease in maize ranges between 40-100% (Anaso *et al.*, 1989). Yield losses in individual fields are frequently 40–60%, but under favourable conditions, may be as high as 80–100% (Exconde & Raymundo, 1974; Bonde, 1982). In the 1974-75 seasons national yield losses in the Philippines were estimated at 8%, with a value of US $23,000,000 (Exconde, 1976). In general, yield losses correlate with percentage of plants infected (Exconde, 1975; Bonde, 1982).

Symptoms

Symptoms of Philippines downy mildew occur on the leaves and stems of maize plants. However, it is the effects of the disease on the production of viable cobs that is most detrimental to the commercial value of the crop. The severity of the disease in individual corn plants varies with environmental conditions and developmental stage of plants at the time of infection (Weston, 1920). When the disease infects young seedlings, there is usually a complete failure of the plant to develop and produce fertile ears. When infection attacks older plants, the plant may mature and produce stunted, malformed cobs with fewer grains (Weston, 1920).

Downy mildew on maize leaves is characterised by elongated chlorotic streaks with a downy growth of conidia and conidiophores (Fig 2.4). Symptoms first appear 3–6 days after infection as pale yellow to whitish discolourations on the leaf blade. Tassels may be deformed, and ears may be aborted. When the disease is severe, the infected plants are stunted and weakened, and may die within a month. When the attack is moderate, infected plants usually reach maturity but produce small, deformed ears (Weston, 1920; Dalmacio & Raymundo, 1972).

Fig. 2.4: Symptoms of downy mildew on maize leaves and plants respectively

By contrast, infection of corn leaves by *P. sacchari* produces irregular, discontinuous discolourations which do not tend to develop into the long streaks characteristic of *P. philippinensis*. Symptoms and sporulation take longer to appear than in infections with *P. philippinensis*. Although called downy mildew, infection by *Sclerophthora macrospora* causes excessive tillering, dwarfing and replacement of the tassel by a mass of leaf-like growth. *S. macrospore* produces large numbers of oospores in the thickened leaves.

Pathogen's Character

Peronosclerospora philippinensis and *sclerospora sacchari* are morphologically identical but differ from each other in that the former does not produce oospores and does not infect sugarcane. However, Exconde (1970) had claimed that he found infection of sugarcane also in the Phillipines and also observed oospore production on leaves that touched or were partially covered by soil.

The conidiophores of *P. Philippinensis* are reported to be 140-420 µm long. The maximum width is 20-25 µm. Conidia of this species are mostly barrel-shaped but variable on different hosts. The pathogen produces hyaline, upright, short, stout, branched sporangiophores bearing spherical sporangia.

Disease cycle

S. philippinensis is morphologically more or less similar to *S. sacchari*. It normally does not produce oospores in maize leaves and does not infect sugarcane. Though oospores in maize leaves have been observed in the Philippines but their role in recurrence of disease is not yet clear. Besides maize, this pathogen attacks several other plants like *Saccharum spontaneum* (Kans grass), *Sorghum* bicolor, *halepense,* and *Sorghum propinaquum.* These may serves as collateral hosts and source of primary infection of maize. The sporangiophores are 150-400 µm long with maximum width of 20-25 µm at the point of initial branching. The basal cell is 80-100 x 9-11 µm in size. The sporangia are mostly long, cylindrical-ovoid, though variable in shape on different hosts, and 14-44 x 11-27 µm in size.

Like other downy mildew fungi, these pathogens of maize survive mostly as oospores in plant debris in soil. Some of these, besides maize are also known to occur on several collateral weed hosts. *Sclerospora sacchari* was earlier known to produce oospores in only sugarcane and was considered to alternate between sugarcane *and* maize. In India, oospores production by this fungus on maize was reported in 1968 and thus it is supposed to survive as oospores in maize debris. *S. philippinensis* is known to infect several collateral hosts, mainly wild Sorghum spp. *(Sorghum bicolor, s. halepense) and Kans grass (Saccharum spontaneaum).*

Weather parameter

The disease is spread mainly via airborne sporangia that require high relative humidity (R.H) >90% and drizzling with Low temperature (21-33°C) for germination. Young plants are highly susceptible.

Disease management

Cultural practices

Alternate hosts and diseased crop debris must be destroyed and long rotations should be followed. For valuable seed plots fungicidal sprays are recommended.

Resistant varieties

Some varieties such as Phil, DMR 1 are resistant to downy mildew (Sharma and Payak, 1985). No commonly cultivated variety in India, is however, resistant.

Chemical fungicides

Use of Dithane M-45 at 0.3% (4-6 sprays starting from 10 days after sowing at 7 days interval). Sharma, *et al.*(1981) found good control of the Phillippine downy mildew with 3 sprays of Dithane Z-78 (0.3%) or 4 sprays of Dithane M-45 (0.3%).

2.4. Brown stripe Downy mildew of maize

Pathogen: *Sclerophthora rayssiae* var. zeae

Host plant

The primary host is *Zea mays* L. (maize). The pathogen can also infect *Digitaria sanguinalis* (L.) Scop. (large or hairy crabgrass), which may act as a source of inoculum for early sown maize under favorable environmental conditions. *Digitaria bicornus* (Lam.) Roem.& Schult. i.e Southern crabgrass in Thailand was found infected with *S. rayssiae* var. *zeae*, but this strain of the pathogen was not capable of infecting maize. Susceptibility of other grasses via natural infection has not been reported.

Geographic Distribution

This disease was first observed in several maize growing areas of India in 1962, and since then it has spread throughout India. It has also been reported from Myanmar, Nepal, Pakistan, and Thailand.

Epidemics and losses

Brown stripe downy mildew incidence is greatest in regions of high rainfall. In India the worst epidemics have been in areas that receive 100 to 200 cm of rainfall annually. Disease also occurred in regions where average annual rainfall was roughly 50 to 70 cm, but incidence was light. Moderate disease incidence has been recorded in India where average annual rainfall was 60 to 100 cm. The disease was observed in India in 1962, but not described until 1967 (Payak and Renfro, 1967). It is a common and destructive disease in India, with losses ranging from 20 to 90% (Payak, 1975). It tends to be most severe in areas with 100 to 200 cm rainfall, declining in severity as rainfall declines (Frederiksen and Renfro, 1977). The losses that the disease may inflict on a crop in a given area varies in a complex way. In general, if three-quarters or more of the foliage is affected prior to flowering, then the loss may be total, ear formation is either totally suppressed or markedly attenuated. Grain yield reductions may vary from 20 to 90%. Losses above 70% occur only in highly susceptible cultivars grown in conditions favourable for the disease (Payak, 1975). Yield losses of up to 63% have been recorded in the tarai area of Uttar Pradesh (Sharma *et al.*, 1993). McGregor (1978) established the parameter designated as Expected Economic Impact (EEI), which formed the basis for listing' 49 top-ranking exotic pathogens' from a total of 551 that pose a significant threat to agriculture in the USA. *S. rayssiae* var. zeae was ranked 43rd with the EEI standing at US$ 53 million at 1978 prices.

In India, where the Brown Stripe Downy Mildew (BSDM) pathogen was first discovered, yield losses from BSDM range from 20-70%. If BSDM were to cause comparable damage in the U.S., it would translate to a \$4.6-16.1 billion financial loss.

Disease Symptoms

The characteristic feature of brown stripe downy mildew, as its name suggests, is the vein-limited striping of the foliage (Fig 2.5). Other parts of the plant including leaf sheaths, husk leaves, ears or tassels, do not show symptoms even though all of the leaves including the flag leaf may be affected by the disease.The early zoosporic infections appear as vein-limited chlorotic flecks which enlarge lengthwise and coalesce. The merger of the flecks leads to the formation of rudimentary stripes in the inter-veinal areas. The stripes range in width from 3 to 7 mm. in length, however, they may extend to the full length of the lamina and also on either side of the midrib. At first the stripes are chlorotic or yellowish but with age they turn yellowish-tan to purple ferrugineous and necrotic. In some maize genotypes, the pathogen induces stripes, which have reddish to reddish-purple borders with bleached centres. This process of stripes becoming necrotic coincides with the development of the teleomorphic stage (antheridia and oogonia) and heralds the cessation of sporangial prodution.The disease first appears on the lower most leaves, which are proximate to ground level. These leaves show the highest level of striping; as a result they present a pale-brown, burnt appearance and severely affected leaves may be shed prematurely. Leaves around the ear shoot show a lesser amount of striping and the leaves above it (near the tassel) show the least striping. Occasional infections through localized rain-splashing or a similar kind of disturbance may alter this pattern of disease appearance to some extent. Infections leading to severe striping result in blotching of extensive areas of leaf laminae. In the early stages, zoosporic infections en masse in large patches lead to rapid coalescenced, which produces the blotching effect. The damage becomes more severe and premature defoliation may result. As the veins are not affected, laminar shredding is uncommon. However, when severe infection occurs, leaves tear apart near the apices and hang in tatters.Vegetative or floral malformations of any kind are completely absent. What differentiates this disease from crazy top, caused by Sclerophthora macrospora, is the absence of malformation, stunting, leaf thickening, leaves becoming strap-like, etc. Greatest damage due to the brown stripe disease occurs when the disease severity is high in the pre-flowering stage. It suppresses emergence of the ear or at best a rudimentary ear of no consequence may deep out of the sheath of the ear shoot. On the under surfaces of the chlorotic stripes, a greyish-white, downy growth develops, which has a fine granular rather than fibrous appearance. It is not restricted to the adaxial surfaces of the stripes on the leaf though it is more commonly found there. On the upper leaf surfaces a downy growth is also quite often encountered. Thus, in a strict sense, it is amphigenous rather than hypophyllous. The downy

Fig. 2.5: Symptoms of brown stripe downy mildew of maize leaves and plant

growth has been observed even in the afternoons and so it is not so evanescent as is the case with other downy mildews affecting poaceous crops (Payak and Renfro, 1967; Payak *et al.*, 1970). As the stripes lose their chlorotic appearance and turn necrotic, the downy growth disappears. Oospores appear to be produced only in necrotic tissues. Striping of maize leaves can occur due to a variety of causes: genetic, nutritional or pathological. However, the presence of granular downy growth on the undersides of vein-limited stripes without any malformation is a diagnostic feature of brown stripe downy mildew.

Disease cycle

Primary inoculum comes from oospores overseasoning in the soil or plant debris or from mycelium in infected seed. Oospores in air dried leaf tissue remained viable for four years, although infected seed dried to 14 % moisture or less and stored for four or more weeks will not be capable of transmitting the disease. Warm soil temperatures (28°-32.5°C) are required for disease development. Germinating oospores produce sporangiophores bearing sporangia, and secondary spread occurs with dispersal of sporangia in wind and water splash, or from an infected plant to a healthy plant via physical contact. The pathogen apparently does not systemically infect the plant.

Weather parameter

Downy mildews are very significant disease in tropical regions of Africa and Asia, where prolonged periods of leaf wetness and cultivation of alternate hosts are prevalent during the growing season. Cool, wet and humid conditions are optimal for disease development. In favorable conditions, disease cycles are rapid, leading to severe infection and spread of disease.

Disease forecasting

1) Deliver surveillance and monitoring network to provide timely information of the incidence and severity of corn downy mildew in the United States and off-shore source areas such as the Caribbean Basin.

2) Provide a web-based system (USDA Corn Downy Mildew Monitoring and Prediction System) for information management of monitoring observations, forecasts, and decision criteria to stakeholders.

3) Prediction modeling

Disease management

As the disease was recorded in the 1960s, the need for control through host resistance was deemed to be a priority for the All India Coordinated Maize Improvement Project. In the hybrid development programme a large number of inbred lines were evaluated to this disease and many promising lines were soon identified so that it was possible to release a resistant hybrid as early as 1968. This was one of the fastest answers to a disease problem identified in 1965. Subsequently, through cooperative work in the Asian region under the aegis of the Inter Asian Corn Programme, very promising resistance to all the prevailing downy mildews of maize in the region was identified in Philippine germplasm; Taiwan had already released hybrids resistant to sugarcane downy mildew (*Peronosclerospora sacchari*). Materials identified as resistant in the Philippines, Taiwan and Thailand to the prevailing downy mildews have also shown a high level of resistance to S. rayssiae var.zeae. Coordinated work at a national level has been carried out to determine dosage, formulations, etc., for the systemic fungicide metalaxyl, which is specifically active as a seed treatment against oomycetous fungi including the downy mildews. Lal *et al.* (1980) found that seed treatment with the fungicide controlled brown stripe downy mildew for up to 30 days after planting. Resistant cultivars are available from foreign sources. Drying seed to < 14% moisture will prevent seed transmission of the disease.

Seed treatment with apron (metalaxyl) is highly effective in reducing disease severity. Seed treatment with metalaxyl at 4 g/ kg controlled brown stripe downy mildew *(Sclerophthora rayssiae* var. *zeae)* of maize up to 30 days after planting. When combined with seed treatment, one foliar application of metalaxyl (225 ppm) on the 30th day after planting result in excellent disease control and increased grain yield and 1,000 kernel weight. Apron has also been found to be compatible with bio-agents such as *Ps. fluorescens*. Single application of calcium hypochlorite @ 25 kg/ha, successfully reduces the disease and has also been found to be compatible with *P. fluorescens*.

2.5. Downy Mildew of Wheat and Minor millets

Pathogen: *Sclerophthora macrospora*

Host plant

Wheat including Durum Wheat, Barley, Cereal Rye, Maize, Oats, Sorghum, Triticale and many grasses including most cool season turf grasses.

Geographic Distribution

The disease occurs all over the world. It is widespread but sporadic in Illinois.

In June 1985, Lloyd durum wheat and Azure barley plants showing stunting, leathery leaves, excessive tillering and deformed heads were observed in wet areas of two fields near Casselton, North Dakota (20-50% incidence). Similar symptoms were observed in wet areas in oat breeding nurseries in Fargo and Prosper. This was the first report of downy mildew on these cereals in North Dakota (Jons *et al.*, 1986), despite the fact that the pathogen was widespread on grasses in the Dakotas region (Semeniuk and Mankin, 1964). Yellow tuft caused by *S. macrospora* was observed in Kentucky bluegrass [Poa pratensis] sod fields in Illinois, USA in 1992. The sod was 18 months old and growing on soil used continuously for sod production during the previous 5 years. Disease tufts (5-15 cm diameter) were observed at three field locations that collectively represented about 3 ha; about 10% of that turf was affected. Excessive soil moisture for 4 weeks and daily air temperatures ranging from 15 to 26°C had preceded disease development (Wilkinson and Pedersen, 1993).

Epidemics and losses

Yield losses of up to 25% can occur depending on the time of attack and the course of the epidemic. In Denmark the yield loss through mildew attacks in barley and wheat is often 5-10% in the field trials. In addition to the cereals, most wild and cultivated grasses can be attacked. No infection can occur between the plant species but occur between winter and spring forms of the same plant species; for example from winter barley to spring barley. Among the cereals, attacks occur especially in barley and wheat, followed by in rye and to a lesser extent in oats.

Fig. 2.6: Symptoms of downy mildew on wheat plant

Disease Symptoms

Yellowing of leaves, stunting, excessive tillering and occasionally early death of young plants occur. Older plants often have twisted, leathery and thickened leaves. Heads on surviving plants may be distorted, twisted and thickened ("Crazy Top") with little grain (Fig 2.6). Usually observed in plants growing in or very near to standing water.

In turf grasses there is initial stunting, followed by circular yellow patches that are 1-10 cm in diameter. The plants in these patches have stunted roots and may be easily pulled up. White, downy mycelia may cover the leaves during cool wet weather. It is often difficult to see in mown turf. Symptoms may vary depending on the species of grass.

Disease cycle

The downy mildew fungus produces a large number of round, pale yellow, smooth walled resting spores (oospores) within infected, senescing leaf, glume, and culm tissue. The fungal structures are best seen with a light microscope in decolorized leaf tissues stained with acid fuchsin. The thick-walled oospores may persist in dead host tissue for several years and are released into the soil when diseased tissues decay. Oospores are carried from one field to another in diseased plant residues and soil, in seed grain, and by the wind, surface runoff water, and tillage equipment. The oospores germinate in water or saturated soil to produce lemon-shaped sporangia (conidia). Within an hour or two the sporangia usually liberate 30 to 90 motile zoospores. The zoospores are capable of swimming short distances in water before settling down and forming slender germ tubes that penetrate the host tissues of mostly seedling plants. Following infection, the downy mildew fungus develops systemically within the plant, becoming most abundant in actively growing tissue. Sporangia are formed and infection occurs over a wide range of soil temperatures (6 to 31° C; optimum 10 to 25° C). The causal fungus is incapable of reproducing in the absence of a host plant. The fungus must infect living plants of small grains, grasses, corn, sorghums, or rice each season; however, oospores persist and remain viable for several years to cause primary infection. Primary inoculum comes from oospores overseasoning in the soil or plant debris , or from mycelium in infected seed. Oospores in air dried leaf tissue remained viable for four years, although infected seed dried to 14% moisture or less and stored for four or more weeks will not be capable of transmitting the disease. Warm soil temperatures (28°-32.5° C) are required for disease development. Germinating oospores produce sporangiophores bearing sporangia, and secondary spread occurs with dispersal of sporangia in wind and water splash, or from an infected plant to a healthy plant via physical contact. The pathogen apparently does not systemically infect the plant.

Control measures

1. Early sowing of winter crops and late sowing of spring crops increase the risk of attacks.

2. Avoid high nitrogen level, remove volunteers and use mildew resistant cultivars.

3. Cultivar mixtures consisting of cultivars with different resistance to mildew can also be used.

4. Reduce waterlogging and inundation.

5. Control grasses that may be a source of infection.

3

Downy Mildew of Pulses

3.1. Downy mildew of peas

It is an important disease of pea crop particularly in those areas where cool and wet climate is prevalent during its cultivation. The disease during favourable weather conditions causes considerable yield losses.

Pathogen: *Peronospora viciae* (formerly *P. pisi*)

Host range

Vigna faba, Vigna sativa. Pisum arvensis, P.sativum, Lathyrus sativus, *Vicia hirsuta, Vicia sativa, Vicia villosa* (Dixon, 1981).

Geographical distribution

Reports of the pathogen are from different countries on different crop host.

On Pea- Berkeley in England first discovered the disease in 1846 (Chupp and Sherf, 1960) and since then it has been found in all parts of the world where the crop is grown intensively. The occurrence of the disease has been reported from Argentina, Australia, Canada, Chile, Denmark, France, Germany, Kenya, Netherland, New Zealand, Rhodesia, Spain, Tasmania, USA, Ukrain, Italy, Poland and Australia (Dixon,1981). In India it is prevalent in Indogangetic plains including Uttar Pradesh and Punjab (Bains and Dhaliwal, 1986).

On Broad bean- England, Italy, Rumania, Tunisia.

On Lathyrus sativus- India, UK, Union of Soviet Socialist republics.

On Pisum arvense- United states of America.

On Vetch- Argentina, India, UK, Union of Soviet Socialist Republics

Losses

In Wisconsin, some fields had 62-85 per cent diseased pods during downy mildew epidemic which occurred in 1973 (Hagedorn, 1974). It was also identified as one of the major contributor to the pea yield decline syndrome in South Australia (Davidson and Ramsey, 2000). Though, in general, P. viciae does not cause high crop losses (Pegg and Mence, 1972; von Heydendorff, 1977; Bugiani and Giovoni, 1996), these may occur in soils heavily infested with oospores. Crop losses of 30 and 45% in Sweden and in the UK, respectively, have been reported (Oloffson, 1966; Biddle *et al.* 1988). Local infection of leaves does not usually lead to yield losses (Pegg and Mence, 1972), however; infection of the pods may lead to considerable yield losses. Diseased pods produced few peas which are of a poor quality.

The most damaging epidemics of downy mildew of faba bean,caused by *Peronospora viciae* f. sp. *fabae* (Berk.) Caspary, occur in northwestern Europe, though the disease has been reported in most regions where faba bean is grown.

Symptoms

The disease generally starts from the lower leaves and then progresses upward. Two types of infection i.e. local and systemic, occurs on pea plants. In local infection, round or less frequently elongated, yellow to brown *spots* appear on the upper surface of leaves, stipules, and petioles and sometimes on stems (Fig 3.1). The patches may be of small size or may occupy a larger area of the leaf lamina. During cool and moist weather, the corresponding under surface is covered with a white to grayish violet downy growth of the fungus, which is composed of sporangiophores and sporangia. Systemic infection, results in stunting and distortion of the plants, usually killing them before flowering. On pods, the patches are pale green; more or less elliptical which later turns brown. Seeds developing under these patches remain undersized and aborted. Later in the season, oospore develops in the senescent tissues and can survive in soil for up to 15 years (Van der Gaag and Prinking, 1997).

Fig. 3.1: Downy mildew symptoms on leaves of pea and pods.

Differences in symptoms also occur in different host plants. For example, the characteristic stunting of oospore infected seedlings usually occurs in pea but may be less common in faba bean. Van der Gaag and Frinking (1997a) did not observe stunting of oospore infected faba bean seedlings but Jellis *et al.* (1998) indicated that soilborne infection results in stunted and distorted seedlings. Stunted growth of faba bean may also occur after conidial infection above soil. Leaf lesions on faba bean become brown and necrotic and shot holes may be formed which do not occur in pea. The fungal growth appears on infected pea stems at high humidity but not on faba bean stems.

The Pathogen

Fungal mycelium is aseptate, hyaline, intercellular and branched with finger shaped haustoria. Sporangiophores first appear as simple elongating hyphae from stomata on abaxial leaflet surface, then branching from a single axis (monopodially) to produce multiple, terminal sporangia, which developed synchronously and polyblastically on each sporangiophore . Sporangia had smooth surface during development but finely echinulated when mature and are delimited by septa at the ends of terminal sporangiophore branches.

Gametangia (oogonia and anthredia) developed extensively on inner surface of field grown pea pods from smooth, bulbous hyphae, adhering to the host epidermis. Each oogonium is surrounded by several anthredia. Oospores within pod tissue of field grown plants are enclosed by oogonial membranes. Each oospore had a heavily reticulated outer wall enclosing cytoplasm and liquid possibly lipid. The pathogen also shows some variation, and atleast six races have been reported (Hubbeling, 1975). There are now two strains of the downy mildew fungus. They are the Parafield strain and the Kaspa strain. The Kaspa strain is detected in Australia.

Disease cycle

The pathogen perenates in soil through oospores lying in soil . Sometimes, it is seed borne also, both under the seed coat or as a contaminant with seed lots. These oospores cause primary infection in the ensuing season. Secondary spread of the disease occurs through sporangia disseminated by moist wind.

Weather Parameter

Disease occurrence is usually associated with unusually wet weather and long periods of cool overcast weather (Hagedorn, 1974). Sporangial production requires high humidity for atleast 12 hrs. The sporangia germinate best at low temperatures between 4 and 8°C and require rain for dispersal. Sporangia germinate readily in free moisture by germtube. The longevity of sporangia decreases significantly as temperature increases. At 5°C no oospores are produced while 20°C is optimum (Van der Daag *et al;* 1996). Proximity of pea crops to each other also results in severe outbreak of the disease.Manganese concentration in pea plants is positively correlated with the severity of downy mildew (Davidson and Ramsey, 2000).

Disease Management

Cultural practices

Cultural practices like destruction of infected plant debris, atleast **3** year or longer crop rotation, wider spacing, use of healthy seed (may be produced in dry areas), deep ploughing and well-drained fields are useful in reducing the disease.

Host resistance

Complete immunity among pea cultivars is lacking against downy mildew although some degree of resistance/ tolerance have been reported in some cultivars/ lines like Gastro, Starcovert, Starnain, Puget, Cobri, 119, 21 and Mukta (Hubbeling, 1975; Stegmark, 1988; 1990 and Annonymous, 1999).

Chemical fungicides

The pathogen is known to overwinter in the form of oospores in the infected plant tissues and in the seed. Therefore, seed treatment with various systemic fungicides like metalaxyl and A9873 (mixture of metalaxyl, cymoxanil and fludioxanil) has been reported effective against this disease. Foliar application of metalaxyl (Singh and Dickinson, 1980) or cymoxanil + copper (Bugiani and Govoni, 1996) after the appearance of initial symptoms have been advocated against this disease. However, ineffectiveness of seed treatments with metalaxyl and oxadixyl in New Zealand has been reported (Falloon *et al.,* 2000) due to development of fungicide resistant to these molecules, while cymoxanil or fosetyl-Al protected seedlings from infection by metalaxyl resistant strains. To avoid buildup of fungicide resistant strains, it is always better to rotate the fungicides and apply in mixture with non-systemics.

3.2. Downy mildew of lima bean

Downy mildew has been a major disease of lima bean grown in the humid eastern United States since it was first reported and described by Prof R. Thaxter (Thaxter 1889). Phytophthora phaseoli is a unique organism that is pathogenic to a single plant species, *Phaseolus lunatus* or lima bean.

Pathogen: *Phytophthora phaseoli*

Losses

A serious outbreak of downy mildew on lima bean was reported in Bergen County, New Jersey in 1897, and further reports from the New Jersey Agricultural Experiment Station, in 1914, show that the disease was present in the state since then and probably almost every year. The pathogen was first found infecting lima bean in Delaware by Dr. C 0. Smith in 1904. These early reports were for occurrence on Fordhook lima bean. In 1958, it was reported that growers in Delaware suffered an estimated loss of 1,139 tons of lima bean valued at $175,762 (App, 1959). The disease caused losses of about thirty-five percent of the lima bean crop equivalent to 5,715.26 tons of lima bean with a value of $2,268,000, threatening the future of the lima bean industry in Delaware. Race E of *P. phaseoli* was identified as responsible for this downy mildew epiphytotic. In the same year, the new race F was also identified (Evans et al., 2002)

Symptoms

The pathogen attacks and destroys pods in all stages of development The first observable symptoms are white downy mold patches on pods (Fig 3.2), flower racemes and petioles, and more rarely, leaves (Walker, 1952). A dark reddish border surrounding the infected tissue on pods is also common. Pods become dry and shriveled following the infection, remain attached to the plant, and results in a loss of marketable beans. The damage to pods prevents bean development or discolors the beans. The extent of discoloration depends on the stage of the pod at infection. Infection and disease development occur most rapidly under conditions of cool nights and warm days during periods of wet weather and heavy dews (Hyre, 1958). Disease severity also can be increased by higher plant density, which increases the level of moisture in the canopy, facilitating infection of the fungus.

Fig. 3.2: Symptoms of Downy mildew of lima bean

Forecasting

Two Forecasting systems for *Phytophthora phaseoli,* predicting downy mildew of lima bean exist. The first one was developed by Dr. R.A. Hyre of the University at Delaware and was the most widely used system to predict downy mildew in the last fifty years (Hyre, 1958). This forecasting system is based on temperature and rainfall, and is based on the same factors as those for to predict potato leaf blight caused by *P. infestans* (Hyre, 1954). According to Hyre's model, downy mildew will occur if both rainfall and temperature are favourable. Rainfall is considered favorable when the 10-day total is 3.05 cm (1.2 in) or more while temperature is considered favorable when the 5-day mean is less than 26.1°C, and greater than 7.2° C. Once eight consecutive downy mildew favorable days are recorded, the disease is forecasted to appear in 7-14 days, the cycle was broken if temperatures exceeded 32.2° C (Hyre, 1958). This forecasting system was based on the conditions favorable for the growth of the secondary inoculum of the pathogen.

Scarpa and Ranicrc in 1964 developed the second forecasting system to predict downy mildew on lima bean caused by P. *phaseoli,* based on consecutive hours of dew accumulation. This system warned growers to spray fungicides if dew points were 20.6 °C , or greater for twelve hours, and expected to continue for more than twenty-eight hours This predictive system was reported to have success in New Jersey and Philadelphia, but there are no records about its use in other geographic areas.

Disease Management

Cultural Practices

Several cultural practices can minimize downy mildew. Beneficial practices include planting in fields that have good drainage, plowing under all residue of the lima bean crop at the end of the season and crop rotation. Planting lima bean crops earlier in the season also helps to avoid the early fall when the weather is cool and favorable for downy mildew. Resistant cultivars also have provided effective control through the years, however, losses continue because of the changing race structure of the population and that same cultivars lack resistance.

Chemical fungicides

The use of fungicides has been the most effective way to control P. *phascoli* since 1897 when Bordeaux mixture was recommended for control (Slurgis, 1897). However, lima bean is sensitive to copper fungicides, and phytotoxicity became such a serious problem that recommendations warned against using copper based fungicides. From 1946 to 1954 research on management of downy mildew with Tri-basic copper sulfate and maneb fungicides as sprays and dust, showed good control and lower toxicity (frossan *et al.,* 1957). Copper fungicides were the only labeled fungicides for downy mildew until 1998, and during this time resistant cultivars were available and downy mildew was not an important problem. In 1998 a new race of P. *phaseoli* had developed, genetic resistance was no longer effective and downy mildew was once again a problem. Fungicide testing was again initiated to look at new fungicides for control of downy mildew. Copper hydroxide was determined to be more effective than tri-basic copper sulfate and became the only additional labeled fungicide that could be used to control downy mildew. Other fungicides such as Kocide 200 DP and Champ DP have been approved for downy mildew based on successful field trials (Dominiak and Mulrooney. 2000). Davey *et al., (*2004) reported that the systemic fungicides Ridomil Gold/Copper WP 2.0 lb, and Phostrol at all rates tested (0.95,1.42, and 1. 89 liters) were the best fungicides for management of downy mildew. These products reduced disease incidence and severity, increased total number of pods, and consequently increased bean yield. The high level of control was consistent for two growing seasons of testing, making these fungicides the most reliable and effective products labeled for downy mildew control (Davey. 2005).

4

Downy Mildew of Oilseed Crops

4.1. Downy mildew of sunflower

Pathogen: *Plasmopara halstedii* (Fort.) Berl.& de Tom.

Hosts

Over 100 host species from a wide range of genera in the family Asteraceae have been reported susceptible to this pathogen, including wild and cultivated species of *Helianthus*, e.g. sunflowers, which is the principal economic host. Wild Asteraceae hosts (e.g. species of the genera *Helianthus*, *Artemisia*, *Xanthium*, etc.) may also occur widely in the corresponding areas but their potential as reservoirs of the pest is not yet known (Virányi, 1984). For lists of hosts see Leppik (1966) and Novotel'nova (1966).

Geographical distribution

EPPO region: Present wherever sunflowers are grown, in Albania, Austria, Bulgaria, Czech Republic, Egypt, Estonia, France, Germany, Hungary, Italy, Moldova, Morocco, Poland (unconfirmed), Romania, Slovakia, Spain, Switzerland, Turkey, Russia (European, Siberia), Ukraine and Yugoslavia.

Asia: Azerbaijan, China, Georgia, India, Iran, Iraq, Israel, Japan, Kazakhstan, Pakistan, Russia (Siberia), Turkey.

Africa: Egypt, Ethiopia, Kenya, Morocco, Zimbabwe, Uganda.

North America: Canada (widespread), USA (California, Kansas, Minnesota, North Dakota, South Dakota).

Central America and Caribbean: Dominican Republic.

South America: Argentina, Brazil, Chile, Uruguay, Paraguay.

Oceania : Australia (not recorded on sunflower, but on cape weed, Arctotheca calendula, in New South Wales and South Australia), New Zealand (Hall, 1989; unconfirmed).

Europe: Present.

Downy mildew on sunflower is widespread in all sunflower-growing countries with the exception of Australia. According to Tikhonov (1975), it was first discovered on sunflowers in the United States in 1883, and in 1892 it was found on *Helianthus tuberosus* in Russia. As the sunflower expanded to other countries, the disease followed it closely, especially after the World War II. The rapid expansion of the disease may be explained by its transfer with infected sunflower seed. In the former Yugoslavia, it was discovered by Perišiæ (1949) and described by Nikoliæ (1952).

Disease severity and losses

Disease severity may vary considerably according to region, year and growing conditions. The incidence of downy mildewed sunflowers in a field may range from traces to near 50% or even up to 95% (Sackston, 1981). In Europe, after its first appearance in 1941, the disease increased rapidly and by 1977 it was rated a "major disease" in all sunflower-producing countries of Europe (Sackston, 1981).

Symptoms

Symptoms of the disease become evident in different forms depending on the stage of crop growth and occurrence of infection (Fig. 4.1 to 4.3).

(a) **Damping off:** Seedlings are killed before or soon after the emergence due to subterranean infection by the downy mildew fungus. Affected plants dry and become wind blown.

(b) **Systemic symptoms:** Sunflower plants carrying systemic infection are severely stunted and the upper leaves become entirely chlorotic at flowering. The stem becomes brittle. Flower heads of infected plants remain erect, become small in size, remain sterile and produce no seeds or only few seeds on such heads.

(c) **Local foliar lesions:** Small angular greenish-yellow spots appear on leaves as a result of secondary infection. The spots may enlarge and coalesce to infect a large part of the leaf. The fungal growth becomes visible at lower surface of the diseased area.

(d) **Basal root or stem galls:** The root infection may result in formation of galls at the base of the plants on primary roots. Such plants are less vigorous and subject to lodging.

Fig. 4.1: Chlorotic symptom

Fig. 4.2: White cottony growth of downy mildew fungus

Fig. 4.3: Downy Mildew affected plot and sunflower plants

Pathogen

Plasmopara halstedii is an obligate biotrophic oomycete pathogen that attacks annual Helianthus species and cultivated sunflower, Helianthus annuus. The fungus causing downy mildew of cultivated sunflowers is known in the world literature under two scientific names: (1) *Plasmopara halstedii*, used in many parts of the world to refer to a closely related group of fungi, the "*P. halstedii* complex" (Leppik, 1966), attacking cultivated sunflowers, other annual and perennial *Helianthus* species, as well as a number of additional composites; (2) *Plasmopara helianthi*, a name introduced by Novotel'nova (1966) in Russia, referring to the fungus thought to be confined to members of the genus *Helianthus* with further specialization on intrageneric taxa as *formae speciales*, that confined to *Helianthus annuus* probably being *Plasmopara helianthi* f.sp. *helianthi*.

Races of downy mildew pathogen

Seven races of P. halstedii have been recorded so far, and four of these exist in Europe (Gulya *et al.,* 1991). Whether races of the fungus other than race 1 have been introduced by seed or have arisen locally remains to be determined.

Disease cycle

The oospores in the residues of the preceding sunflower crop or oospores in or on seeds from the systemically infected plants serve as primary source of infection. In spring, germination of overwintered sexual oospores leads to sunflower infection. Some oospores have been reported to remain dormant upto 14 years. Intercellular hyphae are responsible for systemic plant colonization and the induction of disease symptoms. Under humid and fresh conditions, dissemination structures are produced by the pathogen on all plant organs to release asexual zoosporangia. These zoosporangia play an important role in pathogen dissemination, as they release motile zoospores that are responsible for leaf infections on neighbouring plants.

Local spread of the fungus is mainly by wind-borne sporangia from infected cultivated or volunteer sunflowers and, within a field, by soil particles (e.g. during tillage) as well. Long-distance spread is by man transporting infected seed. A single oospore that germinates gives rise to a single sporangium with zoospore differentiation(Novotel'nova, 1966; Delanoë, 1972), and release follow. In the presence of free water, the zoospore swarms rapidly and, if a host tissue (root, root-hair, stem or less commonly leaf) is available, settles on an infection site where encystment and subsequent germination take place. Penetration of the host is in a direct manner through the epidermis (Virányi, 1988a). Once established, the fungus grows intercellularly and, in a compatible host/pathogen combination, it starts with systemic colonization towards the plant apex. Systemic mycelium may be present in all plant tissues except meristems (Novotel'nova, 1966). When conditions are favourable, asexual sporulation takes place by means of sporangiophores arising primarily through stomata or other openings on the invaded tissue. Oospores are also produced in infected plant parts, primarily in roots and stem (Virányi, 1988b).

P. halstedii is a soil-borne pathogen, its oospores serving as primary inoculum for young sunflower seedlings. It may also be wind-borne, causing secondary, usually localized, infection of above-ground plant parts by dispersed sporangia, or even seed-borne when seed produced by infected plants carries mycelium and/or oospores of the pathogen. The significance of wind-borne sporangia in disease initiation is usually low. However, secondary infection is considered as an important factor in the spread of the disease in certain regions under favourable environmental conditions (Zimmer & Hoes, 1978). In addition, secondary infection by sporangia is found to incite latent infection in plants with no disease symptoms during the season but producing seeds that might carry the fungus in a latent form (Sackston, 1981).

Since seed-borne inoculum is extremely rare and results in a very low percentage of systemically infected plants, it is unlikely that dramatic outbreaks of the disease could be attributed to such inoculum. Instead, soil-borne oospores from a previous sunflower crop or even from volunteer sunflowers are the usual sources of severe attack of downy mildew in the field.

Environmental factors

Moisture and temperature are the most important environmental factors affecting infection and spread. Zoospores, originating from either sexual or asexual sporulation, require free water to retain viability and move towards infection sites. Consequently, rainfall or intensive irrigation will be a prerequisite for the initiation of primary infection, particularly during the critical first 2-3 weeks after sowing (Zimmer & Hoes, 1978; Kolte, 1985). Plant age and host tissue are also of significance in determining susceptibility of sunflower to systemic infection by *P. halstedii* (Sackston, 1981). From a practical point of view it can be stated that the earlier the infection occurs in the season, the more severe the disease will be in the plant.

Control Measures

   i)   Early planting at the onset of the rainy season decreased disease incidence.

   ii) Treat the seeds with metalaxyl compound (Apron 35 SD) @ 5 g/kg of seed.

   iii) Growing resistant varieties: LSH 1 and LSH 3 are downy mildew resistant hybrids.

   iv) Follow clean cultivation by removal of crop debris.

   v)   Avoid continuous sunflower cultivation.

   vi) Follow deep summer ploughing.

   vii) Avoid irrigation or rain water from one field to other.

   viii) Uproot and burn the downy mildew affected plants.

Resistant varieties

Commercial sunflower lines have been converted into resistant hybrids to all downy mildew races. The sunflower hybrids viz. Velja, Kazanova and Rimi-PR are resistant to downy mildew.

Co-dominant CAPS markers for *Pl6* gene of resistance to downy mildew can be used for efficient identification of plants resistant to downy mildew races.

Table 4.1: Sunflower hybrids/varieties identified in India by ICAR for release based on yield and downy mildew resistance reaction (1996-2004)

Sr.No.	Sunflower hybrid/variety	Year	Remark
1	Sungene-85	1996	This variety was released during AICRP workshop held at JNKVV, Jabalpur in April 1996.
2	LS-11	1998	This variety was released for varietal identification at committee meeting held at TNAU, Coimbatore, in April, 1998.
3	MSFH-47	2000	This hybrid was highly resistant to downy mildew and it was released by ICAR for variety identification at committee meeting held at PAU, Ludhiana, in April, 2000.
4	Pro-009 (Prosun-09)	2003	This hybrid was released for variety identification at committee meeting held at TNAU, Coimbatore, in April, 2003.
5	SH-416	2003	This hybrid was identified for variety release at committee held at TNAU, Coimbatore, in April, 2003.
6	DRSF-108	2003	This variety was identified at ICAR's variety release committee held at TNAU, Coimbatore, in April, 2003.
7	PCSH-243	2004	This hybrid was identified for release in the variety release committee meeting held at ANGRAU, Hyderabad, on May 18-20, 2004.
8	PRO-011	2004	This hybrid was identified for release in the variety release committee meeting held at ANGRAU, Hyderabad, on May 18-20, 2004.
9	SCH-35(Maruti)	2004	Released at Maharashtra State variety release committee meeting held at Bombay in Feb, 2003.

4.2. Downy mildew of soybean

Pathogen: Peronospora manshurica

Geographical distribution

North America: Canada, USA (California, Kansas, Minnesota, North Dakota, South Dakota).

Asia: China,

EPPO region: Italy

Downy mildew of soybeans is rarely of economic importance from a yield perspective, however when pods and seed become infected, results in a reduction in seed quality and marketability.

Symptoms

Small, irregular spots on upper leaf surfaces are initially pale yellow in appearance, later becoming gray brown with a yellowish margin. On the underside of the leaves, the spots have a gray, fuzzy appearance due to the presence of the pathogen. Symptoms frequently occur at low levels throughout the crop canopy. Early leaf spots are non-descript and are commonly confused with leaf spots and pustules caused by soybean rust.

The first symptoms of systemic infection usually are lighter green areas on the lamina near the petiole . The infection quickly spread along the midvein and into the lamina until most of the leaf surface is involved.

The signs and symptoms of systemic infection usually are evident on one or the other of the cotyledons or unifoliate leaves, but occasionally both cotyledons and both unifoliate leaves are involved. Some times the cotyledons are infected but symptoms did not appear in any other parts of the plant.Conversely, symptoms of the disease sometimes appeared in the Early and advanced stages.The development of symptoms of systemic infection by *Peronospora manshurica* on unifoliate leaves of soybean seedlings can occure without the cotyledons showing signs of infection.Symptoms of the disease appeares on the trifoliate leaves as they develope. Many of these leaves developes symptoms similar to those described for the unifoliate leaves; however, on some, symptoms appeared as a mottled, gray-green color. All of the infected trifoliate leaves are conspicuous by their dwarfed appearance. Visible leaf symptoms do not develop on all infected seedlings. However, infection became evident on some 12-14 days old plants when sporulation occurred on the surfaces of one or both of the cotyledons after the plants are exposed to high humidity. Sporulation could be induced seven to ten days after the first visible evidence of infection by placing the plants in a moist chamber for 12 hours.

In local infection the damage to a soybean plant as a whole is slight. Systemic infection causes a marked stunting of all aerial organs. Stems are spindly, internodes are shorter, and leaves are smaller and narrower than normal. Small, irregular spots on upper leaf surfaces are initially pale yellow in appearance, later becoming gray-brown with a yellowish margin. On the underside of the leaves, the spots have a gray, fuzzy appearance due to the presence of the pathogen (Fig.4.4.).

When pods are infected, an encrusted mass of fungal-like growth is visible inside the pods. Infected seed has a dull white appearance and is partially or completely encrusted with the pathogen.

Fig. 4.4: Downy mildew infection on soyabean leaf and seeds

Losses

Although this disease is rarely of economic importance from a yield perspective, when pods and seeds get infected, results in a reduction in seed quality and marketability.

The reduction of soybean yields in Canada was 0.3 thousand metric tons, in china 182.3 thousand metric tons, in Indonesia 5.0 thousand metric tons, in Italy 1.4 thousand metric tons, and in United States 24.1 thousand metric tons.

Disease cycle

Primary infection is initiated by oospores which may be seed-borne as a milky white encrustation on the surface of the seed (Wolf and Lehman 1924; Johnsonand Lefebvre 1942) or soil-borne with infected leaf and pod debris (Hildebrand and Koch 1951). Systemic infection of soybean seedlings may occur when oospore-encrusted seeds are planted (Koch and Hildebrand 1946, Jones and Torrie 1946). The fungus sporulates profusely from the lower surfaces of leaves of such systemically infected plants (Hildebrand and Koch 1945) which serve as a source of initial inoculum for spread of the pathogen in the field after a primary outbreak of downy mildew in the field of soybean. The seconday spread of disease depends on a number of environmental conditions. These conditions, such as light, temperature, humidity, rainfall, dew, and wind, are all involved in sporulation, spore discharge, dispersal, germination, and reinfection.

The downy mildew fungus overwinters in the field as thick-walled resting spores (oospores) in leaf debris and on seed. The oospores serve as the initial source of inoculum; however, wind dissemination of spores(sporangiospores) produced on the lower surface of leaf lesions is the most important means of spread within the field and from field to field. Typically, infection requires the presence of extended periods of dew and temperatures between 10° C and 26.6° C.

Disease Management

In most cases, control of this disease is not warranted and there are no "rescue" treatments available. Seed quality, for sowing in the field, is most important as preventative downy mildew control measures. Following control measures should be adopted for prevention of this disease.

Certified seed: Use certified seed for the sowing

Plant clean seed: Do not plant seed from infected fields. If contaminated seed is planted the next season, the fungus can infect the seedling systemically and cause stunting and mottling of the leaves.

Crop rotation and tillage: The downy mildew fungus survives in crop residue and on the surface of seed. Therefore, crop rotation or deep burial of infested crop residue is an effective way to reduce inoculum.

Chemical control: Metalaxyl effectively controlled downy mildew race 12

Resistant varieties: Use resistant varieties like Beeson, Callend and century.

4.3. Downy mildew of Mustard

Pathogen: *Peronospora parasitica*

Host range

The pathogen infects a wide range of plants including broccoli, brussel's sprouts, cauliflower, cabbage, kale, chinese cabbage, chinese broccoli, chinese mustard, radish, turnip, kohlrabi, swede, water cress, shepherds purse, and many more. A different race of *P. parasitica* causes downy mildew on stocks. The pathogen on brassicas will not infect stocks and vice versa.

Symptoms

The first observable symptom is small, light green-yellow lesions on the upper leaf surface, while corresponding lower surface became faint green yellow. Lesions are angular and variable in size, but are often bounded by large veins. On the bottom sides of leaves a fluffy or downy fungal growth with sporulation appears during cool, moist conditions (Fig. 4.5). Old lesions become necrotic and translucent after invasion by secondary saprophytes. Seedlings when infected may be killed or develop dark brown vascular systems due to severe infections, but older plants are rarely killed. The disease can be devastating under cool, moist environmental conditions.Young seedlings are more susceptible than adult plants and can die from downy mildew if infected while young. The spots turn yellow or light orang as they enlarge. Later the leaf may become papery and die.

On fall-planted seed crops, downy mildew can create tiny leaf spots during the winter months or other times where small clusters of conidia and conidiophores are produced on the underside of the leaf surface; rarely in large enough quantity to be seen without magnification. During warmer rain breaks, individual lesions may rapidly enlarge and then quickly collapse with a secondary soft rot during rainy periods. This can be confused for cold-damaged tissues with subsequent secondary leaf rot, which are also common in areas of Oregon and Washington seed production during certain winters.

Downy mildew infection on an inflorescence can lead to stag head formation, abnormal shape and development of the inflorescence. White rust and downy mildew are commonly found together on leaf colonizing on brassicas in western Oregon and elsewhere.

Fig. 4.5: Symptoms of downy mildew infection on mustard leaves

Peronospora parasitica overwinters in roots or infected plant debris, affects nearly all cultivated and weed plants in the Crucifer family. Strains of the mildew pathogen are generally more aggressive on the crop plant from which they were obtained. For example, the radish strain causes little damage to cabbage. Spores may be spread as contaminants on seeds or by wind or water droplets.

Environmental conditions

The disease is favoured by cool temperatures with an optimum range of 8-16°C. The vegetative spores need water to germinate and can infect seedlings within three hours of contact with a leaf. Atmospheric temperature in the range of 10 -20 °C and relative humidity>90% RH favours disease development.

High humidity, fog, drizzling rains, and heavy dew favor disease development and spread. Optimum conditions for disease development are night temperatures of 7.7°C to 19.4°C for 4 or more successive nights, and day temperature of about 23.8°C or lower. In areas with mild wet winters, such as western Oregon or Washington, downy mildew can continue infecting fall-planted seed crops during the winter months, although only small amount of the total leaf surface area may be colonized on individual leaves, but this allows for a build-up within a field, potentially leading to severe increase in infections of spring growth if spring conditions are wet and mild. Fall and spring downy-mildew management is important in areas of fall-planted seed fields to avoid extensive staghead formation, where an inflorescence became abnormal in shape and development.

Disease Cycle

Downy mildew pathogen *Peronospora parasitica*, can attack broccoli, cauliflower, cabbage, and brussel sprouts. Infection occurs when soil borne resting structures called oospores germinate and produce sporangia under moist, cool conditions. Sporangia are produced on the underside of leaves in the evening and are released during the day as leaves dry. The wind blown sporangia land on leaves and directly penetrate leaves and flowers. The pathogen survives between cole crops in the soil as oospores.

Oospores (sexually produced or survival spores) of the fungus are formed in ageing and dying leaves, and survive in debris in the soil from one crop to the next. The fungus is known to survive on seed and on several species of crucifer weeds. In other cruciferous crops the spores of the downy mildew fungus are the major means of dispersal of the disease, especially once it is established within seedlings crops. Spores are produced overnight and released the following morning in such crops as the air dries out. They are dispersed by wind and rain splash.

Control Measures

Cultural control

- Eradicate cruciferous weeds (wild mustards, etc.) that may harbor the fungus.

- Manage irrigation to reduce periods of high humidity. Subsurface drip irrigation may provide some disease suppression as compared to furrow or sprinkler irrigation. If possible, time irrigations to end before dusk to prevent extended periods of leaf wetness.

- Spring-planted, summer-harvested crops have fewer problems than fall-harvested.

Chemical fungicides

In the downy mildew endemic areas, apply first spray as soon as seedlings appear; repeat three times a week until plants are set out in field. In other areas apply first spray as soon as the disease appears and next sprays at an interval of 10 to 15 days as per requirement. Spraying for mildew requires completely covering the plant. In general, high-gallonage ground application has been more effective than aerial application.

- Actigard at 7.8 to 14.17g/A on 7-day intervals for up to four applications can suppress the development of downy mildew by inducing host-resistance pathways. Begin applications 7 to 10 days after thinning, before disease is present. If disease is present in the area, tank-mix the first application with another fungicide registered for downy mildew. Preharvest interval is 7 days.

- CAA-fungicide (Carboxylic Acid Amides) formulations (Group 40) in combination with another fungicide that has a different mode of action. Do not apply more than once before alternating to a different mode of action

  - Forum at 1680 ml/A on 7-day intervals. Do not apply within 7 days of harvest. 12-hr reentry.

  - Revus at 2240 ml/A on 7- to 10-day intervals. Preharvest interval is 1 day. 4-hr reentry.

- Copper products are not recommended as stand-alone materials.

  - Champ Formula 2 at 156 to 312 ml/A. 48-hr reentry.

  - Cueva at 1.89 L to 7.4 L/378 L water on 7- to 10-day intervals. May be applied on the day of harvest. 4-hr reentry.

Resistant varieties

Resistance is being incorporated into commercial cultivars, but at present no resistant commercial cultivars are available.

5
Downy Mildew of Fiber Crop

5.1 Downy Mildew of hemp

Pathogen: *Pseudoperonospora cannabina*

Geographical distribution

It's spread has been limited to Asia and Europe.

Symptoms

Downy mildew begins as yellow leaf spots of irregular size and angular shape, limited by leaf veins. Opposite the spots, on the underside of leaves, the fungus emerges from stomata to sporulate. Mycelial growth on the underside of leaves is best seen in early morning when dew turns the mycelium a lustrous violet-grey colour (Fig 5.1). Lesions enlarge quickly and affected leaves become contorted. Leaves soon become necrotic and fall off. Whole plants and entire fields may follow this pattern.

Fig. 5.1: Symptoms of downy mildew on hemp leaves and downy mildew growth on dorsal side of leaf

Losses

Economic impact is greatest in warm temperate regions, such as southern Europe, Italy and southern France. The disease is particularly destructive in regions where hemp is continuously cropped without rotation. Reports of disease intensity range from 7 to 10% of the leaves per plant.

Control Measures

Cultural and Mechanical practices

Sanitation: Destroy infected crop debris, and remove residues for burning. Soil solarization under sheets of polyethylene plastic for 4 to 6 weeks may be effective (Barloy and Pelhate, 1962).

Avoid excess atmospheric humidity. Excess humidity is a common problem and permits sporangia and sporangiophores to flourish. Humidity always increases after canopy closure, which is when leaves of adjacent plants touch each other and shade the soil. Canopy closure is unavoidable in fibre crops, but can be avoided in seed crops and seedling beds. Properly spaced plants keep the canopy open, increase air circulation and reduce humidity and leaf wetness. Plant rows in the direction of prevailing winds, or in an east-west or northeast-southwest orientation to promote solar drying. Air circulates better on sloped hillsides or high points than on low or flat ground. Avoid overhead irrigation during flowering.

Proper ventilation is critical so humidity can be reduced. Avoid dew at all costs during flowering. Do not irrigate plants late in the day or at night.

Rotate with other crops for minimum three years. McCain and Noviello (1985) reported two Italian hemp cultivars, Superfibra and Carmagnola Selezionata, as resistance to a strain of *P. cannabina* that destroyed all cultivars.

Biological Control

A unique strain of *Bacillus subtilis* is sold as a foliar spray for controlling downy mildew.

Chemical fungicides

Ferraris in 1935 controlled *P. cannabina* epidemics in Italy with copper sulphate. Hewitt (1998) reported control of *P. humuli* with Bordeaux mixture or copper oxychloride. The undersides of leaves must be sprayed.

6

Downy Mildew of Cash Crops

6.1. Downy mildew of Sugarcane

Pathogen: Sugarcane downy mildew can be caused by four species:

1) *Peronosclerospora sacchari*

2) *P. miscanthi*

3) *P. spontanea and*

4) *P. philippinensis*

Major Host

Hosts in the *Saccharum* genus include: *S. officinarum, S. robustum, S. spontaneum and S. edule.*

Minor Host

Other hosts include *Zea mays* (maize), teosinte (*Euchlaena mexicana*), *Sorghum halapense* (Johnson grass), *S. sudanense* (Sudan grass), *Tripsacum dactyloides* (Gama grass), *Sorghum bicolor* (broom corn). Other studies suggest there are 18 species that are systemically infected in the sub-family Panicoideae; these include species of Andropogon, Bothriochloa, Eulalia, Schizachyrium and Sorghum (Bonde & Peterson 1981)

Geographic Distribution

The disease is restricted to the Pacific, South Asia and South East Asian regions viz Fiji, India, Indonesia, Japan, Papua New Guinea, Philippines, Taiwan and Thailand (Suma & Magarey 2000).

Disease Symptoms

Typical symptoms of downy mildew are leaf streaks of 1-3 mm in width which are separated by normal green leaf tissue. Streaks vary in length and may be just a few centimetres or run the length of the leaf blade. There may be up to 30-40 leaf streaks on individual leaves, though in some varieties there may be relatively few streaks of much wider dimension (Fig 6.1). Streaks can occur on the lower surface of the mid-rib but not usually on the leaf sheath (Hughes & Robinson 1961, Leu & Egan 1989, Suma & Magarey 2000). Cooler conditions may lead to a narrowing of leaf streaks. Initially, leaf streaks are pale creamy-yellow in colour but with age streaks turn yellow and then to a mottled brick-red colour (Fig 6.2) (Hughes & Robinson 1961; Leu & Egan 1989). In some varieties, streaks may be a consistent red colour. When symptom development is intense, the whole shoot may appear mottled-red and discoloured. On warm humid nights, the pathogen produces a soft white velvety growth on the underside of leaves (Fig.6.3) (Hughes & Robinson 1961; Leu & Egan 1989). With time, the downy growth may appear like a grey powder as the conidiophores and conidia dry and age. Initial symptoms in young systemically-infected plant cane are a mottled paleness of the young spindle leaf (Leu & Egan 1989). Surviving stalks and stools are characterised by narrow discoloured leaves with upright habit, and abnormally thin stalks with varying degrees of stunting (Hughes & Robinson, 1961; Leu & Egan 1989). Mature stalks may develop side-shoots in autumn and early winter leading to a witch's broom effect (Leu & Egan 1989). Not all lateral buds will side-shoot and it appears that it is the infected buds that germinate. Oospores (sexual spores of the fungus) develop as winter approaches and as these form they may lead to leaf splitting; as the spores cause a sideways force on the interveinal leaf tissue causing the tissues to separate. Leaf splitting can be a quite spectacular symptom. Often the production of oospores is also accompanied by the rapid elongation of infected stalks which leads to thin, brittle stalks referred to as 'jump ups' (Hughes & Robinson 1961). These weakened stalks may bend but even when they do, they are often higher than the remaining healthy stalks within the canopy. The stunting associated with systemic infection of susceptible varieties can cause losses of up to 40% or more (Rauka *et al.* 2005b; Suma & Pais 1996).

Fig. 6.1: Classic leaf streak symptom in the early stages

Fig. 6.2: Brick red leaf symptoms seen as the leaf ages

Fig. 6.3: White downy growth on the underside of leaves

Epidemics and losses

Periodic epidemics of sugarcane downy mildew have occurred in Taiwan. Between 1960 and 1964, 70% of the fields of one variety were affected (Payak, 1975; Sun *et al.*, 1976). It is not as severe in other south-east Asian countries or in Australia. In Fiji, sugarcane downy mildew at one time caused major losses in sugarcane growing areas, but these were reduced to very low levels by the release of resistant cultivars coupled with an intensive system of disease control (Daniels *et al.*, 1972). In 1965-78, *P. sacchari* infection of sugarcane and maize and host factors were compared in Taichung (nonepidemic), Huwey (sporadic), Hasinying (main damage) and Pingtung (marginal damage) geographical regions of Taiwan.

In Nepal, there was a large epidemic in 1967, followed by a couple more in 1970 and 1973.

Pathogen

Conidiophores of *P.sacchari* emerge through stomata singly or in groups of 2, rarely 3. They are massive and wedge shaped, short, stout structures widening gradually towards the upper portion and dichotomously branching at the tip 2-3 times. Conidia of *P. sacchari* may vary in size but are usually between 25-55 μm in length and 15-25 μm in width (Sivanesan & Waller 1986). Conidia are elliptical or oblong, rounded at the apex, and slightly apiculate at the base with a thin, smooth, hyaline wall. The conidiophores arise singly or in groups from stomata and are erect, thin, smooth, have a hyaline wall and the apex is branched several times (Leu & Egan 1989). Oospores are embedded in the interveinal leaf tissue, are globular and yellow with a wall thickness of 3.8-5 μm. Their diameter is 40-59 μm (Leu & Egan 1989). Sivanesan & Waller (1986) reported haustoria to be bulbous, Conidiophores hypophyllous, 125-190 x 18-25 μm, dichotomously branched twice or thrice at the apex with 2-4 sterigmata on ultimate branches. Conidia ellipsoid to ovoid, base pendunculate or rounded, 25-55 x 15-25 μm. Oospores yellow to brown, average 50 μm diameter with a wall 3.5-5 μm thick.

Disease cycle

The pathogen has two forms of sporulation; asexual conidia produced under high humidity on leaf surfaces and sexual oospores which are produced in infected leaves during the cooler months (Hughes & Robinson 1961; Leu & Egan 1989). Conidia are the primary infective propagule and constitute one of the important transmission mechanisms. The conidia are very fragile and not long lasting. Conidia are produced on humid nights but they loose viability within a matter of hours after sunrise (Hughes & Robinson, 1961). Studies suggest taht the conidia are dispersed by air currents and that they rarely move more than 400 m from an infested crop (Hughes & Robinson 1961). Oospores are not released from leaf tissue until the leaves begin to decay on the soil surface. Except for *P. miscanthi,* no one has shown oospores to have a role in downy mildew infection so their role in the disease cycle remains unclear (Suma & Magarey 2000).The diseases are most prevalent in warm, humid regions. Moist soils favor oospore germination and therefore damp soil as a result of irrigation or reduced tillage techniques will encourage disease development.

Disease management

Cultural method

- Select healthy planting materials.
- Avoid ratooning in infected field.
- Rouging of diseased stools.
- Dig out diseased stools at the early stage of the disease.Careful rouging at frequent interval of each stool, as the first symptom of the disease appears, is recommended.
- Do not plant downy mildew susceptible corm near sugarcane field. Corn is an important alternate host of downy mildew. Susceptible corn can carry the disease from one season to another or allow the causal organism to survive when there are no canes available.

Chemical fungicides

The pathogen may be eliminated from systemically infected planting material by hot-water treatment at 50°C for 2 hours (Hughes and Robinson, 1961, Leu and Egan, 1989).

Metalaxyl may also be applied to setts in commercial cane planters and to both the soil and foliage of infested crops. In ratoon crops, metalaxyl has been applied at 1.8-2.0 kg (a.i.) / ha using a granular formulation (Malein 1993; Eastwood & Malein 1998). The fungicide has a prophylactic effect in the growing crop, protecting the crop from infection for up to 20 weeks post application (Eastwood & Malein 1998).

Disinfect setts before planting. Soaking seedpieces in metalaxyl solution (Ridomil) at 50g a.i/200lit of water for 10 minutes effectively control the downy mildew.

6.2. Downy mildew or Blue mold of Tobacco

Pathogen: *Peronospora tabacina, Peronospora hyoscyami* f.sp. *tabacina,*

Host plant

Nicotiana species are the only known hosts.

Geographic Distribution

The first appearance of blue mould on tobacco was reported by Cooke in Australia in 1891, where it is considered to be the most important disease affecting tobacco both in seedbeds and in the field. The disease was first reported in the United States in 1921, but some people (Stevenson and Archer, 1940) believe that Harkness in 1885 had already observed the parasite and that *Nicotiana repanda*a wild, spontaneous and susceptible species contributed to the spread of the fungus (Godfrey, 1941; Wolf, 1947).

Blue mould was also reported in Canada and Brazil in 1938, in Argentine in 1939, in Chile in 1953, in Cuba in 1957 and in Mexico in 1964.

The pathogen was unknown in Europe until 1958, when it was introduced accidentally into England and spread very rapidly. It was found in the Netherlands and the German Federal Republic in 1959, and in 4 years had colonized the whole of Europe, North Africa and the Near East (Corbaz, 1964). Its progress towards southern Africa was probably stopped by the Sahara, whereas it progressed eastwards and was reported to have reached Cambodia by 1969 (Anon., 1969).

Symptoms

Once established, blue mold is fairly easy to identify, although symptoms vary with plant age. On beds of seedlings with leaves up to 4 cm in diameter, blue mold is first seen as circular, yellow areas on diseased seedlings. Plants in the center of the affected area may have distinctly cupped leaves. Some of these leaves exhibit a gray or bluish downy mold on the lower surface; hence the name *blue mold*. The upper surfaces of infected leaves remain almost normal in appearance for 1-2 days before the plants begin to die and turn light brown. Diseased leaves often become twisted so that the lower surfaces turn upward. Blue mold can affect plants in the field throughout the growing season. Single or groups of yellow spots (lesions) appear on the older, shaded leaves (Fig 6.4). Often the spots grow together to form light brown, necrotic (dead) areas. Leaves become puckered and distorted, large portions disintegrate, and the entire leaf may fall apart. Under continuous favorable weather conditions, blue mold can destroy all leaves at any growth stage. Lesions may occur on buds, flowers, and capsules.

In its early stages, blue mold can easily be confused with cold injury, malnutrition, or damping-off. However, the presence of the characteristic downy gray-blue spores on the undersurface of leaves quickly identifies the disease as blue mold and distinguishes it from other problems.

In severe situations, blue mold may also cause systemic stem infections, resulting in partial or overall stunting of the plant, with narrow, mottled leaves. Discoloration (brown streaks) can be found inside these stems. The plants often lodge or snap off if systemic infections occur near the base of weakened stems.

Fig. 6.4: Symptoms of downy mildew on tobacco leaves

In general, burley tobacco is much more susceptible to blue mold than flue-cured varieties.

Epidemics and losses

Since 1979, blue mold has become epidemic in the field in the United States in some years. In Western Europe, the tobacco crop yield decreased by *27,* 500 tons (CORESTA, 1960) and the financial loss amounted to US $50 million (Todd, 1961). In 1961 losses again accounted for 75% of the crop in Algeria and 65% in Italy (CORESTA, 1961b).

Weather parameters

Extremely wet and cool weather provide favorable conditions for fungus development, spread, and infection. Spore production can occur from 7.7-30°C. Temperatures above 30°C or below 7.7°C restrict spore production. The time from infection to sporulation is typically from 4 to 15 days, but can be considerably longer depending upon day and night temperatures, variety, and strain of the fungus. Night temperatures from 10 to 18.3°C and day time temperatures from 21.1-29.4°C are ideal for disease development. It is important to remember that during most years in Florida ideal temperatures for blue mold occur during much of the early part of the tobacco growing season.Thus, rainfall and irrigation tend to strongly influence blue mold.

Disease Cycle

Once blue mold is present, its development depends on weather conditions. Spores require wet leaves for germination and infection. Cloudy weather increases susceptibility, but sunlight is fatal to spores and stops the production of new spores. Therefore, blue mold is most severe and can develop rapidly during periods of cloudy, wet weather, but stops developing during sunny, dry weather. A 5- to 7-day, symptom-free incubation period takes place before the appearance of the first visible symptoms (yellow lesions). Incubation becomes longer with less than ideal conditions and with the age of the tobacco plants. The latent period for sporulation is generally 5-7 days. Sporulation can occur on the day symptoms first appear, but it usually occurs the following night. Under favorable conditions, a second set of spores is usually produced 7-10 days after initial infection; without chemical control, this cycle may be repeated several times during the growing season, creating a much larger epidemic.

The pathogen is not known to overwinter in the more temperate zones; and therefore it is assumed that inoculum is introduced each year into the U.S. Sporangiospores can be dispersed thousands of kilometers by weather events and are the primary source of inoculum for epidemics. Likely sources of yearly blue mold epidemics are windblown spores from tobacco crops in Mexico and the Caribbean that move northward, or from wild tobacco in the southwestern U.S. It is unclear whether the pathogen is capable of overwintering in infected debris and the role of oospores in disease is not clearly understood.

Another common way blue mold spreads is by the distribution of infected transplants. In some cases, transplants that appear healthy may actually be infected. Farmers periodically buy transplants from distant growers and run the risk of buying diseased plants and introducing blue mold into their region.

Disease management

Cultural methods

(1) Use transplants produced in open non-shaded areas.

(2) Do not seed the transplant bed prior to January 10. The use of a plastic cover maintains heat for accelerated plant growth, and temperatures above 30°C inhibit blue mold development.

(3) Avoid use of excessive amounts of nitrogen.

(4) Irrigate beds when needed. Excessive irrigation will create favorable conditions for blue mold especially coupled with high rainfall amounts.

(5) Inspect fields and beds routinely. Inspect near high trees, hedge rows and low areas as blue mold usually begins in such areas first.

(6) Destroy old plants in the bed immediately after transplanting is complete. Cut and plow down stalks in the field immediately after harvest. Plough down reduces inoculum of several diseases for the following season.

7

Downy Mildew of Fruit Crop

7.1 Downy mildew of grape
Pathogen: *Plasmopara viticola*

History and Epidemics
In France downy mildew is called as "le mildiou" while in Germany it is called as "falscher Meltau". Downy mildew is undoubtedly of American origin and was probably indigenous on the wild grapes of the eastern United States, from which it spread to the cultivated vineyards when these were established by the colonists. It was not known to occur outside America until 1870, although it was first described from this country as early as 1834. In the early part of the last century various horticultural writers referred to the destructive action of the mildew on cultivated grapes.

Its history in Europe dates back to the time when French vine-yards were suffering from the grape *Phylloxera*, an aphid insect pest that was also introduced into Europe from America. It was found that rootstocks of the American species were more resistant to this aphid than was the European vinifera grape, and therefore consequently around 1870 importations of the American Vines were made into Europe, and somewhere around 1875 the downy mildew fungus evidently was also introduced with these importation and became established, and by 1878 it was prevalent enough to cause alarm. Since the European vinifera grapes were far more susceptible to this disease than were the American species, it soon became a real menace to grape culture throughout the humid sections of Europe.

The serious threat to the grape industry in Europe gave rise to a series of investigations by French research workers. The contributions of Cornu (1882), Millardet (1883), Cuboni (1887), and Viala (1893) cleared up many details in the life history of the pathogen, but probably the most significant result was the accidental discovery of the value of a mixture of lime and copper salts in the control of the mildew. Millardet, in studying the disease, noticed that many vines along the roadside at Merloc had retained their foliage, while those farther from the road were completely defoliated. On inquiry he found that the growers were in the habit of sprinkling these vines with a mixture of verdlgris (copper acetate) and lime or with copper sulfate and lime in order to make the grapes unattractive to passers-by who were in the habbit of picking the fruit. Following up this clue he tried various combinations of copper salt and lime and thus introduced Bordeaux mixture.

In America, Farlow in 1876 gave an excellent botanical description of the fungus, including a detailed study of the germination of the conidia and the formation of swarm spores. Between 1886 and 1889 Scribner of US Department of Agriculture issued a series of reports on mildew and black rot. These reports not only recorded numerous experiments in various parts of the country on the control of the disease, but also contained translations of a number of French and Italian articles and many points of historic interest. He followed very closely the results of the French trials of copper compounds, and it is significant that within 5 years after the discovery of Bordeaux mixture in France, many grape-growers in America started using this material for the control of both downy mildew and black rot.

Geographic Distribution

Downy mildew is now found in most regions of the world where grapes are grown under humid weather conditions. Originating in America, it was introduced into France during the seventies and soon spread over the Continent and to northern Africa. It was reported from South Africa in 1907, from Australia in 1917, and from New Zealand in 1926. It is also known to occur in several countries including South America, Japan, China, India, Asia Minor, and Russia.

In the United States it occurs in all regions east of the Rocky Mountains, but is most destructive in the humid regions in the northern section of the Mississippi Valley, along the Atlantic sea board, and about the Great Lakes. It has never become established in the great California grape-growing areas, in spite of the fact that the vinifera grape which is especially susceptible is extensively grown. This is probably due to the dry conditions, under which the fungus does not thrive.

Economic Importance

Loss from downy mildew is more likely to result from the effect on the vine than from direct rotting of the fruit, although the grower is not likely to realize this fact fully. If the disease occurs early in the season, the young bunches of berries may be entirely killed by the attack of the stems as well as of the young fruit by the fungus. If the disease appears later in the season, the berries are not so likely to become infected and the main loss comes through defoliation and killing of leaf tissues, which results in the failure of the fruit to ripen properly.

Fruit from seriously diseased vines has low sugar content. Another serious loss results from the attack of young shoots, which are consequently dwarfed and do not make the canes necessary for bearing the crop. Under very favorable conditions half-grown or older berries may be infected, resulting in the production of unsightly bunches.

In France, before the control of the disease was worked out, it was not uncommon for entire vineyards to be so completely defoliated as to result in a loss of 5%" to 35 % of the crop. Ordinarily the loss from downy mildew is not serious in the regions where rainfall is light and humidity is low. The loss over the period of years is not likely to be great since epidemics occur only during those summers when rainfall is unusually heavy at critical periods. In such situations 50 to 75 per cent of the crop may be destroyed where no protection is given, provided weather conditions are favorable.

Symptoms

The disease may appear on any succulent part of the vine. It is commonly found on the leaves, young shoots, and immature fruit clusters (Fig 7.1). On the leaf it first appears as pale yellow, rather indefinite spot on the upper surface. The translucent character of the young stages of the spot has given rise to the name "oily spots." If the variety attacked has a hairy under-surface it is difficult to detect affected area on this surface, but on some wild species having few hairs the white aerial growth of the fungus can be seen as soon as, or before, the yellow area is evident on the upper surface. The downy appearance of the fungus beneath the corresponding oily upper spots gave rise to the name downy mildew for the disease. Later the leaf tissue is gradually killed and the necrotic lesions become light to dark brown and irregular in outline. The white color of the downy fungal growth turns to a dirty gray. Frequently, when the spots are numerous or the fungus especially vigorous, the dead areas merge into each other so that a large section of the leaf between the main veins may die and turn brown.

Fig. 7.1: Downy mildew symptoms on grape leaves and berries

Young shoots attacked by the fungus are decidedly shortened but are thicker than normal shoots owing to hypertrophy. The whole shoot, including tendril and young leaves, often becomes covered with the white, muting structure of the fungus. After a period of growth in which the whole shoot may become decidedly distorted, the tissues turn brown and die, preceded by the collapse and disappearance of the downy growth on the surface. If the attack is localized on a portion of the shoot or the variety is somewhat resistant, the cane may survive, but it frequently shows the effect of the attack in the distorted appearance of the growth.

Flowers and young fruit clusters may be attacked, in which case usually the entire cluster is invaded, and after the usual appearance of the downy fruiting masses, the entire cluster dies. In some cases only a portion of the cluster is attacked and the remainder may remain alive and develop. If infection occurs on the fruit when young, it is usually covered with the downy growth, but when the attack occurs after the berries are half-grown, the rotting takes on an entirely different pattern. In this case the growth of the fungus is largely internal; the berry turns to a dull green, then to a brown, and finally wrinkles somewhat but does not take the deeply shrunken appearance and the black color of fruit is due to attack by the black rot fungus. This type of fruit rot is often called "brown rot." Berries attacked by the downy mildew inclined to "shell" and are rarely mummified on the vine.

Hosts

Plasmopara viticola attacks all common species of wild and cultivated grapes (*Vitis* spp.). In addition it has been reported on the Virginia creeper (*Parthenocissus quinquefolia*) and on Boston ivy (*Parthenociuus tricuspidata*). The European grape (*Vitus vinifera*) is more susceptible than the cultivated American grapes which are derived from native species. In some hybrids of the American and European species the lack of resistance to downy mildew is evident.

Pathogen

According to Wilson (1907), the first collection of the fungus was made by Schweinitz in America in 1834. In 1848 it was named *Botrylis viticola* by Berkeley and Curtis. De Bary in 1863 redescribed it as Peronospora viticola. In 1886 Peronospora was separated into two genera by Schroeter, and the grape mildew pathogen was referred to the genus Plasmopara by Berlese and De Toni in 1888. Wilson in 1907 subdivided the genus Plasmopara into Rhysolheca and Plasmopara, but this division has not been generally accepted and the pathogen is reffered as Plasmopara viticola.

The mycelium of the fungus is non septate, except occasionally in the older hyphae (Fig 7.2). In the host tissue, where it is intercellular, the hyphae are very irregular, both in a diameter and branching. Often the threads conform to the general outline of the spaces in the cellular structure traversed by the fungus. For this reason the diameter of the hyphae may vary from 1 to 2 μ to as much as 50 to 60 μ within a short space. The mycelial wall is thin, the contents hyaline and granular. Older portion of hyphal growth may be entirely empty of contents, while the advancing young threads are densely filled with the granular protoplasm.

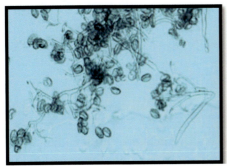 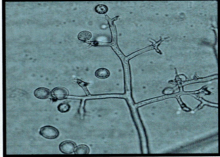

Fig. 7.2: Microscopic photograph of *Plasmopara viticola*

Haustoria are produced in abundance. These enter the cell through a very small opening in the cell wall made by a small tube arising from the mycelium growing alongside the cell. Inside the cell wall this tube expands into a globose structure which presses back the plasma membrane but does not pierce it. Where the haustorial tube enters the cell, the cell wall is decidedly thickened and a collar, composed of remnants of the invaginated cell wall, may be seen around the bane of the swollen haustorium.

A cushion of swollen mycelial segments is formed beneath the stomata preparatory to the production of the conidiophores. From these swollen bodies arise many smaller conidiophores which push upward through the stomata. 1 to 20 conidiophores may arise through a single stomatal opening. The base of the conidiophore is constricted where it passes between the guard cells of the stomata, but swells again at the surface. In some cases the conidiophores may be pushed directly through the epidermis, and in the young fruit they arise from lenticels. Conidiophores are produced most readily in subdued light or at night under humid conditions. On the leaf, with rare exceptions, they are formed on the under surface only. This is due in part to the absence or rare occurrence of stomata on the upper surface.

The conidiophores are 300 to 500 µm long and 7 to 9 µm in diameter when measured above the swollen base. The branching of the conidiophores is characteristic of the species. 4 to 6 monopodial branches arise at rather definite intervals along the upper portion of the conidiophore, the final branch being formed by the terminal portion. These primary branches in turn form 2 to 3 secondary branches. The lower branches may in turn form short tertiary branches. On the final branch 2 or 3 short, blunt sterigmata arise on which the conidia are borne.

The conidium is formed by the swelling of the end of the sterigma, into which the protoplasm from the conidiophore flows until the spore attains about normalize, when it is cut off by a septum.

The hyaline, thin-walled conidium is usually ovoid with a small papilla at the tip which gives it a lemon shaped outline. The conidia are 11 to 18 µ X 15 to 30 µ, but variations both in shape and size are common.

Germination commonly occurs by the formation of zoospores inside the conidium (sporangium). Two other types have been described, one the formation of a germ tube, the other the emission of the entire contents of the conidium, from the extruded mass of protoplasm. When the zoospores are formed within the conidium the protoplasm separates into indefinite masses. These masses then round out into definite zoospores which separate and move about within the conidial chamber. The swarm spores escape through an opening in the tip of the conidium. The zoospores are pear-shaped with a depression on one side where the two cilia or flagella arise. They measure about 7X 9 µ, and the flagella are about 30 µ long.

Oospores are formed in the host tissue. Commonly they are produced in the palisade cells or spongy parenchyma of the leaf very near to the principal veins. They have also been found in berries and in the cortex of the shoots. Oospores are not formed abundantly, as in some other species of the downy mildews. The oospores are 25 to 36 µ in diameter surrounded by the thickened, persistent rough oogonial wall.

The antheredium is a long slender thread which partially encircles the oogonium as per reports of Farlow, however; other investigator have described a clavate structure which appears to be an antheredium, but Gregory states that if an antheridium is present in this species it is certainly rarely found, and in no instance has the fertilization of the oogonium or the formation of a definite impregnating tube been observed, as can usually be seen in other species of this genus. Gregory has observed the germination of the oospore during which a single germ tube is produced which functions as conidiophores (sporangiophore). A single sporangium is formed on the end of this tube i.e sporangiophore, and in this sporangium swarm spores are formed and escape as described above. Arens confirmed Gregory's observations, but states that some oospores have been found to require exposure through two winters before they will germinate.

Disease Cycle

Plasmopara viticola is an obligative parasite, means it develops in the living host tissue. So far as known, it does not over winter in the live dormant parts of the plant. The only known method of overwintering is by means of the oospores, although in Brazil where the grape is evergreen the fungus may continue to produce sporangia throughout the year. The oospores are embedded in the dead leaves which disintegrate during the winter, thus freeing the oospores. These must be subjected to low temperatures before they germinate. The production of sporangia and swarm spores from the germinating oospores probably occurs during rainy periods when drops of water carrying them are splashed on the leaves near the ground.

Sporangia are abundantly produced on the lesions of primary infection, and the disease may then spread throughout the vineyard. The sporangia are easily detached from the sterigmata, and on disturbing the leaves, clouds of these are seen to be carried by currents of air to adjoining tissues. These may also be carried by water or mechanical means. The sporangia germinate quickly, producing myriads of swarm spores, so that reinfection takes place very rapidly under favorable weather conditions. The swarm spore produces a germ tube which grows along the surface of the leaf until a stomata is encountered, when it enters the substomatal chamber through the opening. So far as known, infection takes place only through the stomata, and since these are on the undersurface the all infections occur on this surface. After entering the substomatal portion the germ tube swells and sends out branches, invading the tissue for some distance.

According to Gregory the incubation period of the fungus is from 7 to 12 days on susceptible varieties, but may be as much as 20 days on other varities. At the end of the Season oospore are formed in the old leaves, thus completing the cycle.

Favourable Environment

Downy mildew is of economic importance only in localities where the rainfall is high during the summer month and where humid condition prevails. Since infection is dependent upon the presence of water both for germination and for dissemination of the zoospores, free water must be present on the leaves for a considerable period of time. However, heavy dews may be sufficient to form the necessary water film after the dissemination of the sporangia.

There is some difference of opinion as to the range of favorable temperatures for sporangial germination. The optimum according to Gregory is between 10 and 15.5°C and a minimum about 4.4°C. At optimum temperature germination of sporangia may occur with in ½ hour.

Sporangia may germinate best in diffuse light or darkness, but contrary to this opinion of early workers, germination occurs in direct sunlight, provided a low temperature is maintained. Sporangiophores are almost invariably formed at night or in very subdued light.

Control Measures

Since the only known method of primary infection is by means of the oospores which are in the old leaves, sanitary measures are of great importance in the control of this disease. Removal or ploughing under of the leaves and keeping the canes as high above the ground as practical , serve to reduce the chances of spores being splashed on the new growth in the spring.

Free circulation of air by proper spacing and pruning, thus facilitating rapid drying, is also of importance in reducing the chances of secondary infection.

In regions where downy mildew is prevalent, the selection of resistant varieties to this disease is advisable, but in general, if the loss is not sufficient to warrant sacrificing desirable market qualities the disease may be controlled by chemical sprays.

Cultural practices

These are aimed to reduce the size of the primary inoculum, and to prevent the availability of optimum conditions for their germination and infection. It involves:

1. Destruction of the debris from the soil

2. Growing the crop in well-drained soils and in rows to provide good air movement to keep the leaves dry

3. Avoiding overhead irrigation to prevent wetting of the leaves.

Host resistance

It is not of much significance as most of the cultivars that are used for wine making are susceptible to the disease, though some are more susceptible than others. However, interspecific hybrid cultivars of *V. vinefera* and the North American species have good resistance and also good wine-grape qualities.

Chemical control

Sprays of pre-infection (protective) fungicides, Bordeaux mixture and dithiocarbamates, at regular intervals, right from the early seedling stage, provide good control. Post-infection, systemic fungicides are used scarcely as these generate resistance in the pathogen .Disease forecast models help farmers in the use of the fungicides. The recommended systemic fungicides are foestyl aluminium and metalaxyl. Resistance development in *P. viticola* against Metalaxyl has been reported from several countries, and, therefore, these fungicides have to be used judiciously and as per recommendations of the manufacturers.

In regions where both downy mildew and black rot occur, the mildew is likely to be controlled by the sprays applied for black rot. Bordeaux mixture or other copper sprays are most effective in the control of downy mildew and should be applied under conditions where the disease is likely to cause serious loss. The time and number of applications depend upon local conditions. The first application should be to protect the new foliage from primary infection. In Switzerland, Captan 0.5 per cent is reported to be even more effective than Bordeaux in the control of downy mildew.

7.2 Downy Mildew of melons (Watermelon, Muskmel and sweetmelon)

Pathogen: *Pseudoperonospora cubensis*

Geographical distribution, epidemics and losses

P. cubensis is widely distributed in all continents of the north and south hemispheres where cucurbit plants are cultivated. It mainly occurs in warm, temperate, subtropic and tropic areas on field cultivated as well as on protected (glass house, plastic house and shade house)crops, especially in areas with annual precipitation of >300 mm.

Although distributed worldwide, essential differences among geographic areas may be observed in the occurrence of *P. cubensis*, and the damage it causes (80 % to 90 %losses) to various host species. The highest frequency of *P. cubensis* is apparent on the genus Cucumis. It currently occurs in more than 80 countries on C. sativus and in more than 50 countries on C. melo. In the eastern U.S.A., *P. cubensis* has re-surged as a major problem on cucumbers beginning in 2004. In Europe, P. cubensis is originally common in the Mediterranean region. Recently, it has quickly spread to most European countries, reaching Scandinavia.

It is recorded in approximately 40 countries around the world and the main centre of this distribution is Central America and the Caribbean region. In

Europe, the documented occurrence of *P. cubensis* comes from Yugoslavia and former USSR.

The distribution of the pathogen on the genus Citrullus is even more limited, in about 25 countries, with the main centre of occurrence in Central America. The natural occurrence of *P. cubensis* on Citrullus lanatus is common in Florida, USA but not in Europe or the Middle East.

Symptoms

Downy mildew symptoms first appear as small yellow spots or watersoaked lesions on the upper side of older leaves while corresponding lower side give rise to mildew fungal growth. The centre of the lesion eventually turns tan or brown and dies (Fig 7.3). The yellow spots sometimes take on a "greasy" appearance and do not have a distinct border. During prolonged wet periods, the disease may move onto the upper crop canopy.

Fig. 7.3: Symptoms of Downy mildew on water melon crop

In melons crops, the lesions appear irregular shaped, where these are smaller and round on infected leaves. Expanding lesions on melons leaves are often restricted by leaf veins, giving the lesion an angular or square appearance. As the disease progresses, the lesions expand and multiply, causing the field to take on a brown and "crispy" appearance.

Downy mildew lesions on watermelon leaves appear smaller and round than on cucumber or cantaloupe leaves. Under humid conditions, a downy growth develops on the underside of the light yellow lesions observed on the top of the leaf. This downy growth is particularly noticeable in the mornings after a period of wet weather or when conditions favour dew formation. The downy growth on the underside of the lesions frequently speckled with dark purple to black sporangia (spore sacks) that can be observed with a hand lens. Lesions are sometimes invaded by secondary pathogens such as soft rot bacteria or other fungi. Sporangia (spore sacks) formed in the lesions on the underside of leaf surface appear as black specks.

Due to the rapid spread of this disease and because symptoms often do not appear until 4 to12 days after infection, a successful disease management program must be implemented prior to the appearance of the disease symptoms.

Pathogen

The resting sexual spore i.e Oospore formed during previous season germinate by forming zoosporangium. The whole zoosporangium content is cleaved and biflagellate zoospores, 10-13 μm in diameter, are formed and released. However, there are also reports on direct germination of sporangium by germ tubes. Sporangia easily dislodge from the sporangiophore and are distributed by air or rain splash (similar to other *Pseudoperonospora* spp).
Pseudoperonospora spp. has dichotomously-branched sporangiophores with terminal growth; the sporangia of similar age are present at the ends of sterigma.

Disease Cycle and Environmental relation

Downy mildew is favored by cool, wet and humid conditions. The pathogen produces microscopic sac like structures called sporangia over a wide range of temperatures (5°C-30°C). Optimum sporangial production occurs between 5°C-20° C and requires at least 6 hours of high humidity. The sporangia act similar to spores.

They are easily transferred to healthy plant tissue by air currents or splashing rain. Once they land on a susceptible host, they germinate and can directly infect the leaf within one hour. During prolonged cool wet periods, the sporangia can also burst open and release many zoospores. The zoospores swim through of water along the leaf surfaces towards the stomatas. These natural pores are a primary point of entry for pathogen, resulting in multiple infections on the leaf. This disease may progress slowly or stop temporarily when temperatures rise above 30°C during the day. Night time temperatures of 12°C-23°C will promote disease development, especially when accompanied by heavy dews, fog or precipitation. With night time temperatures around 15°C and daytime temperatures around 25°C, downy mildew infections on cucurbits produce more sporangia within 4 days.

Pathogen Survival and Spread

The downy mildew pathogen is an obligate parasite. It requires living green plant tissue to survive. Killing frosts and cold winters effectively prevent spores from overwintering in the field. However, downy mildew can over winter on living cucurbit plant material growing in greenhouses. Furthermore, greenhouse melon crops and transplants are at risk of developing downy mildew from wind borne sources early in spring, before the field crop has been planted.

Downy mildew primarily over winters in the southern U.S. and Mexico where cucurbits are produced year round.In these areas, the inoculum builds up on susceptible hosts in the early spring. Sporangia are carried long distances by storms and may survive for several days. Once the disease arrives and becomes established in a region, sporangia are disseminated by air currents, splashing rains, overhead irrigation, insects, tools, farm equipment, the clothing of workers and through the handling of infected plants.

Forecasting system

The Cucurbit Downy Mildew Forecast is a web based downy mildew forecasting system that follows the movement of downy mildew from the south to north throughout the growing season and alerts growers to potential movement of the disease into a region. Following the movement of the disease throughout growing season and adhering to the regional disease alerts allows growers to make timely fungicide applications.

Forecasting is an efficient aid in control of *P. cubensis*. The parameters used in forecasting and their integration in protection of cucurbits against *P. Cubensis* are monitoring the occurrence and movement of the pathogen which enables the prediction of disease out breaks in specific areas and the application of suitable control measures prior to infection.

Management Strategies

1. Manage downy mildew using cultural practices integrated with registered fungicide applications.

2. If possible, produce fruits and vegetable transplants in greenhouses dedicated solely to transplant production.

3. Do not produce cucurbit transplants in the same greenhouse which was used to mature greenhouse melons plants.

4. When planting cucurbit transplants, ensure that the transplants are free from disease.

5. Apply a fungicide on field planted transplants prior to installing a row cover or tunnel and immediately after the row cover or tunnel is removed.

6. Select fields and manage the crop to promote air movement and reduce humidity levels inside the crop canopy.

7. Avoid excess over head irrigation. Consider irrigating during the late morning to facilitate rapid leaf drying.

Apply a preventative fungicide prior to an overhead irrigation event. If possible, use trickle irrigation.

8. Scout fields for symptoms of the disease every week or more often if possible.

9. Maintain good weed control in the field. Control alternate weed hosts (wild cucumber, golden creeper and volunteer cucumbers) in neighbouring fence rows and field edges.

10. Follow a preventative spray program.

Under wet and humid conditions, apply a fungicide every 5-7days. When dryer weather occurs, the interval between applications can be relaxed to 7-10 day intervals. Always apply fungicides with at least 250-300L of water per hectare (25-30gal/acre). Ensure adequate coverage and spray penetration into the canopy. Rotate between fungicides from different chemical families.

Use both multisite and singlesite mode of action products.Consider washing equipment and tools before moving from one field to another.Ensure field workers wash their hands before moving from one field to another and, if possible, wear freshly laundered clothing each day.If possible, work in diseased fields at the end of the day.

8

Downy Mildew of Vegetables

8.1. Downy mildew of cruciferous vegetables

Cruciferous vegetables are important Kharif vegetable crops, which are grown both for table and seed purpose. This vegetable group constitutes crops like cauliflower, cabbage, radish, turnip, broccoli, brussel's sprouts; knol-khol and rutabaga .The vegetable crops affected by downy mildews pathogen *Peronospora parasitica are* members of the genus *Brassica* and the downy mildew diseases on these crops are widespread in those regions of the world, which have cool and wet climate.

Geographical distribution

The downy mildew disease on cruciferous crops has worldwide distribution. The disease was first recorded in 1883 in USA on *Brassica* spp. (Farlow, 1883). In India, this disease was first reported by Butler in 1918. Subsequently various authors reported this disease from many countries of the world (Channon, 1981) particularly Angola (Serafim and Serafim, 1968), Argentine (Lindquist, 1939), Australia (Samuel, 1925; Anon., 1955), Austria (Glaeser, 1970), Bermuda (Waters-ton, 1940), Brazil (Grillo, 1937), Britain (Moore, 1959), Brunei (Herb. IMI), Canada (Jones, 1944; Downey and Bolton, 1961), Chile (Mujica and Vergara, 1960), China (Porter, 1926; Pai, 1957), Columbia (Orjuela, 1965), Costa Rica (McGuire and Crandall, 1967), Cuba (Fernandez Rosenada, 1973), Cyprus (Herb.IMI), Czechoslovakia (Rydl, 1968), Denmark (Gram and Rostrup, 1924), Dominica (Anon., 1972a), Egypt (Elarosi and Assawah, 1959), Ethiopia (Herb. IMI), Fiji (Anon, 1969), Finland (Herb. IMI), France (Darpoux, 1945), Germany (Neumann, 1955), Greece (Herb.IMI), Guatemala (Muller, 1950), Haiti (Anon., 1972a), Hong-Kong (Johnston, 1963), India (Thind, 1942), Iran (Ershad, 1977), Iraq (Herb.IMI), Israel (Peleg, 1953), Italy (Ciferri, 1961), Jamaica (Leather, 1967), Japan (Hiura and Kanegae, 1934), Kampuchea (Soonthronpoct, 1969), Kenya (Anon., 1957), Korea (Anon., 1972b), Libya (Herb.IMI), Malawi (Peregrine and Siddiqi, 1972), Malaysia (Mcintosh, 1951), Malta (Herb.IMI), Mauritius (Orian, 1951), Mexico (Rodriguez, 1972), Morocco (Herb. IMI) , Mozambique (Decarvalho, 1948), Nepal (Bhatt, 1966), Netherlands (Thung, 1926b), New Zealand (Jafar, 1963), Northern Ireland

(McKee, 1971), Norway (Ramsfjell, 1960), Panama (McGuire and Crandall, 1967), Pakistan (Perwaiz *et al.*, 1969), Papua New Guinea (Herb. IMI), Philippines (Ocfemia, 1925), Poland (Zarzycka, 1970), Portugal (Da Costa and Da Camara, 1954), Puerto Rico (Anon., 1972a), Roumania (Savulescu, 1960), Sabah (Anon., 1962), American Samoa (Firman, 1975), South Africa (Doidge *et al.*, 1953), Spain (Gonzalez Fragroso, 1924), Sri Lanka (Park, 1932), Sweden (Nilsson, 1949), Switzerland (Gaumann, 1923), Taiwan (Lo, 1961), Tanzania (Herb. IMI), Thailand (Chandrasrikul, 1962), Trinidad (Stell, 1922), Turkey (Herb.IMI), Uganda (Herb.IMI), USA—various states (Chupp, 1930; Weber, 1932; Ramsey, 1935; Kadow and Anderson, 1940; Pound, 1946; Foster and Pinckard, 1947; Shaw and Yerkes, 1951; Borders, 1953; Kontaxis and Guerrero, 1978), Uruguay (Koch de Brotos and Boasso, 1955), USSR (Antonov, 1978), Venezuela (Herb.IMI), Vietnam (My, 1966), and Yugoslavia (Sutic and Klijajic, 1954).

Losses

The losses from this disease can be enormous because it causes infection in nursery beds as well as in the field after transplanting. Systemic infection may occur in the curds/heads and flourish in storage. Under favourable weather conditions, *P. parasitica* may infect up to 50-60 per cent of cabbage seed crop and reduce yields by 16-20 per cent (Vasileva, 1976). Downy mildew can significantly affect the yield and developmental characters like size and weight of silique, number of silique/plant, number of seeds/silique and weight of seeds of radish (Achar, 1992). Seed yields loss can be as high as 58 per cent while infection also affects the size and weight of roots. In addition to the damage caused by the downy mildew itself, the affected tissues become susceptible to secondary bacterial *(Xanthomonas* spp.) and fungal *(Alternaria* and *Rhizopus* spp.) invasions that induce more rapid decay of the heads.

Host range

The pathogen attacks several species of the genus Brassica. Other alternate hosts are *Arabidopsis* spp., *Chenopodium album, Uimelina sativa,, Cheiranthus allioni, C. cherrii, Capsdla bursa-pasteris, Coronapus didymus, Eruca sativa, Iberis amara, Lepidium sativum, Malcomia africana, Matthiola incana, Nasturtium officinale, Raphunus sativus, Radicula nasturtium-aquaticum, Rorippa* spp., *Sinapsis alba, Sisymbrium officinale* and *S. irio* (Channon, 1981; Bains and Jhooty, 1983; Verma and Thakur., 1989; Icochea *etal,* 1994).

General Symptoms

An attack by *P. parasitica* on the seedlings of all common brassica hosts results initially in the development of discoloured spots on the surface of the cotyledons. These spots then turn yellow and later shrivel and die. At such an early stage of growth the loss of the cotyledons may be fatal. Before the seedling dies, however, sporulation of the pathogen occurs, in the form of fine loose carpet of sporangiophores and sporangia which develop mainly on the lower surface of the cotyledons and hypocotyls.

Later infection of hosts such as cabbage, cauliflower, Brussels sprout, turnip, swede and radish appears first as small discoloured spots and yellowing on the upper surface of the true leaves, with a sparse patchy growth of mycelium on the lower surface (Fig.8.1). While sporulation is often profuse on the mycelium on the undersurface of the leaves, it is more delicate and "wefty" than that of the powdery mildews. In slight attacks the infected areas remain discrete, as angular patches on the leaves, but under moist conditions favourable for mildew development, the infection may spread and cause the leaves to shrivel and die. In many situations, infection and partial or complete destruction of some of the leaves may be the total expression of the disease in the field.

Fig. 8.1: Downy mildew symptoms on cruciferous vegtables

Symptoms on Cauliflower

This disease more seriously attacks young plants than the adult plants. The characteristic symptoms appear as angular translucid spots in the inter-veinal spaces of the leaves which are purplish brown on the underside while corresponding upper surface is tan to yellow in colour. In wet weather, the downy growth of the fungus bearing sporangiosphores and sporangia also appears on the under surface of the leaves. In many situations, partial or complete infection of the leaves may be the total expression of the disease in the field. Infection of cauliflower curds has also been observed in fields (Bains *et al.,* 1981). The affected curds look brownish at the top, which later turn dark brown to black, also the secondary rotting of the tissues by bacteria results in destruction of the curd.

Symptoms on Cabbage

When the fungus enters the stalk at the leaf base of an old head of cabbage, a greyish-black discolouration occurs. In some storage lots of cabbage this discolouration has been found extending up through the stalk to the innermost bud leaves. On cabbage heads, the pathogen may cause numerous sunken black spots, varying in size from minute spots to an inch or more in diameter (Sherf and Mcnab, 1986). Extensive greyish-black discolouration spreads through the heads and even penetrates the parenchyma of the stem. The affected cabbage tissues are very susceptible to attack by secondary bacteria and fungi, e.g. *Erwinia, Alternaria* and *Rhizopus* spp.

Symptoms on Broccoli

Downy mildew appears first on the lower leaves of the broccoli plants. The first symptoms of leaf infection are small water-soaked spots surrounded by a halo of light green tissue on the under surface of the leaf. Under conditions favourable for development of infection, the spots enlarge to form indefinite yellow areas. Later, the tissues within these infected areas collapse and become light brown and parchment-like. The mildew lesions vary in shape and size. The largest lesions are usually bounded by leaf veins. The initial spots of infection may also be localized. The tissues of the spot collapse to form a small brown lesion. Systemic infections are usually confined to the upper portion of the main stalk and to the branches leading to the florets of the head. Infected lesion develops brown to black netted lesions, and in others as long strands of discoloured tissues. In some plants systemic invasion of the head can be detected by diffuse blue to purple areas on the stalk and branches of the head.

Symptoms on Radish

Infection of the stem and inflorescences of radish may also occur (Butler and Jones, 1949). The roots of radish may also be attack by *P. parasitica* (Ramsey *et al.*,1954), causing a brown to black epidermal blotch or streak extending around the circumference of the "roots", accompanied by slight russeting and cracking. The internal tissues are extensively explored by the fungus, and though remaining firm, are discoloured grayish brown or black.

Symptoms on Turnip

The pathogen also parasitizes the "roots" of turnip (Gardener, 1920). The stored turnips develop dark discoloured zones spreading down from the crown into the stealer regions of the "roots". The affected tissues vary in colour from light brown to black, and though initially firm, later become predispose to rotting by *Rhizoctonia* and soft rotting bacteria.

The Pathogen

The fungus responsible for this disease is *Peronospora parasitica* (Pers.) Fr. which is considered to be a composite species consisting of many smaller morphological and biological species. The species occurring on *Brassica campestris* was often called *Perenospora brassicae.* More recently, the fungus attacking crucifers has been named *Hyaloperenospora parasitica* on the basis of hyaline sporangiophores, recurved sporangiophores branch tips and rDNA sequence comparisons (Constantinescu and Fatehi, 2002).

The fungus is an obligate parasite. The mycelium of the fungus is strictly intercellular with large, finger-shaped intracellular haustoria, which become clavate and branched and nearly fill the cell cavity. Bains *et al. in* 1981 reported the size of haustoria from cauliflower infected plant as 19.5 - 51.7 x 9.9 - 27.0 μm and are mostly bulbolus, rarely lobbed. The sporangiophores are produced during darkness, initially as unbranched hyphae from aggregations of intercellular mycelium beneath the host epidermis. They are divided into primary and secondary branches (dichotomously branched), which ultimately bifurcate to form sterigmata which bear the single sporangium at the tip. The sporangia are broadly oval, ellipsoidal and hyaline and they fall off mainly by hygroscopic twisting of the sporangiophores in response to change in atmospheric humidity. In cauliflower leaves, sporangia measure 25.6-33.8 x 22.5-29.5 μm.

Oospore formation is favoured by conditions which induce senescence of the host tissues. There is evidence of both homothalism and heterothalism in *P. parasitica.* Sheriff and Lucas in 1989 studied the cytogenetics of heterothalic and homothalic isolates of this fungus and reported that self fertility of the isolates may be due to the presence of a third mating type allele on the 5th chromosome, a condition known as secondary homothalism. The oospores are globose measuring 26 to 43 μm in diameter. They are enclosed in crest like folds and appear pale yellow in colour and germinate by germtube.

Variability in the Pathogen

Specificity in the downy mildew fungus on crucifers is very complex, since it occurs on a wide range of wild hosts as well as cultivated species. *P. parasitica* is highly specialized and seldom occurs in the same biological form on more than one crucifer. A turnip isolate is able to infect seedlings of turnip but not rutabaga or radish.

Earlier *P. parasitica* was divided into three biological strains, viz.1). f.sp. *brassicae,* which attack *B. oleracea, B. napus, B. rapa, B. nigra, B.juncea, B. tournefortii* and *B. fructiculosa; 2).* f.sp. *sinapidis,* which attack S. *arvensis* and *S. alba; and 3)*. f.sp. *raphani,* which attack and cause disease in *R. raphanistrum* and *R. sativus.*

Felton and Walker (1946) differentiated races of *P. parasitica* found on *R. sativus* and *B. oleracea* on the basis of their host specificity. Natti *et al.* (1967) reported that the predominant physiologic races of *P. parasitica* pathogenic to broccoli and other types of *B. oleracea* grown commercially in New York were race 1 and race 2. The latter race was pathogenic to plants resistant to race 1. Bains and Jhooty (1983) reported that *P. parasitica* isolates from different hosts i.e. from *Brassica, Raphanus, Eruca* and *Sisymbrium* vary and were not cross infective. Mehta and Saharan (1994) studied stagheads of 6 host species on 17 host differentials and classified into two distinct pathotypes, one from cauliflower and the other from oilseed *Brassica.* There was no significant difference between the isolates in percentage of spore germination.

Disease cycle

The downy mildew pathogen perennates as oospores in senescenced host tissues and in *Albugo Candida* induced malformed inflorescence. It also survives as sporangia on leaves and inflorescence and as latent systemic mycelium in seeds or infected plant debris (Sahran *et al.,* 1997). Infections are favoured by low temperatures and high atmospheric humidity following rain or dew. The penetration is usually direct but occasionally also occurs through stomata. Primary infection occurs due to soil borne oospores while sporangia released from sporangiophores found on the cotyledons or hypocotyls favour secondary spread. Water droplets also helps pathogen dispersal over short distances.

Weather Parameter

This is a polycyclic disease, which initiates from the low level of initial inoculum and then increases exponentially through successive cycles on the host during the growing season. Among the factors which markedly influence the development of downy mildew and Epidemics of this disease are governed by the prevalent environmental factors like air temperature and relative humidity. At 15°C, sporangial germination, appressoria formation and penetration of the host take place in 4, 6 and 12 h, respectively. Felton and Walker (1946) found that in pre-penetration stage, sporangial germination was most rapid at 8-12°C while penetration of the host by infection hyphae and formation of the haustoria occurred most rapidly at 16° and 20-24°C, respectively. For disease development a similar temperature optima of 8-14°C has also been reported by d' Ercole (1975). Saharan *et al.,*(1997) observed that lower temperature of 15-16°C results in slower growth of both the host and the pathogen, less damage to the host,while temperature above 18⁰C favours more prolific sporulation, more reinfection and consequently more profuse disease development.

The discharge of sporangia from diseased leaves of Chinese cabbage showed a periodic cycle each day (Lin, 1981). Sporangial release increases steadily after 2 a.m. each day and reaches a peak between 6-8 a.m., decreasing rapidly after 8 a.m. Few sporangia can be detected from noon to 10 p.m. The discharge of sporangia is favoured by temperature of 18°C and RH < 75 per cent. Shao *et al.* (1990) reported that in favourable weather conditions (temperature and RH), sporangial release is three times higher in the morning than afternoon.

Cauliflower plants, which are deficient in potash, suffer most from this disease. Cabbage plants grown in soil fertilized with less potash and more phosphorus are more prone to the attack of downy mildew than in unfertilized soil. However, no consistent effect of fertilizers on development of downy mildew on crucifers has also been observed by Felton and Walker (1946).

Management

Since the fungus survives in the form of oospores or on perennial hosts, the management can be attempted through integration of cultural practices with fungicidal cover and host resistance.

Cultural practices

Destruction of crop debris and perennial weed hosts, crop rotation with non-cruciferous crops, avoidance of dense planting to reduce relative humidity around plants, use of clean seed (without fragments of pods which may contain oospores) are some of the cultural practices recommended for the control of this disease (Sherf and Macnab, 1986; Singh, 1987).

Host resistance

Clear evidence has been obtained on the physiologic specialization (pathogenic variability) of *P. parasitica* at the generic and species levels of the host. A list of cultivars/lines of different crucifer vegetables reported resistant is as under:

In Cauliflower

Cultivars/lines like Igloo, Snowball Y, Dok Eglon, RS-355, PI 181860, PI 188562, PI 204765, PI 204768, PI 181860, PI 188562, PI 204779, PI 241612, PI 261656, 291567, PI 373906, KPS-1. PI 231210, PI 189028 PI 208474, CC 3-5-1-1 PI 246077, Early Winters White Head, Pusa Hybrid -2, and Pusa Kartik Sanskar (Kontaxis *et al.* 1979; Thomas and Jourdam, 1990; Sharma *et al.* 1991; Hoser *et al,* 1995; Mahajan *et al.* 1995; Sharma *et al.* 1995, Anonymous, 2012) have been reported as resistant to this disease. Lines Kuwari-17, Kuwari-8, Kuwari-4 and First Early Laxmi exhibited moderately resistant reaction (Pandey *et al.,* 2001).

In Cabbage

Cabbage Cultivars/lines like January King, Balkhan, Spitz Kool, PI 246063, Copenhagen Market Express, PI 263054, PI 263057, Pi 357374, PI 418784, PI 418985, PI 418986, PI 418987, PI 418988, PI 246077, PI 245013, Tromchuda cabbage, Algarvia, PI 245015, and Geneva 145-1 (Elenkov, 1979; Sherf and Macnab, 1986; Hoser *et al.* 1991; Carvalho and Monteiro, 1996) have been reported resistant to this disease.

In Broccoli

Cultivars/lines viz., Calabrese, Grand Central, PI 231210, Italian Green sprouting, Hybrid 1230, Green surf, 2804, Hyb. 2805. Hyb. 2803, GSV 82-4310, XPH 1117, Hyb 288, AVX 7361, PI 263056, PI 263057, PI 3573, PI 3574, PI 418984, PI 418985, PI 418986, PI 418987, PI 418988, OSU CR, Citation, Excalibur, Nancy, and Ganesh broccoli (Laemmlen and Mayberry, 1984; Sherf and Macnab, 1986; Hoser *et al.,* 1991; Anonymous, 2012), have been reported resistant to this disease.

In Radish

Cultivars like Okura, Tokinishi (All season), Baniba, NoirLon, d'Horloge, and Rave a Forcer (Bonnet and Blancard, 1987; Silvae *et al.,* 1996) have been reported resistant to this disease.

Chemical control

(i) Seed treatment

Fungicidal seed treatment followed by foliar sprays is a common practice to control this disease. Metalaxyl seed treatment @ 0.3 - 0.6g a.i. /kg reduced downy mildew infection on broccolli (Paulus and Nelson, 1977) and a significant yield increase was observed when plants raised from such treated seed were sprayed once or twice with the same compound. Seed treatment with Apron SD (35%metalaxyl) controlled downy mildew of cauliflower for more than 2 weeks after sowing (Crute, 1984). White *et al.* (1984) observed that seed treatment with Apron SD (15g metalaxyl/kg seed) gave complete control of this disease on cauliflower when inoculated 10 days after sowing. Following seed treatment, metalaxyl was detectable in the cotyledons, true leaves and roots of cabbage seedlings upto 4 weeks after sowing.

In summer cauliflowers, good control was achieved by drenching the compost with propamocarb, fosetyl-aluminium foliar sprays and by applying a dichlofluanid foliar spray programme (Davies and Wafford, 1987).

Foliar sprays

During early period (1920 to 1960's) reliance for control of this disease rested on frequent application of spray/ dust of chemicals such as chloranil, copper based material (Bordeaux mixture), zineb, spergon, dichlofluanid and propineb (Whitwell and Griffin, 1967). Later other non-systemic fungicides like captafol, mancozeb, copper oxychloride, have been found to be superior to other fungicides on a large number of crucifers at several locations.

The introduction of systemic fungicides like metalaxyl during 1980's provided effective control of this disease either alone or in combination with non-systemics. Ryan *et al.* (1977) reported the efficacy of fosetyl- aluminium, metalaxyl + mancozeb, cyprofuram and propamocarb in the management of this disease on cauliflowers. Later oxadixyl plus mancozeb, cupric hydroxide and chlorothalonil gave significantly better protection from this disease than mancozeb alone (Brophy and Liang, 1992). Neutralized phosphoric acid (2 applications of 2.4 kg a.i./ha, 21 and 7 days before harvest) sprays applied on to the cauliflower in the field within 3 weeks of harvest reduce curd infection of downy mildew under storage conditions. There was no effect of phosphoric acid on crop appearance and maturity. The maximum phosphonate residue in curds at harvest was 12μg/g which was considered a safe limit (Mckay *et al.,* 1992).

Three sprays of Ridomil 25 WP @ 2kg/ha at weekly intervals begining from 28 days after transplanting controlled this disease effectively on Chinese cabbage and provided 65 per cent more marketable yields (Yang *et al.,* 1983). Due to risk of resistance development, metalaxyl in prepacked mixture with mancozeb (Ridomil MZ) is preferred for the control of downy mildew on various crops. Four sprays with Daconil (0.1%), Dithane M-45 (0.2%), Ridomil MZ (0.2%) or Alliette (0.3%) at 8-10 day intervals were highly effective in controlling this disease on radish and root yield was significantly higher in sprayed plots besides significant reduction in the apparent infection rate (r) and the basic infection rate (R) of downy mildew in treated plots (Sharma and Sohi, 1982; Sharma, 1983). Gupta and Shyam (1994) studied the erradicant activity of eight fungicides and found metalaxyl + mancozeb (0.25%) and cymoxanil (0.03%) + mancozeb (0.2%) more effective over a 14 days period on cabbage seedlings.

8.2. Downy midew of cucurbits

Cucurbits, form an important group of vegetable crops in the family cucurbitaceae, and is cultivated in different parts of the world. The family cucurbitaceae contains at least 9 genera and of them 16 species are cultivated as vegetables. In India, cucurbits like ash gourd, bitter gourd, bottle gourd, cucumber, sponge gourd, long melon, or snake cucumber, kundru, muskmelon, pointed gourd, pumpkin, ridge gourd, snake gourd, squash, round gourd and watermelon are cultivated round the year in one or the other region of the country.

Downy mildew is an important disease of cucurbitaceous crops particularly in those areas of the world having adequate moisture, due to intermittent rains and moderate temperatures during the crop period. The pathogen is confined to cucurbitaceous crops only and during favourable weather conditions, this disease can cause considerable yield losses. It occurs practically on all the cultivated cucrbits in open field and protected crops under glasshouse/ plastic houses (Palti and Cohen, 1980).

Pathogen: *Pseudoperonospora cubensis*

Geographical distribution

The disease was first recorded in Cuba during 1868 by Berkeley and Curtis (Chupp and Sherf, 1960). Subsequently it was reported from Japan, U.S.A., Germany, Massachusetts, Czechoslovakia, Ceylon, South Africa, Kenya, Porto Rico, Queensland, Palestine, Sierra Leonne, Mozambique, Israel, Serbia and Yugoslavia, Nicaragua, Rhodesia, Turkey, South Australia, Alta Canada, Bulgaria and during 1984-85 in Poland (Robak, 1995).

Losses

Losses are directly related to the length of the lag period between planting and onset of the disease. If the outbreak is earlier, the losses may be enormous. Prevalence of unfavourable weather conditions for plant growth may also increase losses due to this disease. Overhead sprinkling irrigation is usually associated with higher losses than those encountered where trickle irrigation is used (Cohen, 1980). In North Carolina cucumber production was reduced by 30 per cent due to the infection of downy mildew (Ellis and Cox, 1948) while Mahrishi and Siradhana (1988) reported 79.06 per cent reduction in muskmelon fruit yield in Rajasthan when no sprays of Dithane M-45 were applied. They further reported that the losses are directly proportionate to disease intensity.

Symptoms

Downy mildew affects plants of all ages. Generally the symptoms appear as pale green areas on the upper surface of the leaves, which are separated by islands of darker green colour. Soon the colour of infected areas changes from pale green to yellow angular spots bounded by leaf veins (Fig 8.2). During moist weather, the corresponding lower surface is covered with a faint purplish to greyish black fruiting layer of the fungus containing sporangiophores and sporangia. Lesions on the upper leaf surface turn necrotic brown from the center outwards. Mostly the leaves, which are located in the centre of the vine, are attacked first and then disease progresses upward and downward until the entire vine is killed. Also it causes reduction in photosynthetic activity early in plant development, resulting in stunted plants and yield reduction, especially in cucumber. Premature defoliation may also result and the fruits may get sunscald due to over exposure to direct sunlight. Symptoms of downy mildew infection exhibit themselves differently on the various cucurbit crops.

Downy mildew causes a variety of symptoms depending on cucurbit type. On cucumber, during periods of leaf wetness from dew, irrigation or rainfall, incipient lesions is formed underside of the leaf which become conspicuously water-soaked. This is the earliest symptom produced by the disease, but will disappear as moisture dissipates. Yellow, irregularly shaped lesions confined by the small leaf veins appear soon after on the top of the leaf. . These lesions then turn brown and may drop out of the leaf. The "checkerboard" arrangement of lesions is characteristic of cucumber downy mildew. The underside of an infected leaf reveals downy, fuzzy growth, arising within the leaf veins. The hue of the sporulation ranges from colorless to gray-brown to deep purple. The color depends on density and age of the sporangia that darken with age. Severe infection results in leaves that are completely dead and curled up. This symptom has been described as "wildfire" as the leaves appear to be burned.

Fig. 8.2: Downy mildew symptoms on cucurbits

Primary lesions on cucumber are angular due to their limitation by leaf veins but on melon (*Cucumis melo*) and other cucurbits no such limitation of lesion growth is seen. The fruits are rarely affected but heavy disease pressure prevents fruit maturation and reduces flavour and quality.

The Pathogen

The fungus responsible for causing cucurbit downy mildew is Pseudoperonospora cubensis (Berk, and Curt.) Rostow. The mycelium is hyaline, coenocytic and intercellular, developing abundantly within the leaf mesophyll layer but also penetrating to the palisade layer. Individual hypha have an oval shape, with intercellular haustoria formed at finger-like branches; hyphal size is 5.4- 7.2 µm in diameter. Sporangiophore develops in groups of between 1 and 5, produced through individual stomata. These are 180-400 µm long and 5-7 µm wide with a bulbous base and are dichotomously branched in the upper third of their growth. Sporangia form singly on subacutely shaped sporuliferous tips, are pale gray to purple in colour, ovoid to elliptical in shape, possess a thin wall with a papilla at the distal end and measure 20-40 x 14-25 µm. They germinate to produce biflagellate zoospores 10-13 µm in diameter. Rarely the sporangia of *P. cubensis* germinate directly by forming a germ tube. Oospores are produced rarely and under some specific environmental conditions. The oospores are spherical (19-22 µm diameter), rarely obovoid to ellipsoid, light yellow to pale yellow with a smooth wall of 1.5- 3.5 µm thickness (Mahrishi and Siradhana, 1984).

Host range and host specificity

Various workers from different parts of the world reported that *P. cubensis* infecting cucurbits differed in its host range (Ellis, 1951; Hughes and Van Haltern, 1952). The host range includes *Bennincasa hispida, Bryonopsis laciniosa* var. *erythrocarpa, Citrullus vulgaris, C. vulgaris* var. *fistulosus, Coccinea indica, Cucumis angulatus, C. anguria, C. callosus, C. dipsaceus, C. melo, C. melo.* var. *dudaim, C. melo* var. *utilissimus, C. sativus, C. sativus.* var. *angulicus, Cucurbita maxima, C. moschata, C. pepo, C. Pepo.* var. *ovifera, Echinocystis lobata, Lagenaria siceraria,var. vulgaris, Luffa aegyptiaca, L. acutangula, L. cylindrica, Melothria maderaspatana, Melothria scabra, Momordica balsamina, M. charantia, Sicyos angulatus, Trichosanthes anguina* and *T. cucumerina* (Hoerner, 1940; Bains and Jhooty, 1976b; Mahrishi and Siradhana, 1988a).

Bains and Jhooty (1976b) from Punjab reported pathological specialization in the downy mildew pathogen, as the muskmelon *(Cucumis melo* L.) isolate of *P. cubensis* did not infect ashgourd *(Bennincasa hispida)* and pumpkin *(Cucurbita moschata* Dush.), while the reverse was possible. Round gourd *(Citrullus vulgaris* var. *fistulosus)* was not observed to be infected under natural conditions of Rajasthan but infection occurred on artificial inoculation with sporangial suspension taken from muskmelon (*C. melo)* and ridgegourd *(Luffa acutangula)*

Bains and Jhooty (1976b) reported that Watermelon and ashgourd were not infected by the muskmelon isolate of *P. cubensis* under Punjab conditions but they were potential hosts in Rajasthan (Mahrishi and Siradhana, 1988a), while watermelon is a host in Mysore.

Differences in the severity of infection on different crops indicate presence of physiological races in the *P. cubensis*. Bittergourd, a non-susceptible crop in other regions is heavily damaged in Assam. Five pathotypes infecting watermelon, muskmelon, bottlegourd, *Cucurbita pepo* and *Luffa aegyptica* were distinguished based on high compatibility with specific hosts (Bains and Parkash, 1985). Inoculation of 26 cultivars representing 13 species and subspecies within 7 genera of Cucurbitaceae in Israel, USA and Japan revealed distinct physiological specialization among pathogen's isolates (Thomas *et al.,* 1987a). Five isolates of *P. cubensis* could be distinguished based on high compatibility with specific hosts and all the five pathotypes were highly compatible with *Cucumis sativus* and *C. melo* var. *reticulum.*

Disease cycle

The pathogen mainly survives as mycelium and sporangiophores in the off-season on living hosts, as the oospores are not common. The pathogen can survive on cultivated and wild cucurbits throughout the year and between cropping periods in vegetative state and sporangia on cucurbits that survive cold weather (Sherf and Macnab, 1986). In Denmark, in the absence of oospores, zoosporangia have been reported to survive below freezing temperature (-18°C) for 3-4 months and may serve as the overwintering structure in northern latitudes (Lange *et al.,* 1989). Wherever the oospores are produced, they may explain the mode of perenation but their role in initiating the disease on cultivated cucurbits has not been elucidated (Cohen, 1980). Oospores have been found in cucumber in Japan (Hiura and Kawada, 1933) and China and in pumpkin, bottle gourd and sponge gourd from Rajsthan, India (Maharishi and Siradhana, 1984; Singh and Sokhi, 1989). These have also been recorded in wild cucurbits from India (Khosla *et al.,* 1973; Bains *et al.,* 1977) and in older infected leaves of greenhouse cucumbers at the end of the season in Austria (Bedlan, 1989). However, their role in acting as primary source of infection for the next season has not been elucidated. The incidence of downy mildew increased five folds over the years when the crop was continuously grown in the same area year after year (St. Anand and Weliner, 1991).

In India, Butler (1918) reported the possibility of perpetuation of *P. Cubensis* in wild cucurbits in regions with suitable climatic conditions. Similarly, Bains and Jhooty (1976a) also proposed that in Punjab and possibly in North India this pathogen survives as active mycelium on plants of *Luffa aegyptiaca*, since

90   The Plant Mildews

during their earlier studies they could not observe oospores on any of cultivated or wild cucurbit, a host of the fungus. However, Khosla *et al.* (1973) and later Bains *et al.* (1977) were able to record oospores on parwal *(T. diocia)* from Madhya Pradesh and on *Melothria maderaspatana* (a weed) from Punjab, respectively. The oospore may serve as a mode of overseasoning of the fungus in those regions where they are produced but yet to be proved and confirmed.

In the presence of free moisture on leaf surface, sporangia germinate, by producing zoospores. Zoospores become encysted after swimming for some time in film of water. After germination, the germ tube penetrates the stomatal openings and this is succeeded by the production of intercellular hyphae and haustoria. Symptoms appear 3-7days after infection on which secondary inoculum is produced which is blown by wind to the neighbouring plants/ fields.

Weather Parameter

The pathogen might be overwintering in mycelial form in wild and self sown cultivated hosts under protected conditions. Under favourable weather conditions sporangia are produced which are blown by wind to cause infection in spring and early summer. Sporangium concentration in the air seems to be the most important biotic factor determining disease onset. Disease appeared at cotyledonary stage i.e. 8 days after sowing in plants growing near by infected fields (Cohen and Rotem, 1971). When no such inoculum source was available, disease occurrence was noticed 7 weeks after sowing. Dew is another important factor, which provides free moisture on the leaf, there by providing ideal conditions for sporangial germination and infection.

Bains and Jhooty (1978a) studied behaviour of *P. cubensis* on muskmelon in Punjab. They reported that the pathogen started producing sporangia before midnight and the sporangia mature by 3 A.M. Maximum dispersal of sporangia occured between 6 and 10 A.M. Maximum sporulation occurred at $20 \pm 2°C$ in the moist saturated atmosphere and optimum temperature for germination of sporangia and infection was 20°C. Temperature above 35°C arrested infection as well as sporulation and lesions become brown, necrotic. Zoospores are motile for longer periods at 10 to 15°C than at 30 to 35°C (Bains and Jhooty, 1976a, 1978a). Relative humidity above 75 per cent was conducive for disease development. Maximum distribution of sporangia was recorded at a height of 25 cm above the ground level and decreased with the increase in the height from the ground (Sharma *et al*, 2003). Apparent infection rate (r) showed positive correlation with temperature and moisture conditions (Maharishi and Siradhana, 1988c). Ullasaand Amin (1988) observed that day temperature of 25-30°C, night temperture of 15-21°C and RH > 75 per cent favoured the infection of *Luffa arutangula* by *P. cubensis.* Singh *et al.,*(1990) observed the influence of post -infection high temperature on disease development in muskmelon cv. Punjab Hybrid and found that continuous exposure of plants at 35 ± 1°C for more than 48 h or even a short exposure of 12 h at 40 ± 1'°C was lethal as it eliminated already established infection. Alternation of high temperature with low temperature had less effect on survival of *P. cubensis.* Maximum sporulation was recorded in cucumber leaves when exposed to 20: 15°C day: night thermo periods. Exposure to higher temperatures decreased sporulation because of rapid necrotization of the infected tissues (Cohen and Rotem, 1970; Singh and Singh, 1994). Singh *et al.* (1996) reported maximum temperature and wetness duration during SMW 21 to 24 as key limiting factors in disease development. A function of these two parameters, HTR (humid-thermal relation) has been used in developing prediction system for muskmelon downy mildew in Punjab. Moderate temperature (23°C) coupled with high rainfall and relative humidity (>80%) helped significantly in disease development on cucumber (Sharma *et al.,* 2003).

Low nutritional status of the muskmelon plants pre-disposes them to infection by *P. cubensis* while disease development was less on plants grown in high P, low K and high N nutrition solution (Bains and Jhooty, 1978b). Maharishi and Siradhana (1988b) also observed reduced incidence of disease in plants supplied with higher doses of P, but its lower concentration with N and K increased disease intensity. Trace elements like Zn and Cu also reduced the disease significantly.

Disease Forecasting

Two types of forecasting systems have been developed for cucurbit downy mildew diseases. One uses data on environmental conditions within a crop to schedule fungicide applications when conditions are favorable for disease development. This approach is typical of most disease forecasters. Another uses reports of disease occurrence and models pathogen movement from these areas to other locations. This system differs from the previous one in that it uses macro-scale weather patterns, rather than site-specific weather, and it predicts pathogen dispersal.

Regional occurrence of downy mildew in the eastern United States is being forecasted based on disease observations and modeling long-distance movement of the pathogen with large-scale weather systems (http://www.apsnetorg/ online/ feature/forecast/top.htm). The pathogen, *Pseudoperonospora cubensis,* is believed to survive winters south of the 30th latitude and from there is reintroduced to the north each year. Disease reports are provided by a network of approximately 40 plant pathologists and horticulturists representing Mexico

92   The Plant Mildews

and various states. Weather fore-casting models and pathogen epidemiology are then used to determine the likelihood of disease outbreaks along weather trajectories. The model utilizes factors important to sporulation at the source, survivability during transport, and deposition from rainout and washout. The HY-SPLIT trajectory model from NOAA's Air Resources Laboratory is used to calculate spore transport in the atmosphere. Twice weekly forecasts are made of the likelihood of inoculum spread and disease risk 48 hours into the future. Risk is described as high, moderate, or low. Disease reports and forecasts are placed on the World Wide Web (http://www.ces.ncsu.edu/depts/pp/cucurbit/; Gerald J. Holmes, North Carolina State University, Raleigh, NC). This system is related to the Tobacco Blue Mold forecaster developed previously for a biologically similar pathogen.

A forecasting system has been developed for downy mildew in Ukraine. This system uses temperature, leaf wetness duration, relative humidity and concentration of sporangia in the air to schedule fungicide applications. This system is recommended for use as a component of a management program that also includes resistant cultivars and highly effective systemic and contact fungicides. With this program, the amount of fungicide can be halved for managing downy mildew, which has been one of the most widespread and economically important diseases of cucumber in Ukraine since 1985 (Naqvi, 2004).

Management

The disease can be kept under check by integration of various management methods like cultural, chemical and host resistance.

Cultural practices

Cultural methods include destruction of plant debris of previous crop and cucurbitaceous weeds, separating mature and newly sown crops, avoiding high crop population densities for preventing the buildup of moist microclimates within leaf canopy and use of irrigation from overhead, which leads to moisture being retained on the leaves.

Host resistance

Cvs/lines/ strains/hybrids like Shipper, Poinsette, Calypso, Addis, Sampson, Israeli lines 17 and 22, Sadao, Russiu, Heiva-K-2841, Schimo- schirazu S 2, Chimochirazu S 2, Belair, County Fair, Nanet, Ninghuang 2, Konkurent, Pasad, Salyut Fi249, Progress X169-3, 176 x 182, lines 1144-74, Gibrid 2, Line 22, 371, 196, Levina Fi, Kamon Fi, 6501, 6502, Belcanto Pi, 88-2, 88-7 and 88-8, Ningyang Spiny, Rodnichok Fi, Bngadnyi Fi, Mover Fi, Edinstvo 31, Feniks,

Wonye 501 and Wonye 502, Jinchun 4, 86604065-90, S440S, Slicer Astrea F, Nastasja F, Regal F, Xia Qing 4, Zao Qing 2 are reported to be resistant to downy mildew from different parts of the world (Cohen, 1976; Pivovarov *et al.,*1977; Jiang, 1981; Meyer and Ziesenis, 1982; Yakimenko, 1983; Pivovarov, 1984; Lisitisin and Pluzhnikova, 1990; Wu, 1991; Lai and Zhang, 1992; Stryapkova *et al.,*1992; Om *et al.,*1992; Lu *et al.,* 1994; Angelov and Georgiev, 1995; Komnenic *et al.,*1995 and Zuo *et al.,*1995). Singh *et al.* (1996) identified KP2, KP7, KP 9, SP 2, SP 3 and EC 163888 as resistant sources of snapmelon against *P. cubensis. Cucumber c*ultivar Chitradurga is reported as immune to this pathogen under Solan conditions (Raj *et al.,* 2003).

Determination of the resistance in plant is determined by the type of lesions exhibited by the plant. Thomas (1978) observed that plants which produced brown lesions had fewer lesions of infection , while the plant which produce yellow lesions had more lesions of infection. Thomas *et al.* (1987b) introduced the use of reaction types 2,3 and 4 representing levels of resistance compared to the susceptible reaction type fixed by RT of Leaves. A double digit classification system for the RT of leaves 1 and 2 respectively, in green howal inoculations of two leaf stage plants is an excellent predictor of the disease reaction of older plants under field conditions. This provided plant breeders with a technique that will expedite selection for downy mildew resistance in muskmelon. Progeny from test crosses and new accession can be evaluated reliably in the green house-growth chamber for downy mildew resistance as soon as the seed is available.This classification technique for resistance evaluates the specific type of lesion produced rather than estimating foliage damage as per cent chlorotic or necrotic tissue, per cent lesion area, per cent foliage loss etc. The extent of sporulation by *P. Cubensis* on muskmelon leaves exhibiting RT 1, 2, 3 and 4 gives an additional measure of the relative resistance level indicated by each RT. Thus spore production decreases progressively as the level of resistance represented by RT 2, 3 and 4 increased.

Cohen *et al.* (1989) showed that *P. Cubensis* produced 10-15 mm lesion in the susceptible cv. Anaras-Yoknean (A-l) but 1-3 mm water-soaked lesions in the resistant cv. PI 12411 IF (PI). Sporulation on cv. PI was extremely limited compared to the conspicuous sporulation on cv. A-l. Reuvoni *et al.* (1992) and Chauhan *et al.* (1994) used peroxidase activity in uninfected musmelon plants to determine resistance and susceptibility to *P. cubensis.* After infection, peroxidase activity increased with time in both susceptible and resistant plants. The ratio of activity in infected to uninfected leaves increased overtime in the susceptible plant.

94   The Plant Mildews

Chemical control

The disease can be kept under check by spraying or dusting of chemicals. Copper in the form of Bordeaux mixture or fixed copper and mancozeb are the principal fungicides for the control of this disease but care must be taken with copper fungicides since some cultivars exhibit copper phytotoxicity (Dixon, 1981). Jhooty and Munshi (1975) screened nine fungicides for the control of this disease on muskmelon and found Dithane M-45 as most effective. Conventional fungicides like Dithane M-45, Dithane Z-78, Difolatan and Tricop-50 could not eradicate the established infection of P. *Cubensis* but Dithane M-45 significantly reduced sporulation and sprayed leaves were protected for 9 days in case of Dithane M-45, Dithane Z-78 and Difolatan and 5 days in case of Tricop-50 (Bains and Jhooty, 1978c). Mahrishi and Siradhana (1988a) also reported the efficacy of Dithane M-45 in the management of this disease on muskmelon.

These protectant fungicides have to be applied in anticipation of the infection but had little or no therapeutic effect on disease development after infection had established. During the past two decades highly active systemic fungicides have brought in a new era in practical disease control with considerable benefit to the farmers. Phenylamides is one such group of fungicides which is specifically developed for the control of one of the most threatening types of diseases like downy mildews and late blights caused by Oomycetes on various annual and perennial crop plants. Downy mildew of cucurbits is polycyclic in nature causing serious epidemics within short periods of favourable weather conditions. It was difficult to control these diseases earlier with conventional fungicides, which required repeated applications at high rates. Phenylamides have specific activity against Oomycetes and have significantly improved control of these diseases by post -infection curative effect, low application rates, and high activity under high infection pressure. However, development of resistant strains to metalaxyl application has been reported in several countries (Reuveni *et al.*, 1980). To avoid this problem combination products are now available. Thind *et al* (1991) reported that mancozeb (0.3%) provided good control of downy mildew of muskmelon when used as protective treatment but failed to check the established infection even when applied 24 h of inoculation whereas Ridomil MZ (0.25%), Galben M8-65 (0.25%) and Fosetyl-Al (0.3%) were remarkable in their protective as well as eradicant activity both under artificial and natural disease conditions. Ridomil MZ (0.25%), Indofil M-45 (0.25%) and Curzate (0.08%) plus Indofil M-45 (0.25%) also provided complete control of this disease on cucumber in prophylactic spray programme (Gupta *et al.,* 1993; Gupta and Shyam, 1998). Gupta and Shyam (1996) reported the antisporulant activity of Ridomil MZ for longer periods that can be utilized in planning various spray schedules for effective management of this disease. Efficacy of cyanofamid against this disease has also been reported (Mitani *et al.,* 2003). Thind and Mohan (2001) and Sharma *et al.* (2003) studied disease-weather relationship in downy mildew of muskmelon and cucumber, respectively, and reported effective control of the disease by fungicide mixtures, such as Patafol (ofurace + mancozeb), Acrobat MZ (dimethomorph + mancozeb), Ridomil MZ (metalaxyl + mancozeb) and Curzate M-8 (cymoxanil + mancozeb) than individual components under high disease risk period. However, Khetmalas and Memane (2003) recommended the sprays of Bordeaux mixture or Alliete followed by mancozeb for the control of this disease and economic returns.

Synergistic effect of plant activator and mancozeb in reducing this disease has been reported (Baider and Cohen, 2003). In green house studies, a mixture composed of 5 parts of plant activator BABA (DL-3-aminobutyric acid) and 1 part of mancozeb was found highly effective in the management of downy mildew of cucumber. The results also showed enhanced effect of mancozeb in BABA-induced plants, suggesting, therefore, that lower dosages of this fungicide may be sufficient to control downy mildew under field conditions.

Plant extracts

Ethanol extracts of *Hedera helix* and *Paeonia suffruticosa* were found highly effective in the inhibition of downy mildew of cucumber when applied as pre-infectional treatment i.e. 2 days prior to inoculation (Rohner *et al,* 2004). The disease reducing activity was confined to the site of applications within the leaf area and the extracts induced no systemic resistance in higher inserted leaves. Both extracts also exhibited curative effects when applied at 1-4 days after inoculation, sporangial production in *P. cubensis*is drastically reduced.

8.3. Downy mildew of Suger beet
Pathogen: *Peronospora farinose betae.*

Host range

Wild and cultivated *Amaranthaceae*: *Amaranthus, Atriplex, Bassia, Beta, Chenopodium, Halimione, Salsola, Spinacia.*

Sugar beet downy mildew pathogen also attacks other cultivated varieties of *Beta vulgaris* (Leach, 1931, 1945; Singalovsky, 1937; Byford, 1966a)

Symptoms
The fungal pathogen attacks plants in all stages. The leaves of seedlings become extensively yellow and curl downward. An attacks on older leaves result in

more restricted spots, which sometimes are in ring form with darker pigmentation. If dry weather prevails after lesions formation, the spots may become dead and produce few spores. In wet weather the underside of the lesions becomes covered with spore-bearing growth of the fungus. The mycelium develops systemically in the cortex and invades the crown of the plant. Subsequent leaves are infected as they expand, and the entire crown becomes a rosette of small, distorted, mildewed leaves (Fig. 8.3).

When crown-infected roots are planted for seed production in the following spring, the floral stalk is systemically invaded and shows marked symptoms. Growth is severely retarded and distorted. Leaves become curled and thickened. Occasionally adventitious buds develop to give an effect of witches' broom. Infected floral parts are swollen and distorted. Various affected parts may be covered with the downy mildew growth.

Fig. 8.3: Downy mildew symptoms on beet

Pathogen

The downy mildews of beet and spinach were initially described as *Peronospora schachtii* Fuckel but following the taxonomic study of Yerkes and Shaw (1959) these species, together with others found on members of the Chenopodiaceae, were included in the single species *P. farinosa* (Fr.) Fr. However, many workers had shown that these downy mildews are host specific (Hiura, 1929; Leach, 1931; Cook, 1936; Singalovsky, 1937; Richards, 1939; Darpoux and Durgeat, 1962) and, following further cross-inoculation experiments, Byford (1967a) proposed the formae speciales *P. Farinose betae.*

The most detailed study carried out is by Singalovsky (1937), who stated that the germinating conidia penetrate the leaf via the stomata. Coenocytic hyphae 5-6 μm diameters ramify through the intercellular spaces while branched haustoria develop within the host cells. The infected leaf responds hyper-trophically with cells expanding to approximately twice their normal size, and the differentiation between palisade and spongy mesophyll tissue, decreases. Sporangiophores emerge through the stomata in small groups of 3 to 5. Their length varies from 200 to 500 μm and their diameter from 8 to 12 μm. The average size of sporangium is 24-8 x 18-9 μm (Yerkes and Shaw, 1959). Sporangia germinate best at about 4.4° to 7.2°C although germination will occur in a range of 3.5° to 8.5°C.

Oospores are frequently found in infected leaves and measure approximately 35 x 38 μm. They carry the fungus over long periods of unfavorable conditions. The pathogen also is carried in seed which may initiate the disease in the spring. The chief source of primary inoculum, however, especially in seed-growing areas, is the over-wintering roots that were infected in the steckling beds.

Weather Parameter

Downy mildew is serious only in cool weather when there are frequent showers or heavy dew or fog. The environmental requirements of *P. Farinose betae* have been studied by Leach (1931), Singalovsky (1937), Darpoux *et al.* (1960), Darpoux and Durgeat (1962) and Byford (1968) and shown to be as follows:

1. Sporulation

Spores develop between 5 and 22°C, optimum around 12°C. A relative humidity above 70% is needed, and most spores are produced above 85% RH. Spores are usually produced overnight and there is some evidence of a diurnal rhythm. Continuous light checks spore production which occurs most abundantly when leaves that have been exposed to daylight for 6-8 h are placed in the dark.

2. Longevity of spores

Cold prolongs the life of spores; their ability to survive falls off above 10°C and few survive more than one day above 20°C. Differences in relative humidity above 60% scarcely affect spore survival. Under favourable conditions some batches of spores can survive for at least 1 week. Spores that survive are able to infect.

3. Spore germination

Spores can germinate between 0-5 and 30°C, with the optimum between 4 and 10°C. Germination is fastest at 9-10°C.

4. Infection

Infection occurs between 0.5 and 25°C, most readily between 7 and 15°C. It usually requires at least 6 h, preferably 8-10 h, in the presence of free moisture. The growing point is the most susceptible part of the plant, and after the cotyledon stage it is the only part where infection is likely to occur. Seedlings are most susceptible in the cotyledon stage, and become more difficult to infect

The Plant Mildews

98

artificially after 4 weeks growth at a minimum of 10°C. However, plants are not immune to infection after 8 weeks in the glasshouse and in the field some new infections continue to appear for most of the season.

Management

Many workers have reported experiments with protectant fungicides to control beet downy mildew but the control obtained was often poor (Moller-strom, 1955; Byford, 1966b). In particular attempts to control the disease by spraying seed crops during the winter were ineffective (Byford, 1966b).

In consequence most advice on limiting mildew incidence has concentrated on cultural practices of which the most important is separating sugar-beet root crops from their main source of infection, beet seed crops. In the Soviet Union 1000 m is usually recommended. In England most sugar-beet stecklings are raised under cereal cover crops which give some protection from infection (Byford and Hull, 1967). Stecklings not raised under cereal cover must be separated by at least 400 m from other *Beta* seed or root crops. Good crop hygiene is also necessary to limit survival of the fungus on ground keepers in fields and on clamp sites and proper rotation reduces the risk of infection from oospores in the soil. Byford (1967c) reported that increasing the spacing between plants increases the proportion infected with downy mildew. Byford (1967c) also reported that infection was increased by high doses of nitrogen fertilizer.

Sugar beet varieties differ in their susceptibility to downy mildew but in areas where the disease is endemic all varieties probably have some non specific resistance to the disease. Russell (1969) reported an inbred line completely resistant to downy mildew but a race of *P. farinosa betae* able to infect the line was soon found (Russell and Evans, 1972). A "downy mildew nursery" has been established at Trawsgoed in Wales, U.K in 1965 where the damp cool climate favours the disease and there are no commercial beet crops that could be accidentally infected. Here all established and new varieties are grown under heavy natural infection and plant breeders can evaluate material for resistance and the most susceptible can be avoided in areas particularly at risk e.g. near seed crops; secondly it has allowed very susceptible material to be eliminated early in the selection process.

Recently the development of systemic fungicides active against Peronosporaceae has given promise of direct control of beet downy mildew. Broyakovskaya (1975) reported that editon and antracol showed strong systemic action when applied to the soil while metalaxyl applied to the seed protect the young seedlings from infection for up to 3 weeks.

8.4. Downy mildew of Carrot

Pathogen: *Plasmopara umbelliferarum*

Downy mildew occurs on many plants of the umbelliferae family; however, the strains that infect carrot, occur only in Europe. Synonyms for this fungus mentioned in the literature include *Plasmopara crustosa* (Fr.:Fr.) Jorst, *Plasmopara nivea* (Unger) J. Schrfi, and *Peronospora umbelliferarum* Unger. There may be more than one species of *Plasmopara* on various umbelliferous plants since morphology of the fungus varies widely with the host plant (Constatinescu, 1992).

Geographical Distribution

The fungus commonly occurs in Poland and the other regions of the world (Kochman and Majewski 1970).

Symptoms

Symptoms are visible on the upper side of the leaves as chlorotic spots and as whitish sporulation on the corresponding underside of the leaf. Infection first appears on young foliage. During periods of high humidity, sporangiophores emerge in groups through stomata and release air borne sporangia (19 to 22 x 16 to 18 µm in diameter), which germinate to produce motile zoospores that swim in free water on plant surfaces, eventually infecting through stomata. Sexual oospores, which are produced within the tissue, may survive the winter in crop debris or in seed.

Host range

Plasmopara umbelliferarum affects numerous species of the family Umbelliferae, Pasley, Fennel.

Management

Several cultural practices can be adopted to manage downy mildew. Because the fungus may survive in seed, only pathogen-free seed should be planted. Strategies that minimize the duration of leaf wetness, such as decreasing plant density, avoiding the use of excess fertilizers, and managing irrigation and drainage may reduce the incidence of disease. Carrots should be rotated with non-umbelliferous crops to reduce the inoculum load in the environment.

8.5. Downy mildew of spinach

It is probably the most wide spread and potentially destructive disease of this crop worldwide.

Pathogen: *Peronospora farinose* f.sp. *spinaciae*

Geographical distribution

The occurrence of this disease has been reported from U.S.A, Japan, Iran, Netherlands, Italy etc. (Inaba *et al.*, 1983; Correll *et al.*, 1994; Lorenzini and Nali, 1994; Sadravi *et al.*, 2000).

Losses

Though exact figures of losses are not available but the disease under favourable weather conditions is potentially destructive and causes significant losses. Under conditions of prolonged leaf wetness and cool temperatures, epidemics can progress very rapidly and an entire crop can be lost in a short period of time (Correll *et al.*, 1994).

Symptoms

The symptoms of the disease appear as slightly yellow, irregular, chlorotic lesions on leaves. Lesions frequently expand and coalesce and may become necrotic. Heavily infected leaves can appear curled and distorted. Under wet conditions and/ or high relative humidity, blue-greyish sporangia and sporangiophores are produced and can be seen in a mass on the underside of the leaf (Fig 8.4).

Fig. 8.4: Downy mildew symptoms on spinach leaves

Pathogen

The disease is caused by *Peronospora farinose* f.sp. *spinaciae* Byford. The fungus is a heterothallic obligate parasite. It produces asexual sporangiophores, which emerge from stomata either singly or in groups of 2 to 3. Sporangia are hyaline, oval and germinate directly by germ tube. Sexual oospores are formed within the host tissues during cool, moist conditions. They germinate either by germ tube or by production of zoospores. Seven distinct physiological races of *P. farinose* f.sp. *spinaciae* such as race 1, 2, 3, 4, 5, 6 and 7 have been reported from USA and Europe (Brandenberger *et al,* 1991; Irish *et al.*, 2003; Irish *et al.*, 2004). The occurrence of race 4 and race 7 of the pathogen outside USA was first time recorded in Italy and Netherlands, respectively (Lorenzini and Nali, 1994, Irish *et al.*, 2003).

Host range

Besides genus *Spinacia,* the fungus also infects several species of *Chenopodium* (Byford, 1981).

Disease cycle

The pathogen overwinters in the infected seed as oospores and mycelium (Inaba *et al.,* 1983) and the infected seeds have been shown to give rise to infected seedlings. Oospores can also survive in soil and may represent an important source of primary inoculum (Wright and Yerkes, 1950). Wind borne sporangia from surrounding diseased fields may represent an important source of primary inoculum. The oospores/sporangia germinate and cause primary infection. The sporangia produced on primary pustules further cause secondary infections.

Weather parameter

The pathogen survives as oospores/ sporangia from one season to another. The sporangia germinate directly on leaf surfaces within 2-6 h under cool and wet conditions (Richards, 1939). Termination of sporangia and germ tube elongation can occur between 2 and 25°C with an optimum at 10°C (Sadravi *et al.,* 2000). Lesion development is favoured by temperatures of 15-25°C with latent period of 6-12 days. Sporangia are readily dispersed by wind and rain splashes but can rapidly lose viability when desiccated or exposed to sunlight (Richards, 1939; Wright and Yerkes, 1950).

Management

Cultural practices

Practices like use of healthy seed and destruction of plant debris and *Chenopodium* spp. from in and around the field, which can harbour the pathogen, should be followed to keep the disease under check.

Host resistance

Planting cultivars with single-gene resistance to a prevalent race controls the disease (Brandenberger *et al.,* 1991). Varieties Solomon, Samson and G-l possessed good resistance to race 1 or race 2 (Shimazaki and Uchiyama, 1985) while Dynamo, Nores, Vital, Califlay and Medania to race 1, 2 and 3 (Hubbeling and Ester, 1978). Lines such as Bossanova, Bolero and RS 1250 had < 2 per cent diseased area (Koike *et al.,* 1992). Of late, cv. Byford resistant to races 1, 2 and 3 was found susceptible to a new race 4 and the application of commercial formulations of metalaxyl; copper oxychloride and chlorothalonil were ineffective (Lorenzini and Nali, 1994). However, when new races of the pathogen appear, as it happened in California and Italy (Brandenberger *et al.*, 1991; 1992; Lorenzini and Nali, 1994), it may take several years to develop new commercial cultivars with single-gene resistance before these become available.

Chemical control

Several systemic and non-systemic fungicides like maneb, fosetyl, metalaxyl and chlorothalonil can be used to combat the ravages of this disease (Koike *et al.*, 1992).

8.6. Downy mildew of lettuce

This is a major disease of glasshouse and field grown lettuce crops the world over where adequate moisture and moderate to low temperature is prevalent.

Pathogen: *Bremia lactucae* Regel.

Geographical distribution

Downy mildew of lettuce is known to occur in all parts of the world including Western Europe, USA, Israel and South Africa (Grogan *et al.*, 1955; Crute, 1987; Achar, 1996).

Losses

The disease can cause considerable damage to lettuce planted during periods of prolonged wetness from rain, dew, and fog or sprinkler irrigation. Severely infected plants may require excessive trimming during packing, which increases production costs. Invasion of downy mildew lesions by secondary soft rot organisms is another major source of crop loss in the field or in transit and storage (Grogan *et al.*, 1955).

Symptoms

Symptoms of the disease appear as pale green to yellow spots on the upper side of the infected leaf, which may coalesce to affect large areas (Fig 8.5). The downy growth of the fungus appears on the lower surface of the leaf under high humidity and low temperature conditions. Older lesions often become necrotic

Fig. 8.5: Downy mildew symptoms on lettuce leaves

or translucent after invasion of infected tissue by secondary saprophytes. Sometimes the early infection of seedlings may lead to systemic infections that result in dark brown discolouration of vascular stem tissues (Davis *et al.,* 1997).

The Pathogen

The disease is caused by *Bremia lactucae* Regel. 1 to 3 sporangiophores are formed from the mycelium in the sub-stomatal cavities. Branching of the sporangiophores is dichotomous or, rarely, trichotomous. At the tip of each branch sterigmata are formed, each with one sporangium. Sporangiophore length varies from 200 to 1200 µm depending on environmental humidity. The sporangia are hyaline, ovoid-ellipsoidal to globose and are 12-31 µm long and 11-27.5 µm wide. The fungus is mostly heterothallic *i.e* two mating types (B1 and B2) are needed for oospore production, although homothallic isolates have also been found (Davis *et al,* 1997). However, oospores have rarely been observed in field-grown lettuce, lettuce debris or wild hosts. This may be due to the presence of only one of the two mating types. Oospores are abundantly produced in cotyledons or leaves artificially inoculated with two mating types. Oogonia and antheridia are produced about 4 days after inoculation, and oospores mature 10 days later. Oospores are spherical (27-30 µm in dia.) and are surrounded by a thick wall. Germination of oospores has only occasionally been achieved under controlled conditions after 1- 6 months of storage in water and rotting debris. The factors affecting oospore germination have not been described.

A number of races in *B. lactucae* have been reported. There is considerable variation for specific virulence in *B. lactucae* populations and the use of new combinations for R-factors resulted in the increased frequency of pathotypes carrying matching virulences. Based on reaction of differential cultivars to various isolates of *B. lactucae,* 46 different virulence pathotypes possessing 2-12 (mostly 5-6) virulence factors have been reported (Thinggard, 1985). Since *B. lactucae* is predominantly heterothallic, it is likely that sexual recombinations are largely responsible for variations. Fifteen pathotypes of the fungus have been described in Europe .

Disease cycle

The pathogen can overwinter either as oospores in plant debris or in seed or wild *Lactuca* spp. Since both mating types are present in many parts of the world, oospores in lettuce debris may serve as the source of primary inoculum. Seed borne inoculum also forms an important part of primary inoculum. This disease is more severe on cotyledons and the initial infection may lead to

104 The Plant Mildews

systemic infection of the plants. *L.serriola* is the most common wild host that may harbour *B. lactucae*. Sporangia from the infected continuous and over lapping lettuce crops from nearby fields can be another source of primary inoculum. Secondary spread of the pathogen is through air borne sporangia.

Weather Parameter

The pathogen overwinters either in the plant debris as oospores or the inoculum from wild hosts cause primary infections. Sporulation is affected by temperature and humidity. The optimum temperature for sporulation is 15°C. Increase in relative humidity at > 90 per cent increases the sporulation markedly (Su *et al,* 2004). Both RH and wind speed affect the number of plants with sporulation and the number of sporangia produced/ plant. They further reported that sporulation may be affected by stomatal aperture in response to RH, as more closed stomata and correspondingly fewer sporangiophores are present at low RH. Dispersal of sporangia usually starts at sunrise with peaks between 10.00 a.m. and noon. Sporangia survived much longer at 23°C (>12 h) than at 31°C (2 to 5 h), regardless of RH (35 to 76%) (Wu *et al,* 2000). Germination percentage was significantly reduced after exposure to 50 and 100 per cent sunlight. UVB significantly reduced sporangium viability, while fluorescent light and UVA had no effect relative to incubation in the dark. Solar radiation is the dominant factor determining survival of *B. lactucae* sporangia. The infection can be initiated within 3 h under optimum temperatures (10-22°C) in the presence of free water or relative humidity near saturation. The minimum and maximum temperatures for infection are 0.5°C and close to 30°C, respectively. Irrigation practices also play an important role in the development of epidemics of this disease. Generally, more disease is observed under sprinkler and under furrow irrigations than subsurface drip irrigation (Scherm and van Bruggenj 1995) because of the longer leaf wetness duration, increased canopy, air humidity, and/ or enhanced dispersal of inoculum.

Forecasting

Kushalappa (2001) developed a forecasting system "BREMCAST" to predict risk levels of downy mildew of lettuce. The criteria for the forecasting system were derived from the relationships of weather variables to sporulation, dissemination and infection processes of *B. lactucae*. In addition, the presence of inoculum source, or the disease, in the field was taken into consideration. The system predicted the occurrence of downy mildew on 84 and 80 per cent of the days under commercial and field plot conditions, respectively.

Management

Cultural practices

The disease can be kept under check by destruction of infected plant debris, by using healthy seed and removal of wild *Lactuca* spp. from in and around the fields. Subsurface drip irrigation results in less conducive environment for downy mildew than furrow or sprinkler irrigation (Scherm and van Bruggen, 1995).

Host resistance

Lettuce downy mildew can be best managed by the use of resistant varieties. Resistant varieties have either specific host resistance genes (single dominant genes) or moderate level of field resistance. Accessions like 490999, 491204, 491205, 491207 and 401208 of *L.saligna* and 491178 of *L. serriola* (Gustafsson, 1989) have high resistance. High resistance was also observed in lettuce cultivars like Saffier and Mariska. Sixteen commercial lettuce cultivars resistant to all known races of *B. lactucae* were evaluated in greenhouses in Italy for resistance to pathotypes present in the area; and all the cultivars were found resistant to the prevalent pathotypes. However, small necrotic spots caused by the fungus were observed on the leaves of cv. Ninja and Bra 1939 while no symptoms were observed on the other 14 cultivars.

Chemical control

Copper-based fungicides and dithiocarbamates provided effective control of this disease but they had certain limitations. These are surface-acting protectant fungicides and often fail to provide adequate control because it is difficult to achieve a continuous and sufficiently uniform presence of the chemical on all leaf surfaces exposed to spore deposition. Newly produced foliage of a rapidly growing crop such as lettuce is often left unprotected between applications and substantial amounts of chemical can be lost in irrigation or rain. Besides this they have some advantage that insensitive forms of the pathogen have not emerged to negate their effectiveness.

Three distinct groups of chemical fungicides like phenylamide group (metalaxyl, benalaxyl, cyprofuram, oxadixyl and ofurace), fosetyl-Al and propamocarb having genuine systemic activity against B. *lactucae* were introduced during the 1970's. Out of these, metalaxyl provided an excellent control of this disease and this fungicide can be applied to foliage or via roots. These fungicides also have good curative properties and the capability of eradicating established infections. However, with the continuous use of these fungicides, pathotypes of *B. lactucae* have emerged which are insensitive to metalaxyl and fosetyl -Al and have resulted in control failure in many countries of the world (Schettini *et al.*, 1991; Wicks *et al.*, 1994b; Cobelli *et al*, 1998; Brown *et al*, 2004). The metalaxyl resistant strains can be managed by mixtures of either dimethomorph or phosphonic acid with mancozeb effectively (Wicks *et al.*, 1994b). Recently available fungicides like azoxystrobin, Previcur N, and cymoxanil are also found effective in the management of this disease (Paaske, 2000).

8.7. Downy mildew of fenugreek

Pathogen: *Peronospora trigonellae*

Geographical Distribution

The downy mildew of fenugreek is reported from Algeria, India, Pakistan, United Kingdom, USA and Iran. The first report of its occurrence is from California in the United States. In India it commonly occurs during the month of February and March. In India this disease and its pathogen *P. trigonella* was first reported by Uppal *et al. in* 1935 from Bombay.

Symptoms

The upper surfaces of the leaves exhibited small chlorotic spots often at the leaf margins, while the lower surfaces exhibited a grayish violet, felty growth of mycelium (Fig 8.6). The downy mildew infection is more at flowering and pod formation.

Fig. 8.6: Symptoms of downy mildew on fenugreek

Pathogen

The grayish downy fungus growth of *Peronospora trigonellae* on the underside of the leaves consists of sporangiophores which are dichotomously branched 6 to 10 times, are terminally forked and at the tip sporangia are formed on sterigmeta. Sporangiophores measured 280 to 525 ìm (average 420 ìm) with slightly swollen bases (7.5 to 10 ìm broad). Sporangia are slightly pigmented, oblong to ellipsoid, and measured 23 to 33 × 18 to 23 µm (average 27.8 × 20.3 µm). The sporangia spread the disease. Oospores are formed towards maturities which are globose with verruculose walls measured 30 to 40 µm in diameter (average 36.1 µm) and are found embedded in the leaf tissue of older lesions. These are responsible for the recurrence of the disease.

Weather parameter

The effect of weather variability has been extensively studied for downy mildew of fenugreek. The disease is favored by a maximum temperature range of 18-24°C to minimum temperature range of 4-10°C and a relatively high humidity which is more than 80%. Positive significant correlation was observed with morning and evening relative humidity while negative correlations were observed for maximum and minimum temperatures. Light conditions also played an important role in the infection frequency. Under light and dark conditions the percentage of infection differed significantly and showed higher infection points under dark conditions. Elaborate studies of the disease reveal that the inoculums of 12 hours age caused the maximum infection in the plants. The disease was initiated when the host leaves were kept wet for 6 hours and the optimum for maximum infection was found to be 12 hours of wetness period. The flowering stage of the plant showed highest infection. According to Lakra (2002) oospore formation is favoured by 24-27°C at 40% relative humidity. Prakash & Sharan (2001) observed that 12 h old sporangia and 18°C temperature are the best for causing infection.

Host resistance

Malhotra & Vashishta (2008) have recommended Hisar Mukta, as resistant and Hisar Suvarna as moderately resistant to downy mildew. Mehra *et al.* (2002) found fenugreek seeds with green tan seed coat highly resistant to downy mildew.

8.8. Downy mildew of Onion and Garlic

The disease causes wide spread damage on the world wide basis wherever this crop is grown in cool and humid conditions.

Pathogen: *Perenospora destructor*

Geographic distribution

The disease was first reported from England in 1841 by Berkeley (Yarwood, 1943). Subsequently, it was reported from Bermuda Islands, USA, Canada, Venezuella, Brazil, Iraq, Holland, Australia, U.S.S.R., China, Japan, Sweden, Norway, Germany, Israel and Poland (Palti *et al* 1958; Jones and Mann, 1963; Rondomanski, 1967),

In India, the disease was first reported from Kashmir valley during 1974-75 and 1975-76 seasons (Mirakhur *et al.*, 1978). Since then it has been reported from Kurukshetra area of Haryana and Himachal Pradesh (Singh *et al.*, 1987).

Losses

Downy mildew of onion caused by *Peronospora destructor* is an economically important disease and occurs in most onion *Allium cepa* producing regions throughout the world, causing losses in both yield and quality (Lorbeer, Andaloro, 1984; Schwartz, Mohan, 1995; USDA, Crop Profile for Onion, 2003). The pathogen attacks all types of onions, but is more destructive to common onion (*A.cepa* L.) where it is reported to cause losses as high as 60-75 per cent of the crop (Cook, 1932, Develash, Sugha, 1997). These losses mainly result from severe infections in bulb onion crops causing early defoliation, "bottle necked" onion bulbs, reduced bulb sizes, and poor storage quality of bulbs (Rondomanski, 1967, Lorbeer, Andaloro, 1984; Gilles *et al.*, 2004; Gianessi, Reigner, 2005). In salad onions, yield losses can be as high as 100%, with whole crops being discarded as downy mildew symptoms make them unmarketable. Losses to seed production are frequently caused by the collapse of infected seed stalks and poor germination of seeds collected from infected stalks (Schwartz Mohan, 1995).

In India, the losses from this disease range from 12-75 per cent in crop yield depending on the time of disease outbreak and disease severity (Sugha and Singh, 1991).

Symptoms

The first symptom of downy mildew is the velvet-like growth of the pathogen on otherwise green leaves (Fig 8.7). Symptoms differ with the type of infection. Infection develops from three different sources like systemically infected perennating leaves, plants grown from infected onion bulbs or from local lesions resulting from air borne inoculum.

Fig. 8.7: Downy mildew symptoms on onion leaves

Systemic infection occurs when the plants are raised from diseased bulbs or infected seedlings used for transplanting. Plants raised from such bulbs remain stunted, become distorted and light green in colour. In humid weather conditions, sporulation develops on the leaves and covers them with felty whitish to grayish fungal growth. In secondary and local infections, spots with greenish and chlorotic zones appear and at times at the point of infection, the plants fall off. In humid weather the fungus develops as white to purplish downy growth on these spots. In such cases undersized bulbs are produced. Other fungi, especially *Stemphylium botryosum* and *Alternaria porri* may colonize the infected tissue, giving the diseased area a dark appearance. Necks of affected plants remain succulent. Sporulation also occurs on seed stalks, where the affected areas tend to be circular to oval and remain yellow. Infected stalks bend over and break when the seed head enlarges. The pathogen may also invade flowers and seeds.

Pathogen

The disease pathogen *Perenospora destructor* is an obligate parasite. The sporangiophore are aseptate, hyaline swollen at the base, 160 - 485 x 9 - 14 µm in diamension and 3 to 4 times dichotomously branched. The sterigmata are acute to subacute, 9-14 x 2-4 µm and bear pyriform to fusiform sporangia, which measure 36 -69 x 18 -25 µm in size. Asexual sporangia are borne on sporangiophores. Sporangia form on sporangiophores that emerge through the stomata of green host leaves are the source of disease recurrence and spread. Oospores develop late in the season and measure 40-50 µm in diameter. They germinate by germtube and are the source of primary infection.

Host range

Besides common onion *Allium cepa*, the pathogen also attacks other species of *Allium* like *A. Fistulosu.,, A. Ascalonicum., A. Nigrum., A. Porrum., A, ursinum., A. Oleraceum., A.sativum., A cepa var. Multiplicans., A cepa var. bulbifera and A. schoenoprasum* (Yarwood, 1943).

Disease cycle

The carry over of the fungus from one season to another may occur as mycelium in the plant debris, the systemically infected bulbs, seed or soil. The sporulation and infection by *P. destructor* usually takes place overnight while dissemination of the newly formed sporangia occurs during morning hours (Yarwood, 1943; Rondomanski, 1967). The oospores on/in the seed or in plant debris (Yarwood, 1943) and systemically infected onion bulbs are another important source of primary infection, but role of oospores has not been established in induction of disease even under controlled conditions (Viranyi, 1981). The onion seedlings

110   The Plant Mildews

from disease prone areas play significant role in introduction of disease in new localities. Secondary infection is caused by wind borne sporangia disseminated from the plants with primary infection.

Weather Parameter

The pathogen requires cool and moist nights to initiate infection. The temperature of 13°C and >95 per cent relative humidity are optimum for the development of the disease. Presence of dew or rain drops on the leaf surface further enhances the chances of infection. The disease is worse in damp conditions (Devlash and Sugha, 1997b). Germination of sporangia occurs at 7-16°C temperatures, optimum being 10 to 13°C. The infection cycle is characterized by long latent periods of 9 to 16 days and 1 to 2 day periods of sporulation, dispersal and infection.

Forecasting

The first approach to warning systems consisted of the definition of periods favourable for the disease in Bulgaria (Vitanov, 1971). A similar approach was followed by Palti *et al.* (1972) who tried to define mildew-free periods analysing 10-year records of weather conditions in relation to the dates of onion mildew outbreak in Israel.

In 1975, a bioclimatological model, based on microclimatic conditions, was developed in the Netherlands to define potential infection dates (Weille, 1975). In the following year, Stenina (1976) tried to forecast the disease in the Krasnodar region (Russia) using weather data.

The Downy Mildew forecaster DOWNCAST developed by Hildebrand and Sutton (1982) may be used to identify periods when weather favours sporulation and infection by the mildew fungus. DOWNCAST requires weather monitoring in the field and indicates optimal times to look for early signs of mildew and to apply fungicide sprays. The weather data needed for the system are hourly air temperatures for the preceeding day, with relative humidity and leaf wetness duration. In summary, if air temperatures were above 24 °C on one day, no sporulation will occur the following night. For sporulation to occur there must be six hours between midnight and dawn where the relative humidity should be 95% or more, temperatures between 4 and 24 °C and no leaf wetness. If environmental conditions are favourable for sporulation , and conditions for infection occur in the next 24 to 48 hours (6 hours leaf wetness at temperatures between 6 and 25 °C) then there is a high probability that infection will take place (Jesperson and Sutton 1987). DOWNCAST has been implemented in many regions of the world to manage downy mildew on onions. In Australia, DOWNCAST was implemented in 1998 and reduced the number of sprays applied during onion production by 4 or 5. Fungicides application was avoided early in the season when temperatures were too high for downy mildew development. During the growing season, there were several times when sporulation occurred but the leaf wetness required for infection was not present (Jackson *et al.*, 1999).

Battilani *et al.* (1996) developed a preliminary model, ONIMIL to forecast primary infection, which is able to determine for each day the probability of *P.destructor* establishing an infection on onion and its infectivity level. Recently, MILIONCAST, a model was developed based on the data from the controlled environment studies to predict the rate of sporulation in relation to temperature and relative humidity (Gilles *et al.*, 2004).

Management

Various practical measures to preclude sources of infection and to prevent spread of the disease, are of great importance in controlling this disease.

Cultural practices

Cultural practices like use of healthy bulbs for planting, collection and destruction of infected crop debris, cultivation on well drained soils, crop rotation avoiding the *Allium* crops in rotation, application of balanced doses of fertilizers along with potassium as the increase in level of potassium decrease disease severity (Deviash and Sugha, 1997a) or avoidance of frequrent irrigations (Ahmad and Karimullah, 1998) help in the reduction of the initial inoculum and spread of the disease in the field. Widely spaced, inter and mixed cropping of onion with Indian mustard (*Brassica juncea*) alone or in combination with coriander (*Coriandrum sativum*) developed less disease compared with pure crops of seed and bulb onions (Devlash and Sugha, 1997a). Soil solarization for six weeks during July- August retarded symptom development of downy mildew for 20 days and increased the bulb yields by > 97 per cent over that of shaded plots. Seed onion should be grown far away from any bulb crops (Naqvi, 2004).

Host resistance

White skinned varieties are generally more susceptible to the disease than red and lilac skinned onion varieties. Resistance in onion varieties against this disease was correlated to flat leaves and xeromorphic tissue structure (small cell, thick wall, high lignification). Susceptible species were characterized by fistular leaves and swollen flower stalk; cells were larger and thin walled and lignifications less marked (Furst, 1976). The resistance to the pathogen is correlated with the chemical nature of cuticular wax and those species which

112   The Plant Mildews

has resistant had acid content in the wax. Cvs/ lines like IC 48045, IC-32149, IC-38617, IC-49871 and DOP-11 have been reported as good sources of resistance under Indian condition and can be utilized in breeding programme.

Chemical control

Various systemic and non-systemic fungicides like mancozeb or zineb, copper oxychloride, propineb, metalaxyl + thiram, metalaxyl + mancozeb, oxadixyl + mancozeb have been reported to check onion downy mildew effectively (Mir *et al.,* 1987; Jorgensen and Nistrup, 1987; Khokhar and Jaffrey, 2000; Khalid *et al* 2002). The efficacy of different fungicides increased with the addition of adjuvants and Acrobat MZ with Synertrol oil was found effective in both pre and post infection treatment against this disease (Mac-Manus *et al.,* 2008). Combining mancozeb sprays with an alert-based metalaxyl application fortnightly, reduces the total number of fungicide applications by 8 whilst maintaining effective disease control.

8.9. Downy mildew of capsicum

Pathogen: *Peronospora tabacina* Adam.

Host range: Pepper, tomato, eggplant, tobacco

Geographical Distribution

Pepper and Tomato are reported hosts in the USA, Russia, and Argentina, however, some isolates only infect pepper and tobacco and not tomato in other locations (Armstrong and Albert 1933, Gayed 1984, Hindi *et al.,* 1965).

Losses

Although extensive losses in transplants have occurred, the disease is of minor, infrequent importance.

Symptoms

Downy mildew produces its characteristic symptoms on the underside of leaf where the fungus sporulates through the leaf stomata. The sporulation is seen as fuzzy, white to grayish patches (Fig 8.8). On the upper surface of the leaf, pale green to yellow spots may appear opposite the areas of sporulation. Older leaves are affected first and often turn completely brown from infection (Sherf and McNab 1986).

Fig. 8.8: Downy mildew symptoms on capsicum leaf

The pathogen

Sporangiophores are dichotomously branched at acute angles, tapered, and curved with pointed tips bearing sporangia. Sporangia are oval.

Disease Cycle and Epidemiology

Sporangia are wind-dispersed over a long distances. Wind-driven rain or splashing water can transport sporangia at short distances. The fungus survives on infected plant debris in the soil and on related weed species. Environmental conditions for infection are high humidity, leaf wetness, and moderately cool weather (daytime temperatures ranging from 15 to 24° C).

Management

Transplants/seedlings infected with downy mildew should be removed from the transplant house and discarded. Watering of transplants should be performed with a minimum of splashing and done early in the morning to allow foliage to dry. Increased plant spacing and reduced humidity levels should also help to keep foliage dry. Rouging in the field may be useful if a small number of plants are infected. Fields and transplant production areas should be kept free of weeds and isolated from tobacco production. Fungicides labeled for this disease can be used preventively to control the disease.

9

Downy Mildew of Spices

9.1. Downy mildew of Opium poppy

Pathogen: *Peronospora arborescens (*Berk) de Bary

Geographical distribution and losses

This is one of the serious and widespread diseases of opium poppy. The disease was observed epiphytotic in Argentina, Germany, Iran, Hungary, Austria and India. The disease was first recorded in India in 1918 and with the introduction of susceptible varieties, the disease rapidly became serious due to favourable environment in its cultivation area in India. Downy mildew now occurs in most regions in India (Rajasthan, Madhya Padesh and Uttar Pradesh). Downy mildew caused 20-30% loss of yield of opium latex. The secondary infection of downy mildew was found to reduce 11-22% latex.

Symptoms

The infection spread upwards from lower leaves. The entire leaf surface is covered by downy mildew coating, which is composed of sporangiophores and sporangia of the pathogen (Fig 9.1). The stem, branches and even capsules are attacked resulting in the premature death of the plants. The infection start from the moment the seed germinate and persist during the whole period of growth (Yossi fovitch 1929). Khristov (1943) reported that the pathogen often cause abnormal development in opium poppy. The pathogen also causes hypertrophy and curvature of stem and flower stalks.

Fig. 9.1: Downy mildew symptoms on Opium poppy plant

The infection is often systemic. The leaves of infected plants remain small, chlorotic and curl downward. Heavy sporulations occur on entire lower surface of diseased leaves.The capsules, if formed get shrinken, wringled and dried. Severely infected plants show extreme stunting. The infected area on the leaf first become chlorotic and then necrotic with large spots. The disease cause extensive damage to the crop cultivated in most sites.

The Pathogen

The mycelium ramifies between the leaf tissue descends out short and rounded haustoria. The mycelium growths extend to the stomata and produce sporangiophores in groups. The sporangiophores are stout, erect, 300-800×10-12 µm. They are 7-10 times dichotomously branched. Fine branches curved short and diverging nearly at right angles. Sporangia are borne singly on ultimate branchlets of sporangiophores. They are elliptical, oval or globuse, thin walled, smooth, hyaline turning brown with age, 23.1 x19.00 µm (18.4-3ox15.2-26 µm). A wide variation is observed in conidial size. Oospores are 42-48 µm, dark reddish brown with irregular outer wall, width varying from 4 to 10 µm and with irregular ridges over the surface. In early stage each oogonium is found to possess a large round body and occasionally one or more bright yellow bodies resembling oil droplets. Average size oogonium is 33µm diameter.

The abundance of mycelia, sporangiophores, sporangia and oospores on the host suggests that P. arborescens and opium poppy has developed remarkable adaptability. Kothari and Prasad (1970) found that P. arborescens is specific. The oospores play a significant role in initiating the disease and they appear to be the source of annual recurrence of poppy downy mildew. The oospores in the soil constitute a potential source of infection. Severely diseased plants appear throughout the crop produces large number of sporangia, which in turn become a source of infection for the new crop. In addition the disease is also transmitted by seeds.

Control measures

Crop rotation, wide spacing and avoidance of low lying site for growing poppy reduces the incidence of downy mildew. Seed disinfection and spraying seed beds with 0.5% Bordeaux mixture or other copper fungicides is recommended for the control of the disease. Kothari and Verma (1967) reported that spraying with Dithane Z-78, Ferban or bisdithane gave 50% control. Spraying with Dithane M-45 (1.8 kg/ha) or 3 sprays of bisdithane (0.15%) followed by Benlate (0.15%) and crop rotation gave substantial control of the disease. Seed dressing and sprays with Ridomil controlled the disease effectively. Primary infection of downy mildew can be reduced by seed treatment with metalaxyl and further sprays of metalaxyl 25WP (0.2%) at 20 days interval are highly effective in controlling secondary infection. However, Ridomil treatment is the most effective method for the control of downy mildew disease of opium poppy. Fungicidal spray with Gramisan, Germisan or dusting with Thiram, Germisan, Gramisan or Agrosan was found to be useful protective measures against the disease. Application of Dithane Z-78 (0.4%) as a drench treatment of soil was observed to be effective for controlling the infection. However, Captan treatment was not recommended as this fungicide inhibited seed germination of opium poppy.

A higher dose of phosphorus application to the crop was found to decrease severity of downy mildew while nitrogen increased the disease index. Crop rotation of 4 to 5 years with one year fallow is recommended for the control of the disease. The three year gap in cultivation of two consecutive crops in the same field and destruction of infected plants in the field is recommended as a prophylactic measure.

Resistant Varieties

A cultivar Gazipur Local shows high resistance and being used in breeding programme.

Crop Rotation Practices

Measures like crop rotation for three years, removal of infected plants and debris, destruction of a common weed collateral host *Argemon mexicana* , sowing of seeds within first week of November, proper balance between nitrogenous fertilizer (90-12 kg N_2O/ha) and FYM (100-200 q/ha) and avoidance of excess soil moisture significantly reduce the disease incidence

Biological Control

Seed treatment with biofertilizer Azotobacter reduces requirement of nitrogen fertilizer as well as disease incidence. Seed treatment with fungal antagonist, *Trichoderma viride* (10g/kg seeds, 2 × 10' cfu/g) followed by drenching of 15-day old seedlings with water suspension of the antagonist (100/ litre) provide effective control.

Integrated Management

An integrated management strategy consists of improved cultural practices as mentioned above, seed treatment and drenching of 15-day old seedlings with *Trichoderma viride* and sprayings of metalaxyl at 35 and 55 days after sowing provides maximum disease control.

10

Downy Mildew of Medicinal and Aromatic Plants

10.1. Downy mildew of *Plantago psyllium*

Downy mildew is the major and serious disease of Plantago particularly P. ovata. It causes enormous quantitative and qualitative loss and often becomes a limiting factor in the successful cultivation of the crop (Desai and Desai, 1969). Usually the disease appears at the time of spike initiation.

Pathogen: *Peronospora alta* and *P. plantaginis*

Geographical distribution

Plantago ovata under commercial cultivation has been reported to suffer from downy mildew disease. Downy mildew occurs on ripe seeds of Spanish, French and European plantago. The downy mildew is a native of Mediterranean region especially Southern Europe, North Africa and West Pakistan.

The plantago commercially cultivated in Gujarat and to a limited extent in Haryana and Rajasthan provinces in India is observed to be infected with downy mildew.

Symptoms

Initial symptoms consist of small pale yellow patches on the leaves which progressively spread, completely destroying the leaves. The chlorotic areas develop in the upper surface of the leaves and ash white frost like mycelial growth develops on the lower surface of the leaves (Fig.10.1).

Fig. 10.1: Symptoms of downy mildew on plantago leaf.

Subsequently, infected leaves become chlorotic and brownish in colour, followed by curling, crinkling and drying of plants (Desai and Desai, 1969). Occasionally, chlorotic streaks, extending along with the midrib of leaves are observed. In such cases profuse downy growth is observed on the lower surface of the leaves. Ultimately, the whole leaf turns necrotic and in severe cases the entire plants appeared blighted. The abundant oospore formation caused thickening of affected leaves (Kapoor and Chowdhary, 1976). It also infects florets of the plant.

The Pathogen

The disease is reported to be caused by two pathogens (i) *Peronospora alta* Fuckel (Kapoor and Chowdhary, 1976) and (ii) *Peronospora plantaginis* Underwood (Desai and Desai, 1969).

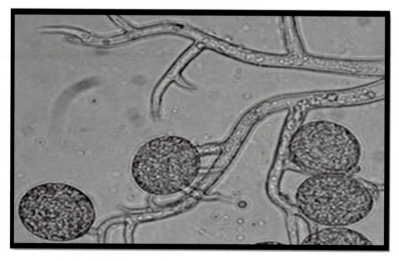

Fig. 10.2: Sporangiophore and Sporangium of *P.alta*

The *Peronospora Alta* mycelium is intercellular, haustoria hyphal with infrequent sporangiosphores measuring 290-450 x 8.0 μm; branching obscurely dichotomous 4-5 times; branch slender, tapering and curved at an angle usually less than 90° (Fig 10.2). Oospores rarely found (Kapoor and Chowdhary, 1976).

The *P. plantaginis* hyphae is intercellular with average breadth of 9.5 μm. Sporangiophores appear slender and tree like with an erect trunk, grey to pale yellow violet in colour emerging singly or in Clusters from stomata on the lower surface and are dichotomously branched. The Sporangia are sub-hyaline, broadly elliptical to subglobose and measuring an average of 35 X 21 μm (Patel 1984).

Disease cycle

Sporangia upon germination produced germ tubes and penetrate through stomata. Sporangia and sporangiophores emerged through stomatal opening before the death of the tissues. Zoospore formation is never observed (Patel, 1984). Under field conditions germination of the sporangia is favoured by temperature around 20°C and relative humidity of 82 Per cent.

Control measures

It is difficult to keep the disease under control once it has affected the crop extensively. The disease is seed as well as as soil borne. Soaking of the seeds in aureofungin (0.75%) solution coupled with two spraying of aureofungin at 15g/ha controlled the disease effectively. The first spray is to be given immediately after the appearance of the disease and is repeated twice at an interval of 8-10 days (Desai and Desai, 1969).

Seed treatment with metalaxyl (5 g / kg seed) coupled with three sprayings of captafol (0.2%) or metalaxyl (0.05%) effectively controlled the disease (Patel1984). The spraying is done first after the appearance of the disease and repeated at 10 days intervals. Fungicidal Sprays with Bordeaux mixture, copper oxycholoride, Dithane M45 or Dithane Z-78 would be useful prophylatic measure for the control of the disease. Studies have shown that high seed rate increased disease severity and therefore should be avoided. No significant source of resistance is available (Desai and Desai, 1969).

10.2. Downy mildew of *Atropa belladonna*

Pathogen: *Peronospora* sp.

Symptoms

The disease appears on the lower surface of the leaves as small mycelial patches, which soon enlarge. On the upper surface yellowish spots are evident after sometime (Fig.10.2). The downy growth of the pathogen on the lower surface consists of sporangia and sporangiophores. As the disease advances, the leaves may wither away.

Fig. 10.2: Downy mildew of belladonna.

Favourable environment

High humidity favours the development of the disease.

Control measures

Proper spacing, balanced irrigation and proper drainage reduce the disease development. Spray with fungicides like blitox, bavistin etc. are effective in reducing disease levels.

10.3. Downy mildew of *Hyoscyamus* sp.

Pathogen: *Peronospora hyoscyami*

Economic losses

The downy mildew disease is known to cause heavy losses in crop yield as well as reduce the alkaloid content significantly.

Symptoms

Yellowish greasy spots appear on the upper surface of leaves with whitish mycelial growth on the corresponding lower leaf surface. The heavily affected leaves curl, become necrotic dry and gives bushy stunted appearance to the plant (Fig.10.3).

Fig. 10.3: Downy mildew of hyoscyamus

Favourable environment

Cool moist weather favours the disease.

Control measures

Spray with blitox, bavistin or metalaxyl have found effective.

10.4. Downy mildew of Basil

Pathogen: *Peronospora belbahrii*

Geographical distribution and epidemics

Downy mildew of basil caused by *Peronospora belbahrii* has been a huge problem for both commercial producers and home growers.

The disease was first reported in Italy in 2004 and subsequently in U.S.A in 2007 and 2008 and has been steadily increasing in prevalence, distribution, and economic importance since then. It was officially identified in Minnesota in 2012. Under the right weather conditions basil downy mildew spread rapidly and results in complete yield loss. It has spread rapidly in the U.S. since, its first detection in South Florida in 2007. In 2008, basil downy mildew was detected in Kansas, Massachusetts, Missouri, New Jersey, New York, North Carolina, and Pennsylvania. The disease was subsequently identified in Alabama, California, Connecticut, Delaware, Georgia, Illinois, Indiana, South Carolina, Vermont, and Virginia in 2009, in Kentucky, Ohio, Texas, and Wisconsin in 2010 and in the Hawaiian Islands in 2011. Sporulation on the abaxial leaf surface and chlorosis of the adaxial leaf surface make basil leaves unacceptable for the market place.

Downy mildew has also been reported in many other countries, such as Switzerland, Italy, France, Belgium, Israel, NewZealand, and South Africa during 2001-2005. Damage from the downy mildew disease has also been reported in Cameroon, Canada, Cuba, and in greenhouse-grown basil in Argentina. Farmers in Europe have experienced the impact of the downy mildew disease as earlier as 2001, but reports of the disease have continued as recently as 2010 from Hungary. This implies the possibility that the disease may already exist in other countries but has neither been detected nor reported. Prior to these outbreaks, downy mildew had only been reported on basil in Uganda in 1933.

Recently, downy mildew was also observed on ornamental plants related to basil, particularly coleus (*Solenostemon* spp.) and salvia (*Salvia* spp.), which belong to the Lamiaceae family and includes basils (*Ocimum* spp.), mints (*Menta* spp.), sages (*Salvia* spp.) and other aromatics.The downy mildew pathogen that infects plants related to basil, such as the ornamental plants coleus and salvia, is present in Florida. However, it is not known whether this is the same downy mildew pathogen and which particular hosts are infected by this strain, although the isolate from coleus is reported to infect basil in greenhouses in New York.

2014 Basil Downy Mildew Outbreak

Basil downy mildew has been confirmed in Chatham and five other North Carolina counties. The disease first appeared in the U.S. in 2007 and has been wide spread throughout the country mostly in the eastern U.S. Many of farmers grow several varieties of both culinary sweet basil and ornamental basil for cut flower bouquets. They often do multiple plantings to extend the harvest. Basil downy mildew is able to devastate a basil crop. One Chatham County grower just had to disc in her entire basil crop due to a severe infection of downy mildew. Even the transplants in the greenhouse which were to be planted became infected. Basil downy mildew has caused a significant economic loss to this grower and many others. Basil downy mildew can be overlooked by growers because the initial symptoms tend to be subtle; the plants exhibit slightly yellowing of leaves as their most noticeable symptom, which may resemble a nutritional deficiency. The dark fuzzy-looking sporulation occurs on the undersides of the leaves. The disease is serious because it renders the leaves unmarketable.

Symptoms

Basil downy mildew causes yellowing, or chlorosis, initially within the leaf veins, on the upper surface of leaves, and purplish, fuzzy growth on the lowerside of leaf (Fig 10.4). Symptoms often start on the lower leaves and moves up the plant, causing defoliation from the bottom to up side.

Fig. 10.4: Symptoms of downy mildew on basil plant

Upon close observation of individual leaves, an irregular-shaped chlorosis or typical vein bounded symptoms are clearly discernible when contrasted to the green foliage of healthy basil leaves. Growers generally do not realize their basil plants are infected with downy mildew disease since the most noticeable symptom on affected basil is leaf yellowing, which is phenotypically similar to the result of a nutritional problem. The discolored area may cover most of the leaf surface. On the underside of leaves, a gray, fuzzy growth may be apparent upon visual inspection. Under high humidity, the chlorotic areas on the leaf turn from dark brown to black very quickly. Sporangia, the reproductive structures of the pathogen are a diagnostic for this disease.

Disease cycle

Germination of sporangia takes place 48 h after inoculation with a single germ tube per sporangium and this is followed by the entry of the germ tube through stomata into the host. Several days after inoculation, hyphae emerged from the stomata of infected leaves. Extensive hyphal growth is observed with the characteristically well-differentiated sporangiophores showing the determinate growth and dichotomous branches ending in slender, curved, right-angled branchlets. In many cases, 2 to 3 hyphae could be seen exiting a single stoma. In addition to hyphae, light brown oval sporangia are observed at the distal ends of the sporangiophore. These sporangia serve as inoculums for secondary infection to occur in the field.

Environmental Factor

The downy mildew pathogen Peronospora belbahrii, can be carried on seed, transplants, or fresh leaves. Infected transplants and leaves may not show symptoms if maintained in cool dry conditions. Spores of *P. belbahrii* can also travel long distances on moist air currents.

Peronospora belbahrii tolerates cool weather and can infect and produce spores in temperatures as low as 15⁰C. The pathogen, however, thrives in warm, humid conditions. As a result the most devastating damage is often seen in late summer.

Peronospora belbahrii needs two different mating types to produce tough resting spores known as oospores. Currently only one mating type has been found in the USA. As a result no oospores are formed and the pathogen may not be able to survive Minnesota's harsh winters. This may change if the second mating type is introduced.

Rapid sporulation and dissemination of the pathogen can be observed during periods of high humidity, mild temperatures, poor air circulation, and duration of leaf wetness (Garibaldi *et al.*, 2005, 2007). The disease can be spread from infected seed or leaves or from wind-dispersed spores which can travel long distances. It thrives in humid summers. Really high temperatures knock it back, which is why the disease was not as severe during the summers of 2011-2012 in the U.S.

Control measures

Few fungicides are currently labeled for downy mildew control on basil. Some phosphorous acid fungicides are effective against downy mildew under herbs on the current label. These fungicides were effective in fungicide efficacy experiments with applications started before or after initial symptoms were found. Actinovate AG is an OMRI-listed fungicide that is labeled for use on herbs and for suppressing foliar diseases, including downy mildew. Other fungicides are expected to be labeled for this use in the future.

Although few fungicides are specifically labeled for this disease, some fungicides that are labeled for basil may be useful in disease management. Reducing the period of leaf wetness by avoiding overhead watering may also be helpful. Heavily infected plants should be discarded. If possible, isolate new plantings to reduce inoculum spread from older plantings. The pathogen is believed to be seed transmitted.

Resistant Varieties

There are no resistant varieties of sweet basil (*Ocimum basilicum*) available. Commercially popular varieties are highly susceptible. Lower disease levels have been observed in red leaf basil varieties (*O.basilicum purpurescens*) and in lemon flavored varieties (*O. citridorum*). Only varieties of *O.americanum*, have shown no symptoms or sporulation when inoculated with downy mildew.

Cultural Control

Peronospora belbahrii is carried on seed. All seedlings and transplants should be monitored closely for yellowing of leaves and gray downy growth on the lower surface of the leaf. If basil downy mildew is identified on any plant, it should be removed and destroyed immediately.

Increase row width and distance between plants to provide good air movement between plants to allow leaves to dry quickly after rain, dew or irrigation. Use drip irrigation if possible. If sprinkler irrigation is the only option, water deeply and infrequently early on a sunny day so leaves dry quickly in the sun.

In greenhouse production, adjust ventilation to reduce humidity. Diseased plants that are past harvest should be promptly tilled under to reduce the spread of the pathogen from one plant to another through spores produced on infected leaves.

Fungicides

Certain fungicides can protect plants from basil downy mildew but sprays must begin before infection occurs to be effective. In one study, extreme periods or rainy wet weather resulted in no control by any fungicide combination. All label instructions should be read and followed whenever a fungicide is applied. Tank mixes and rotation between fungicides should be used to reduce the risk of fungicide resistance.

In Ontario, the fungicides cyazofamid (Ranman and Torrent), mandipropamid (Revus) and phosphorous acid (Confine) are registered for control of downy mildew in commercial field basil. All of these products are preventative, and will have limited effect on the disease once symptoms are widespread in the field. It is important for growers to be aware that once leaves are infected with downy mildew, it takes at least a week for symptoms to develop. Consequently, seemingly healthy leaves at harvest may develop symptoms post-harvest. This is not an issue for dried basil, as long as drying done as soon as possible after harvest.

11

Downy Mildew of Ornamental and Flowering Plants

Downy mildew has been reported on a wide range of ornamentals, especially those in home gardens, but only in a few instances the diseases have been sufficiently severe and widespread .Generally the diseases have been of most concern in a few ornamentals grown as crops either out doors or under glass *or* in nurseries, where many seedlings are raised in close *proximity*, often under humid conditions which favour this disease. Downy mildew having importance on ornamentals and flowering plants are:

11.1. Downy Mildews of Anemone (Windflower)

Two downy mildews are reported on cultivated anemones. The one is caused by *Plasmopara pygmaea* while the other is caused by *Peronospora* sp.

Pathogen .1. : *Plasmopara pygmaea*

It was first recorded in Britain on decaen anemones near Liskeard, Cornwall, in september 1917.

Symptoms

It causes blackening of the foliage of the planting. The sporangiophores emerged from stomata at the periphery of the blackened zone. The fine white mildew covers the under surface of the leaves. Plants become discolored with aborted stamens. Oospores are found in the discolored tissues.

Pathogen .2. :*Peronospora* sp.

Distribution

The first British record is in 1935 from south-west England by Gregory (1950). From 1950 onwards, the fungus became widespread in south-west England (Moore, 1959). It was then found in Jersey (Phillips, 1958) and the Netherlands (Boerema and Silvar, 1959).The Dutch workers considered the disease to be identical to that described in England but cast doubts on the identity of the *Peronospora* sp. The fungus continued to spread south and was reported on anemone varieties in Alpes-Maritimes departments of southern France by

Tramier (1960). The disease meanwhile, continues to concern anemone growers, and the severe outbreaks was reported in Italy (Garibaldi and Gullino, 1972).

Symptoms

The affected leaves have marginal, blackish lesions and tend to curl upwards bringing the sporangiophores, which emerged from stomata on the lower surface, into an erect position. The parasitized leaves in the beginning are typically erect and dull greyish-yellow but the tissues later become necrotic and blackish. When the disease is severe, the leaf development is much reduced, the flower stalks remain short, the flowers sometimes abort and the plant has a stunted, dome-shaped appearance.

Pathogen

The mycelium of the fungus may be found throughout the plant. It consist of intercellular, non septate mycelium, 4-6 μm in diameter with conspicuous digitate haustoria in the parenchymatous cell. The sporangiophores, which emerge from stomato, are 200-250μm long and bear ovoid sporangia The sporangia measure 28-39 x 23-30μm and germinate by zoospores. Gregory referred the fungus provisionally to *Peronospora ficariae. Peronospora sp.* form many brown, thick-walled oospores in the leaf stalks, and also occasionally in the flower stalk which remained infective for 3-4 years. The oospores measures 27-44 μm in diameter, and form abundantly in the dying leaves and flowers. Tramier(1960) considered the spore measurement to be sufficiently different from those of Peronospora ficariae to distinguish the fungus as a new species, confined to Anemone coronaria and A.globosa, which describes as Peronospora anemones(Tramier,1963). Infection may be initiated either by oospore via the root or by sporangia on shoot.

Weather parameter

In the south of France high incidence of mildew is associated with rain in August and September. Maximum sporangial production requires 11 hrs with 100% relative humidity at 12-18°C temperature and for germination to penetration it requires about 8hrs of leaf surface wetness. Tramier (1963) found that sporangial germination was optimal at 11-13°C and that leaves were penetrated directly without the formation of zoospore (Gregory,1950). The sporangia are disseminated particularly at flower picking and also presumably in rain splash.

Disease management

Good control of the disease has been obtained with soil sterilant such as metham-sodium (Tramier,1965), but in many areas this is too costly (Anon,1971), and attention must be given to crop hygiene and rotation.

11.2. Downy mildew of Antirrhinum(Snap-dragon)

Pathogen: *Peronospora antirrhini*

Host plant: *Antirrhinum orontium*

Distribution

Peronospora antirrhini was described by Schroeter as early as in 1874 as the pathogen of cultivated antirrhinum orontium. Until 1936 it devastated nursery stock in several localities and spread to many countries. The disease appeared in Carlow, Ireland in 1937 with an outbreak which destroyed 50000 seedlings (Murphy, 1937). In the following year, the disease was found in Sussex, England (Green,1937, 1938) and by 1952 over 50 outbreaks had been recorded in 25 countries throughout England and Wales (Moore and Moore,1952). During this period the disease had also been reported from Scotland (Wallace,1948), and various European countries, e.g. Denmark, Sweden (Green,1943), Norway (Jorstad,1946), and from Canada (Conners and Savile,1950), several parts of USA (Yarwood, 1947; Jeffers,1952) and from Australia (Anon,1941; Simmonds,1951) and New Zealand (Brien,1946).

In many instances the outbreaks were severe. Yarwood (1947) reported that during 1940-1942 the disease was so destructive in southern California that it was almost impossible to obtain healthy seedlings. P. antirrhini was first observed as a pathogen of cultivated antirrhinum in Norway near Oslo in 1935 but did not reappear until 1944 in the same locality (Jorstad, 1946). The first record in Sweden was in January 1943 from the Weibullsholm Plant Breeding Institute (Green, 1943). Speculation that the fungus is seed-borne has not been substantiated (Yarwood, 1947; Moore and Moore, 1952) but contamination of seed with oospore containing plant debris cannot be ruled out.

Disease Symptoms

The infection on antirrhinums are of two kinds i.e systemic and local. The former are more important and results in the downward curling of the leaves with a marked reduction in growth which results in a rosette of small leaves at the tip of shortened shoot and gives stunted appearance to the affected seedlings (Fig 11.1). In most instances the tip die and affected plants then produce many secondary shoot from the base. The local, non systemic infection commonly appear on leaves as round, pale lesions but are rarely destructive.

Fig. 11.1: Symptoms of downy mildew on Snap dragon

Pathogen

The intercellular mycelium within the tissue give rise to intracellular haustoria, characteristically with 4-8 finger like branches. The dichotomously branched sporangiophores, 350-700 µm long emerge from stomato which bears ovoid shaped conidia, 14-17 x 21-29 µm in size at the tip of sporangiophores. The sporangia are white to yellowish brown. The sporangia germinate by a germ tube, optimally at 13°C in water films. Sporulation is favoured by relatively low temperature and high humidity. In experiments by Yarwood (1947), the sporulation was abundant on systemically infected leaves kept in humid conditions at 13°C and moderate at 19°C. There was no sporulation at or below 7°C or at 22°C and above. Oospores(30-38µm diameter)form abundantly in the petioles, stems and roots of infected plants and are the means of fungal perennation (Mckay,1949).

Weather

The disease is promoted by high relative humidity and low temperatures. The fungus produces spores over the temperature range of 7°C - 22°C with optimum production at 13°C. Spores will only germinate in water. Oospores (sexually produced spores) form in leaves, stems and roots, which later fall to the ground resulting in oospore contamination of the soil. Contaminated seed is suspected to disperse the disease, however, this has never been proven.

Disease management

In glasshouses some control of the disease has been achieved by reducing humidity either through forced ventilation (Yarwood,1947) or by raising the temperature (Pettersson,1954) and by protective spraying with copper or dithiocarbamate fungicides (Green,1938., Jeffers,1952). Sterilization of soil and equipments under glasshouse cultivation may destroy the oospores (Mckay,1949). Reducing relative humidity has been shown to control the disease. Reports of varietal resistance to the disease are contradictory. Breeding snap dragons for resistance to downy mildew was unsuccessful in California.

11.3. Downy mildew of Rose

Pathogen: *Peronospora sparsa*

Distribution

The disease was first reported in 1862 on roses in a cool glasshouse in England by M.J.Berkeley. Subsequently the disease was reported in most European countries, Canada and the USA. It was regarded mainly as a disease of roses in the northern hemisphere. Baker (1953) points out that all records, until 1953, were from north of the Tropic on Cancer expect one from Sao Paulo, Brazil. Since then the fungus has expanded its distribution. It was found in Mauritius in 1965 (Anon,1967), in Isreal in 1968 (Brosh, 1970) and in Australia particularly new south wales in 1971 (Johns, 1972) where it has become widespread (Bertus,1977).

Symptoms

The fungus infects young leaves, stems at the shoot apex, peduncles, calyxes and petals. Frequently, though not invariably, infected leaves and stem develop purplish to black spots, a feature which has merited the name black mildew for the disease (Fig 11.2). Infected leaves often become yellow and drop prematurely, and infected shoots are sometimes killed.

Fig. 11.2: Symptoms of downy mildew on rose leaves

Pathogen

Peronospora sparsa under humid conditions, produces sporangiophores 350 µm long bearing sub-elliptical sporangia, 17-22 x 14-18 µm in size, abundantly through stomata on the lower leaf surface. Under dry condition they are often sparse (hence the species name sparsa) and thus easily overlooked. Sporangia may be produced over longer period from an infected leaf, a feature which assists in the development of epidemics. However, these rarely develop

134   The Plant Mildews

outdoors; the disease occurs mainly in glasshouses and even here outbreaks tend to occur sporadically.

Weather condition

Disease development is favoured by humidities of 90-100% and relatively low temperatures. The sporangia germinate best at 18⁰C and require only 4 hrs in water dews for this, but they do not germinate at all below 4⁰C or above 27⁰C temperature. The formation of oospore seems to be variable but even where they do not form, there is some evidence that the fungus perennates as dormant mycelium in stem (Pickel, 1919., Stahl,1973).

Disease management

Control of the disease can be achieved in glasshouse by temporarily raising the temperature to 27⁰C and ventilating to reduce humidity to less than 90% and by raising temperatures above 27°C (MacLean and Baker, 1951; Jaude, 1959; Brosh, 1970; Gill, 1977). Roses grown at less than 85% RH are not infected. Control measures should include the destruction of infected leaves and stems. The severity of the disease can be reduced by spraying with fungicides of various formulations containing copper and dithiocarbamates (Baker, 1953; Baresi, 1957; Taylor, 1963; Marziano et.al, 1973., Aleksandrova, 1976; Bertus, 1977).

11.4. Downy mildew of Alyssum saxatile

Pathogen: *Peronospora galligena*

Host plant: *Alyssum murale, Alyssum montanum, Alyssum* spp. and several crucifers.

Distribution

It was first described by Blumer in 1938 from collections made in 1937 in the Berne district of Switzerland on *A. saxatile* and its varieties *citrinum* and *compactum.* It was subsequently reported in Germany in 1938 and in England in 1946 (Moore, 1949). A similar type of disease was recorded in west Switzerland, involving malformations of the shoots and inflorescences of *A. saxatile.*

Symptoms

This fungus causes small blisters or gall-like growths on the leaves of its hosts. Profusely branched sporangiophores 400-800 μm long, with ellipsoid or globose sporangia, 16-19 x 14-17 *μm in size* emerge from the disease blistered portions.

11.5. Downy mildew of Chrysanthemum

Pathogen: *Peronospora radii*

Host plant: *C.* cinerariifolium

Symptoms

Peronospora radii form dark, pin head lesions on the outer floret of the American spray variety Yellow Top. The fungus was also found on the varieties White Top, Igloo, Iceberg and Shasta in nearby beds. Some oospore (mean diameter 30μm) are found in infected floret but there are no leaf infection.

Pathogen

The sporangia appeared dark violet en masse and are ovate, 22-36 x 18-25μm in size. Mikuskovic (1968) in Yugoslavia reported the sporangiophores to be 280-630μm (average 423μm) long with ellipsoid ovoid conidia measuring 26-33 x 15-20μm.

11.6. Downy mildew of Cineraria

Two downy mildews pathogens have been reported,

Pathogen: 1) *Plasmopara halstedii* and

2) Bremia lactucae

Host plant

Compositeae, such as Centaurea, Gaillardia (Blumer, 1938b; Viennot-Bourgin, 1954) and Helichrysum (Moore, 1959).

Distribution

Gill (1933) describes an outbreak of P. halstedii on Senecio cruentus in a glasshouse at Long Island, USA.

Symptoms

Pale green spots develop on the upper epidermis of the leaves, and a felt like, whitish mold appears on the undersides of the leaves, corresponding to the spots on the top. The affected leaves shrink and die. It is especially destructive to young plants. On Senecio cruentus, P. halstedii infected spots are up to 3 cm in diameter which are white on the upper surface and slightly brown on the lower surface (Fig.11.3). *B. lactucae* on cinerarias causes irregular leaf spots which are yellow to red in colour and leads to withering and defoliation.

Fig. 11.3: Symptoms of downy mildew on cineraria leaves

Pathogen

Sporangiophores are 247-640 μm long, sporangia 15-30 x 13-23μm in size. These are smaller than those given for this fungus by Wilson (1907).

Control measures

Ferraris (1936) recommends keeping plants in warm but not over damp conditions. Keep the plants as dry as possible. Spray with copper fungicides or Bordeaux mixture. Wittman (1972) reports that the fungus was controlled on Helichrysum bracteatum and other species by spraying early with copper formulations or with zineb, maneb, ziram, captan or daconil.

11.7. Downy mildew of Clarkia(Clarkia elegans)

Pathogen: *Peronospora arthuri (Lewis, 1937; Moore,* 1959)

Distribution

This mildew occured in various localities in Denmark in 1942. It also occurs in Godetia in California.

Symptoms

The first infections are found on the lower leaves as white masses of sporangiophores on the abaxial surfaces which later became grey with faint purplish tings and eventually brownish-grey. On the adaxial surfaces there were pale yellow areas associated with this speculation.

Control measures

Neergaard (1943) suspected that the fungus was seed-borne and recommended a 30 min soak in 0.5% upsulun.

11.8. Downy mildew of Viola (violet, pansy)

Pathogen: *Peronospora violae, Bremia megasperma*

Distribution

Downy mildew on violas is caused by *Peronospora violae* in Europe, Asia and Australia, while in America it is caused by *Bremia megasperma.*

Symptoms

Symptoms consist of light green to yellow patches which appear on the upper leaf surface (Fig 11.4). Masses of mauve coloured spores form directly below on the under leaf surface. There are no reports of the disease being seed borne; however, this may be because nobody has checked.

Spectacular epidemics have been reported after sprinkler watering in dry autumn weather. The spores of *P. violae are trapped* in the mornings, suggesting that the fungus is similar to *P. parasitica* in releasing its spores in the mornings.

Fig. 11.4: Symptoms of downy mildew on Viola (pansy)

Control

Provide adequate ventilation & destroy plants on which the disease appears. Downy mildew develops especially under damp conditions. In France, pansy cultivars were evaluated for their susceptibility to the disease. The white, clear blue and yellow flowering varieties are more susceptible to the disease than the variegated ones.

11.9 Downy mildew of Helleborus (*Christmas rose*)

Pathogen: *Peronospora pulveracea*

Symptoms

The fungus induces precocious flowering, discolouration of the petals, irregular curling of the leaves and general stunting of the plant, and it sometimes seriously affects glasshouse crops of this ornamental.

11.10. Downy mildew of Matthiolae

Pathogen: *Peronospora parasitica, Peronospora matthiolae*

Host plant
Cruciferous plants particularly *M.incana* and *M.bicornis*

Distribution
New Zealand following the first outbreak in1958.

Symptoms
Downy mildew is common *on* the cultivated stock (*Matthiola incana*) especially at the seedling stage. The first symptoms are light green patches on the upper leaf surface with corresponding areas on the lower surface covered with the white downy mold of the massed sporangiophores and sporangia. Infected leaves turn yellow, then become necrotic and often fall prematurely (Fig 11.5). The tender stems & flower parts are also attacked, stunted & dwarfed.

Fig. 11.5: Symptoms of downy mildew on Matthiola

Weather parameter
Infection and sporulation were optimal at 15. 5-21°C, whilst the range is 4.5°C-27°C. The fungus is not active at temperatures outside this range. Under these conditions lesions developes in 5-6 days. The sporangia germinate best at 15.5°C, rather poorly at 10°C or 21°C and not at all at 4.5 and 27°C. No oospores were found.

Control measures
Some control of this disease has been achieved with 0.5% buisol (Anon., 1951), or by spraying weekly with dithiocarbamates (Jafar, 1963) or every 3 to 4 days with a copper oxychloride-zineb mixture (Bertus, 1968).

11.11. Downy mildew of Mecanopsis

Pathogen: *Peronospora arborescens*

Host plant
Several species of Mecanopsis and Papaver particularly the opium poppy, *P. somniferum*

Symptoms
Leaves, buds, calyx, capsules & seed coats are all invaded by this fungus. On the upper sides of the leaves yellowish or light brown blotches occur, which later turn very dark in color. Black spots or blotches on the upper surfaces of leaves are associated with white-grey or faintly mauve masses of sporangiophores emerging from the lower leaf surface. Ornamental garden poppies are sometimes killed at the cotyledon stage whilst older plants become stunted with twisted or swollen flower pedicels and flower buds which fail to open. The fungus is presumed to overwinter as oospores in soil and possibly is distributed as oospores on seed.

Control measures
Spraying seedbeds with copper fungicides has been recommended (Beaumont, 1953). The fungus lives through the winter in the soil & on plant parts that have become more or less buried by debris; all parts of infected plants should be gathered & carted away. Use only seeds from uninfected plants.

11.12. Downy mildew of Papaver (poppy)

Pathogen: *Peronospora arborescens*

Symptoms
This disease often appears as seedling blight in gardens on older plants. Light green to yellow blotches or patches appear on the upper leaf surface (Fig 11.6). Leaves appear distorted and eventually turn brown. Masses of fluffy grey spores develop on the lower leaf surface.

Fig. 11.6: Symptoms of downy mildew on Papaver

Weather Condition

The optimum temperature for spore (vegetative) germination is 19°C, whilst the range is 2°C-23°C. Beyond this range germination is delayed. The fungus, *Peronospora arborescens,* is reported to be seed borne on *Meconopsis* spp. (Chinese poppy) and *Papaver* spp (poppies), especially in the Middle East. Oospores (sexually produced spores) form prolifically in leaves and contaminate soil after leaf fall, thus carrying the fungus over to subsequent crops. The disease is favoured by heavy potting mixes, wet and humid weather.

Control

Spray with a copper fungicide. Several applications may be necessary for control. At the earliest in the spring, destroy infected plants. Collect seeds only from disease free plants. Avoid planting in damp situations.

11.13. Downy mildew of Veronica (Speed well)

Pathogen: *Peronospora grisea*

Symptoms

Large brown lesions develop on leaves and disfigure plants. Downy greyish spores form on the under surfaces of lesions (Fig 11.7). Young infected shoots dry out ,wither and die. Densely planted cuttings are very susceptible to the disease.

Fig. 11.7: Symptoms of downy mildew on Veronica (Speed well)

11.14. Downy mildew of Mimulus
Pathogen: *Peronospora jacksonii*

Distribution
A downy mildew fungus on *M. guttatus* was first described by Shaw (1951) from Oregon, USA as *Peronospora jacksonii*. The same fungus was subsequently found affecting *Mimulus* cuttings at Wisley, England by Mence (1971) and this appears to be the first European record though it was quickly followed by another from Germany (Doppelbaur and Doppelbaur, 1972).

Symptoms
Symptoms consist of light green to yellow patches which appear on the upper leaf surface (Fig 11.4). Masses of mauve coloured spores form directly below on the under leaf surface.

Control measures
Mence (1971) found Orchid to be the most susceptible cultivar and was able to control the disease by spraying with zineb and improving the ventilation.

11.15. Downy mildew of Arabis (Rock cress)
Pathogen: *Peronospora parasitica*

Symptoms
A light gray mildew growth develops on the undersides of the leaves & over the young stems, causing some distoration and swellings of infected parts.

Control
The prompt removal of infected plants is advisable and an application of a fungicide containing copper may be necessary.

11.16. Downy mildew of Argemone (Prickly-poppy)
Pathogen: *Peronospora arborescens*

Symptoms
Leaves, buds and capsules of prickly poppy are affected by this mildew. A light gray mold develops on the undersides of leaves. On the upper surface yellow to brown blotches eventually turn dark.

Control

Remove & destroy infected plants. Use clean seed.

11.17. Downy mildew of Centaurea (Cornflower)

Pathogen: *Bremia lactucae*

Symptoms

The fungus usually causes the development of pale greenish or reddish irregular spots on the upper sides of the leaves, while the undersides may be covered with a soft moldy growth (Fig 11.8). The leaves collapse & the attacked parts die. Young plants are especially susceptible. The downy mildew fungus *Plasmopara halstedii* has been reported from Iowa.

Fig. 11.8: Symptoms of downy mildew on Centaurea leaf

Control

Remove and destroy badly infected plants. Space plants widely apart and provide full light and aeration. An occasional application of Bordeaux mixture or some other copper fungicide will also help to cure downy mildew.

11.18. Downy mildew of Gramineae (Grasses)

Pathogen: *Sclerophthora* spp.

In, 1969, St.Augustine grass in Florida & Texas was found to be affected by this disease with the common downy mildew symptoms on the plants.

Control

The disease does not appear to be serious enough to warrant control measures.

11.19. Downy mildew of Houstonia

Pathogen: *Peronospora calotheca* & *P. seymourii*

These two fungi occur more frequently on wild than cultivated plants with common downy mildew symptoms. Hence no controls are necessary.

11.20. Downy mildew of Lagneria

Pathogen: *Pseudoperonospora cubensis*

This mildew is occasionally serious in the Eastern & Southern United States.

Control

Spray the vines with Bordeaux mixture or other Copper fungicides if the disease is serious.

11.21. Downy mildew of Myosotis (Forget-me-not)

Pathogen: *Peronospora myosotidis*

Symptoms

The lower sides of the leaves usually have a downy fungal growth, with large quantities of spores. Pale spots appear on the upper surfaces of the leaves corresponding with the mold below.

Control

Valuable plantings can be protected from this disease with Bordeaux mixture or any readymade copper fungicide.

11.22. Downy mildew of Paeonia (Peony)

Pathogen: *Phytophthora cactorum*

Symptoms

This disease is similar in appearance to that caused by the Botrytis, except that it forms no characteristic mold & is never seen covering the bases of infected shoots or infected leaves or buds. The entire shoot may turn black & die. Cankers appear along the stems & cause them to fall over. The Botrytis mold seldom invades the crown, but Phytophthora often does & causes a wet rot to develop there, often destroying the entire plant. The variety 'Avalanche' is susceptible to Phytophthora.

Control

Because infections occur in the roots and lower portions of the stems, fungicidal sprays are of no value. Confirmed cases should be lifted out together with adjacent soil & thrown into the trash can. Such material should never be placed in the compost pile. Planting healthy clumps in a new spot where the soil is well drained usually prevents further trouble.

11.23. Downy mildew of Parthenocissus (Boston ivy & Virginia creeper)

Pathogen: *Plasmopara viticola*

Symptoms

Greenish yellow blotches appear on the upper surfaces of Virginia creeper leaves, which later turn brown on the lower surfaces, are covered with white mold (Fig 11.9). Heavily infected leaves may fall in large numbers. The same fungus causes even more serious damage to cultivated grapes.

Fig. 11. 9: Symptoms of Downy mildew on Parthenocissus

11.24. Downy mildew of Rudbeckia (Goldenglow)

Pathogen: *Plasmopara halstedii*

Symptoms

This disease may cause wilting and death of seedlings. Older plants may not die but the foliage is mottled and light yellow (Fig 11.10).

Fig. 11.10: Symptoms of downy mildew on *Rudbeckia* (Goldenglow)

Control

Set plants in clean soil. Spray with Bordeaux mixture during summer to prevent severe outbreaks of the disease.

11.25. Downy mildew of Stelbaria Holostea (Easter bells)

Pathogen: *Peronospora* sp.

Symptoms

The fungus winters in the stems as a mycelium. The inflorescence and stems are distorted & shrunken.

Control

Destroy all infected plants and plant debris at the end of the season. Change the location of the plantings. Spray with Bordeaux mixture or any other copper fungicides during the growing season.

11.26. Downy mildew of Impatiens

Pathogen : *Plasmopara obducens*

Geographical distribution

Impatiens downy mildew caused by *Plasmopara obducens* was first observed on garden impatiens (*Impatiens walleriana*) in the United Kingdom in 2002.

Prior to that, it was identified as early as 1897 on wild impatiens (jewelweed) in Vermont. In 2011, there were outbreaks of this disease in many states of USA including CA, IL, IN, NY, MA, MN, and WI. In January 2012, it was confirmed on plants in the landscape in Florida. During the 2012 growing season, impatiens downy mildew has been widespread in over 30 eastern states including Connecticut. A new and particularly aggressive form of impatiens downy mildew emerged as a major threat to the cultivation of ornamental impatiens in the United States, where they are one of the most popular ornamental plants.

The disease has become a serious issue in the United States, including Wisconsin. Impatiens downy mildew has been so destructive in many areas that it has made impatiens unusable as a garden ornamental. The disease affects garden impatiens (Impatiens walleriana and I. balsamina), as well as native jewelweeds (I. pallida and I. capensis). New Guinea impatiens (I. hawkerii) and its hybrids appear to be either resistant or tolerant to the disease. Other common garden ornamentals are immune to impatiens downy mildew and thus not affected by the disease at all.

Symptoms

Symptoms of impatiens downy mildew often first occur on leaves near the tips of branches (Fig 11.11). Initial symptoms include an irregular yellow-green discoloration of leaves that can be confused with spider mite feeding injury. Affected leaves often curl downwards. Stunting and reduced flowering are other common symptoms. As the disease progresses, leaves and flowers drop off, leaving a bare stem. Eventual death of affected plants can occur. The most distinctive characteristic of impatiens downy mildew is the presence of a fuzzy white fungal growth that develops on stems, buds and particularly the under sides of leaves.

Fig. 11.11: Symptoms of downy mildew on Impatiens

Host Plants

Susceptible hosts include standard garden impatiens, double impatiens and mini-impatiens and any hybrids of *I.walleriana*.including balsam or garden balsam (*Impatiens balsamina*). All varieties of *Impatiens walleriana* and any hybrid with *I. walleriana* in its background are susceptible to impatiens downy mildew. Touch-me-not (*I. balsamina*) and several wild species of impatiens can also be infected.

Pathogen

Plasmopara obducens produces sporangia on the lower surface of infected leaves. Sporangia can be splashed short distances to spread from plant to plant and can also become airborne and travel long distances on moist air currents. *Plasmopara obducens* thrives in cool (17.2-22.7p C) moist conditions. 4 hours of leaf wetness is necessary for sporangia to form. Under hot dry conditions, infected plants may show no symptoms of disease and produce no sporangia on the lower leaf surface.

Favourable Weather

Development and expression of impatiens downy mildew is highly influenced by the weather. Wet foliage, cool temperatures (especially at night), and moist air are ideal conditions for disease development. The disease was noticed late in the season in several locations when night temperatures began to drop into the 50s in early September. Downy mildew likes and requires moisture to sporulate and cause new infections. Plants in heavily shaded locations, where the leave stay wet for extended periods of time will generally have a higher incidence and severity of disease because moisture promotes infection and disease expression.

Disease cycle

The pathogen is commonly introduced into a garden on infected impatiens transplants. It can be spread in the garden by windborne spore-like structures called sporangia. Once established in a garden, *P. obducens* can spread from plant to plant by wind or rain splash. Cool, wet/humid weather favors disease development. *P. obducens* can potentially overwinter in a garden in the form of specialized spores called oospores. These spores can be found in soil and in infested plant debris. Whether P. obducens can be introduced via impatiens seed is unclear. Impatiens downy mildew can be spread short distances by water splashing from infected plants and greater distances by windborne spores from infected plants. Disease tends to be worse in locations where leaves stay wet for extended periods of time or the beds are very dense and the beds receiving overhead sprinkler irrigation, because the foliage does not dry quickly.

Control Measures

Do not plant *Impatiens walleriana* or any hybrid containing *I. walleriana* in previously infected beds. Oospores allow the pathogen to survive from one season to the next. Use Alternate plants for planting which include coleus, caladium, begonia, and New Guinea impatiens.

Use disease free seedlings

Impatiens can be planted into beds with no history of downy mildew but care should be taken to purchase disease free plants. Thoroughly inspect all transplants for yellowing foliage and downy growth on the lower leaf surface. Reject any plants with evidence of infection. Once a plant is infected, it cannot be cured. Choose a supplier that uses a regular fungicide treatment program to protect plants from downy mildew. Purchasing transplants produced in an area where landscape plants are not present during production (i.e. colder northern states) will further minimize the risk of receiving infected plants. Infection of landscape impatiens is still possible from windblown sporangia.

Reduce moisture and humidity

Space plants so that air moves easily between plants and leaves dry quickly. Set sprinkler irrigation for early morning watering and providing deep and infrequent irrigation to reduce leaf moisture. Avoid evening applications of sprinkler irrigation. Use drip irrigation if possible to keep foliage dry.

Destroy infected plant parts

If infection is found, bag and remove the infected plants, any fallen leaves, blossoms and the closest neighbors. Remove the entire plant including roots. Do not compost infected plant material. Fungicides will not cure an infected plant and it is better to remove the plant to reduce spread of the pathogen to other impatiens in the area. At the end of the season completely remove all plant material to prevent overwintering of the pathogen.

Fungicides

Several fungicides will protect plants from infection, but no fungicides will cure the disease once infection has occurred. In a few trials, plants drenched or treated with a granular application of Subdue Maxx at planting have remained disease free for 46 days to 2 months. In another study Adorn plus Heritage or Adorn plus Vital protected landscape impatiens for 5 weeks.

There is a high risk of impatiens downy mildew becoming resistant to certain fungicides if they are over used. Isolates resistant to mefenoxam have been identified in Europe. Rotate between different chemical families of fungicides (each family has a different FRAC Code) or tank mix two fungicides from different chemical families to avoid fungicide resistance developing. All label instructions must be carefully read and followed when applying a fungicide.

Home gardeners should contract with a licensed pesticide applicator to manage impatiens downy mildew fungicide applications.

Disease Resistance

New Guinea impatiens (*Impatiens hawkeri*) types are resistant to this disease.

Literature Cited for Downy Mildew

Abbott, E. V. and Hughes, C.G.1961. Sugarcane Diseases of the World.Vol.1.edited by J.P.Martin., Abbott, E V. and Hughes, C.G. Elsevier Publ.Co., Amsterdam, 141-164.

Achar, P.N. 1992. Yield loss in Raphanus sativus from systemic infection by Peronospora parasitica. Phyton (Buenos Aires). 53: 89-94.

Achar, P.N., 1996. First report of downy mildew disease in lettuce caused by Bremia lactucae in Natal, Southern Africa. Plant Disease, 80(4), p.464.

Agrios, G. N. 2005. Plant Pathology. 5th ed. pp. 427-33.

Ahmad Shabeer and Karimullah. 1998. Relevance of management practices in downy mildew in onion. Sarhad.J.Agric. 14:161-162.

Alcock, N. L.. 1933. Downy mildew of Meconopsis. New Flora and Silva. 5 (4): 279 - 282.

Aleksandrova, I.. 1976. Two new diseases of greenhouse Flowers. Rastitelna. Zasht.24, 10-11.

Amm, K. 1929. Physiological Utersuchungen an Plasmopara viticola, under besondererk Berucksichtingung der Infections bedingungen. Jahr. wiss. Botan., 70:93-157.

Anand, P.C. and Wehner, T.C. 1991. Crop loss to 14 diseases in cucumber in North Carolina for 1983-1988. Rep. Cucurbit Genet. Coop. No. 14pp. 15-17.

Anaso, A. B., Tyagi, P. D., Emechebe, A. M. and S. K. Manzo. 1989. Control of sorghum downy mildew (Peronosclerospora sorghi) of maize by seed treatment in Nigeria. Crop Protection. 8: 82-85.

Angelov, D. and P, Georgiev. 1995. Study of the susceptibility of cucumber accession to downy mildew, Pseudoperonospora cubensis. Rostov .Rasteniev" dni Nauki. 32(5): 245-246.

Anon. 1941. Agric.Gaz. N. S. W. 52, 538.

Anon. 1951. Tidsskr. P. I. A. vol. 55,70.

Anon. 1953. Tasm. J.Agric. 24,366-369.

Anon. 1955. 24th Annual Report NSW Department of Agriculture. Biological Branch, Division of Science Services. 37pp.

Anon. 1957. "Annual Report. Department of Agriculture" Kenya, 1955. Vol. 2,237 pp.

Anon. 1962. Quarterly Report for July-September 1962 Plant Protection Committee for the South-East Asia and Pacific Region. F.A.O. Publication, Bangkok, Thailand.19 pp.

Anon. 1967. Rep. Dep. Agric. Maurii. 1965, 91-97.

Anon.1968. Rev. Appl. Mycol.Plant host-Pathogen .Index to Volumes 1-40 (1922-1961). 820 pp. Comm. Mycol. Inst, Kew, Surrey, England.

Anon. 1969. Rep. Dep. Agric. Fiji 1968,29-30.

ANON. (1971). Guide for the assessment of cereal diseases. Ministry of Agriculture, Fisheries and Food, Plant Pathology Laboratory, Harpenden.

Anon.1972a. Plant pests of importance to the Caribbean. F.A.O. Caribbean Plant Prot Commission. 29 pp.

Anon. 1972b. A list of plant diseases, insect pests, and weeds in Korea. Korean Soc. Plant. Prot, 424 pp.

Anon. 1974. PI. Path. Annual Report for 1974. ADAS Science Service, MAFF180.

152   The Plant Mildews

Anon. 1981. Sunflower downy mildew. Plant Quarantine Leaflet No. 13. Common wealth Department of Health, Canberra, Australia.

Anon. 1996. Proceedings of Annual Group Meeting of Sunflower Research Workers, held at JNKVV, Jablapur, India. April, 13-15, 1996.

Anon. 1998. Proceedings of Annual Group Meeting of Sunflower Research Workers held at TANU, Coimbatore, India. April, 19-22, 1998.

Anon. 1999. Proceedings of Annual Group Meeting of Sunflower Research Workers, held at UAS, Dharwad, India. April, 26-28, 1999.

Anon. 2000. Proceedings of Annual Group Meeting of Sunflower Research Workers, held at PAU, Ludhiana, India. April, 26-29, 2000.

Anon. 2003. Proceedings of Annual Group Meeting of Sunflower Research Workers, held at TANU Coimbatore, India. April, 4-5, 2003.

Anon. 2006. Pflanzeschutz Nachrichten Bayer, 59:2:3 (http://www.bayercropscience.com/BCS Web/ Crop Protection.nsf/id/ENI-Journal)

Anon. 2012. Checklist of commercial varieties of vegetables, Dept. of Agriculture and Co-operation, Ministry of Agril, Govt. of India, pp 6,8,19,56.

Antonov, Y.P. 1978. For protection of cabbage and onion against diseases. Zashch. Rast.4:55.

App, F. 1959. The history and economic importance of lima bean downy mildew disease. Proc. Am. Soc. Hortic. Sci. 33:473-476.

Armstrong, G. M. and Albert, W. B. 1933.Downy mildew of tobacco on pepper, tomato, and eggplant. Phytopathology 23: 837-839.

Arx, J.A. Von & Noordam, D. (1951). ValsemeeldauwiPeronosporapuloeraceaFuckel) op Helleborusniger. Tijdschr. PlZiekt. 57, 32-34.

Babadoost. M., R. A. Weinzierl and J. B.Masiunas. 2004. Identifying and Managing Cucurbit Pests.University of Illinois Extension. p7.

Baider, A. and Cohen. Y. 2003. Synergistic interaction between BABA and mancozeb in controlling Phytophthora infestans in potato and tomato and Pseudoperenospora cubensis in cucumber. Phytoparasitica 31:399-409.

Bains, S.S. and Jhooty, J.S. 1976a. Over wintering of Pseudoperonospora cubensis causing downy mildew of muskmelon. Indian Phytopath. 29 : 213-214.

Bains, S.S. and Jhooty, J.S. 1976b. Host range and possibility of pathological races of Pseudoperonospora cubensis cause of downy mildew of muskmelon. Indian Phytopath. 29:214-216.

Bains, S.S. and Jhooty, J.S. 1978a. Epidemiological studies on downy mildew of muskmelon caused by Pseudoperonospora cubensis. Indian Phyioputh. 31: 42-46.

Bains, S.S. and Jhooty, J.S. 1978b. Relationship between mineral nutrition of muskmelon and development of downy mildew caused by Peronospora cubensis. Plant and Soil. 49:85-90.

Bains, S.S. and Jhooty, J.S. 1978c. Mode of efficacy of four fungitoxicants against Pseudoperonospora cubensis on muskmelon. Indian Phytopath.29: 339-342.

Bains, S.S., Sokhi, S.S. and Jhooty, J S. 1977. Melothria maderaspatana-A new host of Pseudoperonospora cubensis. Indian J. Mycol. Pl Pathol7: 86.

Bains, S.S., Sokhi, S.S. and Jhooty, J.S. 1981. Out-break of Peronospora parasitica on cauliflower curd in Punjab. Indian Phytopath. 34: 389-390.

Bains, S.S. and Jhooty, J.S. 1983. Host range and morphology of Peronospora parasitica from different sources. Indian J. Mycol. Pl. Pathol. 13 : 372-375.

Bains, S.S. and Prakash, V. 1985. Susceptibility of different cucurbits to Pseudoperonospora cubensis under natural and artificial epiphytotic conditions. Indian Phytopath. 38:138-139.

Literature Cited for Downy Mildew   153

Bains,S. S and H.S.Dhaliwal. 1986. Downy mildew of peas in Punjab. Indian Phytopathology. 39(3):493-494.

Baker, K. F. (1953). Recent epidemics of downy mildew of Rose. Plant disease reporter, 37(6).

Barloy, J. And J. Pelhate. 1962. Premieres observations phytopathologiques relatives aux cultures de chanvre en Anjou. Ann. Epiphytes. 13: 117-149.

Bary, A. De. 1863. Recherches sur le developement de quelques champignons parasites, Ann. Sci. Nat. France, IV, 20:5-148.

Battilani, P., Rossi, V., Racca, P., &Giosue, S. (1996). ONIMIL, a forecaster for primary infection of downy mildew of onion 1. EPPO Bulletin, 26(3 4), 567-576.

Beaumont, J. L., Coblentz, B., Maurice I'., Chevalier, H., et LenPgre, J.: Indications et resultats du traitement anticoagulant dam l'angine de poitrine severe B propos de 40 cas, SemainehAp. Paris 28:1926, 1952.

Beaumont,A. 1953. Cyclamen diseases. Gardeners' chronicle. Ser. 3, 133: 240.

Bedlam, G. 1989. First detection of Oospore of Pseudoperonospora cubensis (Berk et Curt) on glasshouse cucumbers in Austria. Pflanzenschutzberichte. 50: 119-120.

Behr, L. 1956 . Der falsche mehltau am mohn(Peronospora arborescens). Untersuchungen zur biologie des erregers. Phytopath. Z. 27:287-334.

Berkeley,M.J, and M. A. Curtis. 1888. Botrytis viticola. Rav. Fungi Carol. Exsz'c. V, no. 90.

Berkeley, M.J. 1862 . Peronospora s parsa.. Gardeners' Chronicle .London. 307-308 .

Berlese, A. N. and G. B. de Toni.1888. Plasmopara viticola. Sylloge Fungorum, 7: 3381.

Bertus,A. L. 1968. Downy mildew of Stocks. Agric. Gaz. N.S.W. 79: 178-179

Bertus, A. L. 1977. Rose downy mildew. Agric. Gaz. N.S.W. 88(4):8-9.

Bhatt, D. D. 1966. Set. Tri-Chandra Coll. Sci. Assoc. 2,13-20.

Biddle, A.J; Knott, C.M and B.J. McKeown. 1988. PGRO Pea Growing Handbook. Processors and Growers Research Organization, Peterborough,UK.

Bigirwa G, Adipala E,and Esele J.P.1998. Occurrence of Peronosclerospora sorghi in Uganda. Plant Disease, 82: 757–760.

Blancard, D., H. Lecoq and M. Pitrat. 2005. A Colour Atlas of Cucurbit Diseases: Observations, Identification and Control. 3rd ed.

Bock, C. H. and Jeger, M. J.1996. Downy mildew of sorghum. International Sorghum and Millets Newsletter 37: 33-51.

Bock. C.H, Jeger M.J, Mughogho L.K, Mtisi E,and K.F.Cardwell.1998. Production of conidia by Peronosclerospora sorghi on sorghum crops in Zimbabwe. Plant Pathology, 47: 243-251.

Bock, C. H., Jeger, M. J., Mughogho, L. K., Cardwell, K. F. and Mtisi, E. 1999. Effect of dew point temperature and conidium age on germination, germ tube growth and infection of maize and sorghum by Peronosclerospora sorghi. Mycological Research 103: 859-864.

Boerema, G. H. and Silver, C. N. 1959. Versl. Meded. Plziektenk. Dienst Wageningen 134: 155-157.

Bonde, M. R., Schmitt, C. G. and Dapper, R. W. 1978. Effects of dew-period temperature on germination of conidia and systemic infection of maize by Sclerospora sorghi. Phytopathology 68: 2 19-222.

Bonde, M. R., and G. L. Peterson.1981. Host range of Taiwanese isolate of Peronosclerospora sacchari. Plant Dis. 65:739-740.

Bonde, M.R. 1982. Epidemiology of downy mildew diseases of maize, sorghum and pearl millet. Tropical Pest Management, 28: 49-60.

Bonde, M. R and G. L. Peterson. 1983. Comparison of host range of Peronosclerospora philippinensis and P. sacchari. Phytopathology.73:875-878.

154   The Plant Mildews

Bonde M. R, Peterson G. L, Duck N. B .1985. Effects of temperature on sporulation, conidial germination,and infection of maize by Peronosclerospora sorghi from different geographical areas. Phytopathology, 75: 122-126.

Bonnet, A. and Blancard, D. 1987. Resistance of radish (Raphanus sativa L) to downy mildew, Peronospora parasitica. Cruciferae News letter.12: 98-99.

Borders, H. I. 1953. Preliminary report on fungicide tests for control of downy mildew of cabbage seedlings in south Georgia. Pl Dis. Reptr.37: 363-364.

Brandenberger, L. P., Correll, J. C., &Morelock, T. E. (1991).Identification of and cultivar reactions to a new race (race 4) of Peronosporafarinosa f. sp. spinaciae on spinach in the United States. Plant Disease, 75, 630–634.

Brien,R.M. 1946 . N. Z. J l. Sci.Technol. A.28: 221-224.

Bwown , S., Koike, S.T., Ochoa, O.E., Laemmlen, F., and R.W. Michemore. 2004. Insensitivity to the fungicide fosetyl-aluminum in California isolates of the lettuce downy mildew pathogen, Bremia lactucae. Plant Dis. 88: 502-508.

Brosh, S.1970. Hassadeh. A survey of the distribution of rose downy mildew. 50(4) :434-436.

Butler, E.J. 1918. Fungi and diseases in plants.Thacker, Spink and Co, Calcutta, pp. 314-315.

Butler, E.J.,and S.G.Jones. 1949. Plant Pathology. McMillan, London.

Byford, W. J., and Hull, R. 1963.Control of sugar beet downy mildew (Peronosporafarinosa) by sprays. Ann. Appl. Biol. 52(23):415-422.

Byford, W. J. 1966a. Experiments on the control of sugar-beet downy mildew with fungicides. Proc. Br. Insect. Fung. Conf. Brighton, 1965, pp. 169-76.

Byford, W. J. 1966b. Rep. Rothamsted Ex@. Sta. for 1965, pp. 270-1.

Byford, W. J. 1967a. Field experiments on sugar beet downy mildew (Peronospora farinosa). Annals of Applied Biology, 60(1), 97-107.

Byford, W. J. 1967b. Rep. Rothamsted Exp. Sta. for 1966, pp. 281-2.

Byford WJ, 1967c. Host specialization of Peronospora farinosa on Beta, Spinacia, and Chenopodium. Transactions of the British Mycological Society 50: 603–607.

Byford, W. J., & Hull, R.1967.Some observations on the economic importance of sugar beet downy mildew in England. Annals of Applied Biology, 60(2), 281-296.

Byford, W.J. 1968. Laboratory experiments on sugar-beet downy mildew(Peronospora farinose). Annals of Applied Biology. Wiley Online Library.

Byford WJ, 1981. Downy mildews of beet and spinach. In: Spencer DM (ed), The Downy Mildews. Academic Press, London, pp. 531–543.

Carvalho, T. and Monteiro, A. 1996. Preliminary study on the inheritance of resistance to downy mildew (Peronospora parasitica (Pers. Ex. Fr.) at cotyledon stage in Tranchuda cabbage "Algaivia" Cruciferae News letter.18:104.

Chandrasrikul, A.1962. Tech. Bull. Dep. Agric. Bangkok.6, 23 pp.

Channon, A. G. 1981. Downy mildew of Brassicas. In :The downy mildews, Ed. D. M. Spencer. Academi Press, London. 321-339.

Chauhan, A., Singh, P.P. and Dhillon, M. 1994. Studies on peroxidase activity in muskmelon genotypes in relation to downy mildew infection. J. Pl. Sci. Res. 10:10-11.

Chen, C. P., Sung, C. C. and Ho, C.C. 1959. A brief report of the discovery of oospores of downy mildew of cucumber (Pseudomonas cubensis (Berk. & Curt). Zhihing Zhishi 3(6): 144-145.

Chupp, C. 1930. Phytopathology. 20:307-318.

Chupp, C. and Sherf, A.F. 1960. Vegetable diseases and their control. Ronald Press Co., New York, 693pp.

Ciferri, R. 1961. Rtv. Patol. Veg.Pavia Ser. 3, I. 333-348.

CMI. 1988. Peronosclerospora sorghi(Weston & Uppal) C.G. Shaw Common wealth Mycological Institute Distribution Maps of Plant Diseases No. 179.

CMI. 1988. Distribution Maps of Plant Diseases No. 286 (edition 5). CAB International, Wallingford, UK.

Cobelli,L., M, Collina., and A, Brunelli. 1998. Occurrence in Italy and characteristics of lettuce downy mildew (Bremia lactucae) resistant to phenylamide fungicide. European J.Plant Pathology. 104(5):449-455.

Cohen, Y. 1976. Quantification of resistance of cucumber and Cantaloups to Pseudoperonospora cubensis. Phytoparasitica. 4(1): 25-31.

Cohen,Y., and J. Rotem. 1970. The relationship of sporulation to photosynthesis in some obligatory and facultative parasites. Phytopathology. 60: 1600-1604.

Cohen, Y. and Rotem, J. 1971. Field and growth chamber approach to epidemiology of Pseudoperonospora cubensis in cucumbers. Phytopathology61: 736-737.

Cohen, Y., and Sherman, Y. 1977.The role of airborne conidia in epiphytotics of Sclerospora sorghi on sweet corn. Phytopathology 67:515-521.

Cohen, Y. 1980. Downy mildew of cucurbits. In: The downy mildews (Ed. Spencer, D.M.). Academic Press, San Francisco, 341-353pp.

Cohen, Y., Eyal., H., Hanamia, J. and Malik, Z. 1989. Ultrastructure of Pseudoperonospora cubensiis in muskmelon genotypes susceptible and resistant to downy mildew. Physiol. Mol. Pl Pathol. 34: 27-40.

Conners, I. L. and Savile, D. B. O. 1950 . Twenty-nine Annual Report of the .Can.Pl.Dis. Survey: 1949.

Constantinescu, O. 1989. Peronospora complex on compositae. Sydowia.41:79-107.

Constantinescu O, 1991. An annotated list of Peronospora names.Thunbergia 15: 1–110.

Constantinescu,O. 1992. The nomenclature of Plasmopara parasitic on Umbelliferae. Mycotaxon. 43:471-477.

Constantinescu,O. 1998. A revision of Basidiophora (Chromista, Peronosporales), Nova Hedwigia. 66:251-265.

Constantinescu, O., and J. Fatehi. 2002. Peronospora –like fungi (Chromista, Peronosporales) parasitic on Brassicaceae and related hosts. Nova Hedwigia. 74(3-4): 291-338.

Cook HT, 1936. Cross-inoculation and morphological studies on the Peronospora species occurring on Chenopodium album and Spinaciaoleracea. Phytopathology 26: 89–90.

Cook, H. T. 1932. Studies on the downy mildew of onion and the causal organism, Peronospora destructor (Berk.)Caspary. New York Agric. Exp. Stn., Ithaca, Mem. 143:1-40.

Corbaz,R., 1964. Evolution de l'epidemie de mildiou du tabac Peronospora tabacina. Journal of Phytopathology, 51(2): 190.

Coresta. 1961. Euro- Mediterranean Blue mold Information Service. http://www.coresta.org.

Correll, J. C., Morelock, T. E., Black, M. C., Koike, S. T., Brandenberger, L. P., &Dainello, F. J. (1994).Economically important diseases of spinach. Plant Disease, 78, 653–660.

Cornu, M. 1882. Etudes sur les Peronosporales, 2:1-99 .

Cotton, A.D. 1929. Gardeners' Chronicle. 85:143-144.

Craig, J. 1986. Sorghum Downy Mildew. In Compendium of sorghum Diseases. (Ed R.A Frederiksen), pp. 25-26. American Phytopathology Society, St. Paul, MN. U.S.A.

Crossan, D. F., Lloyd, P. J., Hyre, R. A and J. W. Heuberger. 1957. Control of downy mildew of lima bean. Plant. Dis.Rep. 41: 156-159.

Crute, I. R. 1984. Downy mildew of Brassicas. 35th Annu. Rep. 1983. Natl. Veg. Res. Sta., Wellesbourne, Warwick, UK, 76 pp.

Crute, I.R., 1987. The occurrence, characteristics, distribution, genetics and control of metalaxyl resistant pathotypes of Bremia lactucae in the United Kingdom. Plant Dis. 71, 763—767.

Da Costa, M. E. A. D.and Da Camera, E. dc S. 1954. Port. Acta Biol. 4:162-176.

Dalmacio, S.C and A.D.Raymundo. 1972. Spore density of Sclerospora philippinensis in relation to field temperature,relative humidity and downy mildew incidence. Philippine Phytopathology. 8:72-77.

156　The Plant Mildews

Daniel, J., Husain, A. A. and P.B. Hutchinson. 1972. The control of sugarcane diseases in Fiji. In: Proceeding of the International Society of Sugar Cane Technologists. 14th Congress 1971. Baton Rough, USA; Franklin Press. 1007-1014.

Darpoux. H .1945. Annls Epiphyt. N.S. 11: 71- 103.

Darpoux, H., Durgbat, L. A. & Lebrun, A. (1960). Contribution to the study of downy mildew of sugar- beet. Paper submitted to the 23rd winter congress of the Institut International de Recherches Betteravikres.

Darpoux, H. & Durgbat, L.A. (1962).Studies on downy mildew of sugar-beet. Paper submitted to the 25th winter congress of the Institut International de Recherches Betteravikres.

Davey, J.F., Mulrooney,R.P., Evans,T.A and R.B.Carroll. 2004. Timing of post-infection fungicide application for the control of downy mildew of baby lima bean, 2003 fungicide and nematicide test (on line) Report 59:106. DOI:10,1094/FN59. The American Phytopathological Society, St.Paul.MN.

Davey, J. F., Mulrooney, R. P., Evans.T. A., and R. B. Carroll. 2005. Evaluation of fungicide for control of downy mildew of baby lima bean, 2004. Fungicide and Nematicide Tests (on line) Report 60:V040. DOI:10.1094/FN60. The American Phytopathological Society, St. Paul, M N.

Davis, R.M., Subbarao, K.V., Raid, R.N and E.A., Kurtz. 1997. Compendium of lettuce Diseases. American Phytopathological Society, St. Paul,MN.

Davison,J. A and M. D.Ramsey. 2000. Pea yield decline syndrome in south Australia: the role of diseases and the impact of agronomic practices. Australian J. Agricultural Research. 51:347-354.

Davies, J. M. L. and Wafford. 1987. Control of downy mildew in module-raised cauliflowers. Mono. Brit. Crop Prot. Counc. 39 : 379-386.

Decarvalho, T. 1948. Coldnia Mozambique, Rep. Agric. Seccao de Micohzia.

Delanoë, D. 1972. Biologie et épidémiologie du mildiou du tournesol (Plasmopara helianthi Novot.). CETIOM Informations Techniques 29, 1-49.

Desai MV, Desai DB (1969) Control of downy mildew of isabgol by aureofungin. Hindustan Antibiotic Bulletin 11, 254–257.

Develash R. K., Sugha S. K. 1997. Incidence of downy mildew and its impact on yield. Indian Phytopathology. 50: 127-129.

Dixon, G. R. 1981. Vegetable Crop Diseases.AVI Publishing Co., Inc., Connecticut, 404 PP.

Doidge, E. M., Bottomley, A.M., Van der Plank, J. E.,and G. D.Pauer. 1953. A revised list of plant diseases in South Africa. South African Department of Agriculture. Science Bulletin. 346.

Dominiak, J. D., and R. P.Mulrooney. 2000. Fungicide evaluation for the control of downy mildew of baby lima bean, 1999. Fungicide and Nematicide Tests 55:141. The American Phytopathological Society, St.Paul.MN.

Doppelbaur, H. and Doppelbaur, Hanna. 1972 .Ber. Bayer. Bot. Ges.43: 145-148.

Downey, R. K. and Bolton, J. L. 1961.Production of rape in western Canada. Publ. Dep. Agric. Canada No. 1021,19 pp.

Eastwood, D and P.J.Malein. 1998. Further observation on the use of fungicides to control Peronosclerospora sacchari in sugarcane in Papua New Guinea International Journal of pest management 44(2): 71-73.

Elarosi, H. and Assawah, M. W .1959. Downy and powdery mildews of certain ornamental plants in Alexandria region. Anexandria. J. Agric. Res. 7:253-268.

Elenkov, E. 1979. Breeding disease resistant varieties of vegetable crops. Rastitelna Zashchita27: 13-16.

Ellis, D.E. 1951. Noteworthy diseases of cucurbits in North Carolina in 1949 and 1950. Plant. Dis. Reptr.35(2): 91- 98.

Literature Cited for Downy Mildew   157

Ellis, D.E., and R.S.Cox. 1948. Dusting cucumber to control downy mildew. Bull. N. C. Agri. Exp. Stn.362:16.

Ershad, D. 1977. Publ. Dept. Bot. 10:277 pp.

Evans, T.A., Davidson, C.R., Dominiak, J.D.,Mulrooney,R.P., Carroll, R.B and S.H. Antonius. 2002. Two new races of Phytophthora phaseoli from lima bean in Delaware. Plant Disease. 86: 813.

Exconde, O.R.1970. Phillipine corn mildew. Indian Phytopath.23:275.

Exconde, O. R, and Raymundo A.D. 1974. Yield loss caused by Philippine corn downy mildew. Philippine Agriculturist,58: 115–120.

Exconde O.R. 1975. Chemical control of maize downy mildew. Tropical Agriculture Research Series No. 8, 157-163.

Exconde, O.R. 1976. Philippine corn downy mildew: assessment of present knowledge and future research needs. Kasetsart Journal, 10: 94–100.

Falloon, R. E., Follas, G. B., Butler, R. C and D. S. Goulden. 2000. Resistance in Peronospora viciae to phenylamide fungicide: reduced efficacy of seed treatments of pea and assessment of alternatives. Crop Protection. 19:313-325.

Farlow, W. G.1876. On the American grape vine mildew. Bussey Inst. Bud, 1:415-425.

Farlow, W.G. 1883. Enumeration of the Peronosporeae of the United States. Bot.Gaz.8: 305-315, 327-337.

Fawcett, H.S. 1909. Cabbage diseases. Florida Agric. Exp. Stn. Ann. Rep., 59-60.

Felton, M.W. and Walker, J.C. 1946. Environmental factors affecting mildew of cabbage. J. Agric. Res. 72: 69- 81.

Fernandez Rosenada, M. 1973. Serie Agricola Academia de Ciencias de Cuba. Institute de Investigaciones Tropicales.No. 27, 78 pp.

Ferraris, T. 1936. Riv. Agric. 32:26-27.

Firman, I. D. 1975. Information Document. South Pacific Commission,No. 38, 73 pp.

Forsberg, J. L. 1963. "Diseases of Ornamental Plants". SpecialPublication no. 33.university of Illinois.

Foster, H. H. and Pinckard, 1947. Phytopathology 37:712-720.

Frederiksen, R.A., Bockholt, A.J., Clark, L.E., Cosper, J.W., Craig, J., Johnson, J.W., Jones, B.L., Matocha, P., Miller, F.R., Reyes, L., Rosenow, D.T., Tuleen, D., and Walker, H.J. 1973.Sorghum downy mildew-a disease of maize and sorghum. Research Monograph no. 2. Texas, USA:Texas Agricultural Experiment Station.

Frederiksen, R. A and B. L. Renfro. 1977. Global status of maize downy mildew. Annual Review of Phytopathology. Vol.14:249-271.

Frederiksen R.A.1980. Sorghum downy mildew in the United States: Overview and outlook. Plant Disease, 64: 903- 908.

Furst, G. (1976). Anatomical and histochemical characters of resistance in onion to downy mildew. Anatomical and histochemical characters of resistance in onion to downy mildew., 51-63.

Gardner, M.W. 1920. Peronospora in turnip roots. Phytopathology. 10: 321-322.

Garibaldi, A.and Gullino, G. 1972. Inftore fitopatol. 22: 9-11.

Garibaldi A, Minuto A, Gullino M-L, 2005. First report of downy mildew caused by Peronospora sp. on basil (Ocimumbasilicum) in France. Plant Disease 89: 683.

Garibaldi, A., D.Bertetti ana M.L.Gullino. 2007. Effect of leaf wetness duration and temperature on infection of downy mildew(Peronospora sp.) of basil. J.Plant Dis.Prot. 114:6-8.

Gaumann, E..1918. Uber die Formen der Peronospora parasitica Ein Beitrag Zur spezies frage bei parasitischen Pilzen. Beitr. Bot. Zentrablatt. 35:395-533.

Gaumann, E. 1923. Contribution towards a monograph of the genus Peronospora corda. Beitrage zur. Kryptogamenflora de schweiz. 5:1-360.

158   The Plant Mildews

Gayed, S. K. (1984).The response of pepper and tomato cultivars to challenge inoculation by Peronosporatabacina. Canadian journal of plant science, 64(1), 225-228.

Geetha, H.M. and H.S. Shetty 2002. Expression of oxidative burst in cultured cells of pearl millet cultivars against Sclerospora graminicola inoculation and elicitor treatment.plant science 163 (3):653.

Gianessi L. P., Reigner N. 2005. The value of fungicides in U S crop production. Crop Life foundation crop protection research institute-Washington, DC, 243 p.

Gill, D. L. 1933. Mycologia 25: 446-447.

GilL,D. L. 1977. Pl. Dis. Reptr. 61: 230-231.

Gilles T., Phelps K., Clarkson J. P., Kennedy R. 2004. Development of MILIONCAST, an improved model for predicting downy mildew sporulation on onions. Plant Disease. vol. 88 (7): 695-702.

Glaeser, G. 1970. The occurrence of important causes of injury to cultivated plants in Austria in 1969. Pfianzenschutzberichte 41:49-62.

Godfrey, G.H., 1941. Noteworthy diseases of economic crops and native plants in lower Rio Grande valley in the spring of 1941. Plant Dis. Reptr. 25: 347-353.

Gonzalez, F., R. 1924. Peronosporaceae at present known to occur in the Iberian flora. Bol. R. Soc. Espan. Hist. Nat. 24:305-312.

Gram, E. and Rostrup, S. 1924. Survey of diseases of agricultural and horticultural cultivated plants in 1923. Tidsskr.for Planteavl. 30:361-414.

Green, D. E. 1937. Gardeners' Chronicle 102: 27-28.

Green, D. E. 1938. J. R. Hort. Soc. 63: 159-165.

Green, S. 1943. Agric. Hort. Genet. 1: 97-98.

Gregory, C.T.1912. Spore germination and infection with Plasmopara viticola. Phytopathology, 53:205-249.

Gregory, P. H. 1950. Trans. Br. Mycol. Soc. 32: 241-245.

Grillo, H. V. S. 1937. Rodriguesia 2:39-96.

Grogan, R.G. and Zink, F.W. 1956. Fertilizer injury and its relationship to several previously described diseases of lettuce. Phytopathology 46, 416–422.

Gulya, T. J., Sackston, W. E.; Virányi, F.; Masirevic, S.; Rashid, K.Y. 1991. New races of the sunflower downy mildew pathogen (Plasmopara halstedii) in Europe and North and South America. Journal of Phytopathology 132: 303-311.

Gupta, S. K., &Shyam, K. R. (1996).Antisporulant activity of some fungicides against Pseudoperonospora cubensis on cucumber. Indian Journal of Mycology and Plant Pathology, 26(3), 293-295.

Gupta, S.K, Shyam, K.R. and Dohroo, N.P. 1993. Effect of fungicides on the severity of Downy mildew and yield of cucumber (Cucumis sativus L.) in Himachal Pradesh. Pestology17(3): 37-39.

Gupta, S.K and Shyam, KR. 1994. Antisporulant activity of fungicides on downy mildew (Peronospora parasitica) on cabbage. Indian J. Agric. Sci. 64: 891-893.

Gupta, S.K and Shyam, K.R. 1998. Protective activity of fungicides against downy mildew of cucumber. Pl. Dis. Res. 13(1): 60-61.

Gustafsson, I., 1989. Potential sources of resistance to lettuce downy mildew (Bremia lactucae) in differentLactuca species. Euphytica, 40(3), pp.227-232.

Gustavsson, A.1959.Op. Bot, Sec. Bot. Ltmd.3(1): 271.

Hagedom,D.J. 1974. Recent pea anthracnose and downy mildew epiphytotics in Wisconsin.. Plant Disease Reporter.

Hall, G. 1989. Plasmopara halstedii. CMI Descriptions of Pathogenic Fungi and Bacteria No. 979. CAB International, Wallingford, UK.

Harkness, H. W. 1885. Fungi of the Pacific coast. Bulletin of the California Academy of Science. 1: 256-271.

Hewitt, H.G. 1998. Fungicides in crop protection. CAB International, Wallingford.

Hildebrand, A.A and L.W. Koch. 1945. Soybean diseases in southwestern Ontario in 1945. Ann.Rep.Can. Pl.Dis. Surv.25: 39-39.

Hildebrand, A. A. and L. W. Koch. 1951. A study of systemic infection of downy mildew of soybean with special reference to symptomatology, economic significance and control. Sci.Agr. 31: 505-518.

Hildebrand, P. D., and Sutton, J. C. 1982.Weather variables in relation to an epidemic of onion downy mildew.Phytopathology 72:219- 224.

Hindi, E., Dishon, I. and Nevo, D. 1965. Observations on tobacco blue mold. Plant Dis. Rep. 49: 154-156.

Hiura,M. 1929. Studies on some downy mildews of agricultural plants. II.Relation of meteorological conditions to the downy mildew of cucumber. Res.Bull.Gifu Imp.Coll.of Agric.6:58.(Japanese)

Hiura, M. and Kawada, S. 1933. On the overwintering of Peronoplasmopara cubensis ((Berk, and Curt.) Rostow. Japanese J. Bot. 6(4): 507-513.

Hiura, M. and Kanegae, H. 1934. Trans. Sapporo Nat. Hist. Soc. 13:125-133.

Hoerner, G.R. 1940. The species of the genus Pseudoperonospora and their recorded hosts. Plant Dis. Reptr. 24(9): 170-173.

Homer, 1000 B.C. The use of sulphur in fumigation and other forms of pest control. In: Plant Pathology, R.S.Mehrotra, 2003. McGraw Hill. Education.846 pp.

Hoser, KJ., Lakowska, R.E. and Antosik, J. 1991. Resistance to some Brassica oleracea L. plant introductions to downy mildew (Peronospora parasitica). Cruciferae New sci. 14/15: 144-145.

Hoser, KJ., Lakowska, R.E. and Antosik, J. 1995. The inheritance of resistance to some Brassica oleracea cultivars and lines to downy mildew. J. App. Genet. 36: 27-33.

Hubbeling, N. 1975. Resistance of peas to downy mildew and distinction of races of Peronospora pisi Syd. Rijksuniversiteit Gent. Vol 40(2):539-543.

Hubbeling, N. and Ester, A., 1978. [Testing spinach for downy mildew resistance]. [Dutch]. Zaadbelangen.

Hughes, C.G. and Robinson, P.E. 1961. Downy mildew disease. In: Sugarcane diseases of the world. Eds: Martin. J.P., Abbott, E.V and Hughes, C.G.. Elsevier.

Hughes, M.B. and Van Haltrern, F. 1952. Two biological forms of Pseudoperonospora cubensis. PL Die. Replr, 36: 365-367.

Hyre, R.A. 1954. Progress in forecasting late blight of potato and tomato. Plant Dis. Rep. 38: 245.

Hyre, R.A. 1958. The development of a method for forecasting downy mildew of lima bean. Plant Dis. Rep. Suppl. 257:179-180.

Icochea, T., Torres, H., and W. Perez. 1994. Downy mildew of maca(Lepidium meyenii): symptoms and identification of the causal agent. Fitopatologia. 29(2): 156-159.

Inaba, T., Tkahashi, K., &Morinaka, T. (1983).Seed transmission of spinach downy mildew. Plant Disease, 67, 1139–1141.

Irish, B. M., Correll, J. C., Koike, S. T., Schafer, J., &Morelock, T. E. (2003). Identification and cultivar reaction to three new races of the spinach downy mildew pathogen from the United States and Europe. Plant Disease, 87, 567–572.

Jafar, H. 1963. Studies on downy mildew of stocks. N.Z. J. Agric. Res.6: 70-82.

Jackson, R. W., Athanassopoulos, E., Tsiamis, G., Mansfield, J. W., Sesma, A., Arnold, D. L., et al. (1999). Identification of a pathogenicity island, which contains genes for virulence and avirulence, on a large native plasmid in the bean pathogen Pseudomonas syringae pathovar phaseolicola. Proceedings of the National Academy of Sciences of the United States of America, 96, 10875–10880.

160   The Plant Mildews

Janke, G.D;Pratt,R.G;Arnold J.D and G.N. Odvody 1983. Effects of deep tillage and rouging of diseased plants on oospore population of peronosclerospora sorghi in soil and on incidence of downy mildew in grain sorghum. Phytopathology. 73:1674-1678.

Jaude. Clotilde.1959. An. Soc. Clem. Argent. 168: 52 - 59.

Jeffers, W. F.1952.Pl. Dis, Reptr. 36: 211.

Jeger M.J,Gilijamse E, Bock C.H, Frinking H.D, 1998.The epidemiology, variability and control of the downy mildews of pearl millet and sorghum,with particular reference to Africa. Plant Pathology, 47(5):544-569.

Jellis, G.J., Bond,D.A.,and R.E. Boulton. 1998. Diseases of faba bean. Pages 371- 410. In: The Pathology of Food and Pasture Legumes. (eds. Allen, D.J and Lenne,J.M), CAB International Wallingford, Oxon,UK.

Jesperson, G.D. and J.C.Sutton. 1987. Evaluation of a forecaster for downy mildew of onion. Crop Prol. 6: 95-103.

Jhooty, J.S. and Munshi, O.D. 1975. Control of downy mildew of muskmelon with fungicides. Indian J. MycoL PL Pathol. 5:105-106.

Jiang, M.X. 1981. Selection and evaluation of the cucumber F1 hybrid Ninghuang 1 and Ninghuang 2. Ningxia Agril. Sci.Techn. Ningxia Nongye Keji. 5:10-11.

Johns, T. H.1972.Rep. Dep. Agric, N.S.W. 1971, 200 - 210.

Johnson, H. W. And C. L. Lefebvre. 1942. Downy mildew on soybean seeds. U.S. Dept. Agr. Pl. Dis. Reptr. 26: 49-50.

Johnston,A. 1963. A preliminary plant disease survey in Hong-Kong. Plant Prod, and Prot. Div. F.A.O. Rome, 32 pp.

Jones,W. 1944. Downy mildew disease of cauliflower seed plants .Sci.Agric.24: 282-284.

Jones, F. R. and J. H. Torrie. 1946. Systemic infection of downy mildew in soybeans and alfalfa. Phytopathology 36s 1057-1059.

Jones HA, Mann LK (1963) Onions and their Allies. Leonard Hill Ltd, London.

Jones B.L, Leeper J.C, Frederiksen R.A. 1972. Sclerospora sorghi in corn: its location in carpellate flowers and mature seeds. Phytopathology, 62: 817–819.

Jorstad, I.1946.Norg. GartForen Tidsskr. 36: 497- 498.

Kadow, K. J. and Anderson, H. W. 1940. A study of horseradish diseases and their control.Illinois Agric.. Agric. Exp. Stn.Bull. 469:531-583.

Kanwar, J. S.~ Brar, K. S.~ Kaur, S.~ Dhiman, J. S. 2015. Screening of germplasm for multiple resistance in fenugreek,(Trigonella corniculata L.). Vegetable Science Vol. 27 (1): 86-87.

Kapoor J.N. &Chowdhary P.N. 1976.Notes on Indian microfungi.Indian Phytopath. 29: 348– 352.

Khalid,P.Akhtar and S.S.Alam. 2002. Assessment keys for some important diseases of mango. Pakistan.J.Biol.Sci.5:246-250.

Khare M. N., S. P. Tiwari , and Y. K. Sharma. 2014. Disease problems in fennel (Foeniculum vulgare Mill) and fenugreek (Trigonella foenumgraceum L.) cultivation and their management for production of quality pathogen free seeds. International J. Seed Spices 4(2):11-17

Khokhar, L. K., & Jaffrey, A. H. (2000).Efficacy of fungicides against downy mildew and yield of onion. Pakistan Journal of Agricultural Research, 16(1), 43-44.

Khosla, H.X, Dave, G.S. and Nema, K.G. 1973, Occurrence of downy mildew of parwal [Triehosanthts dioioa Roxb.) in Madhya Pradesh, India. JNKVV Res. J.7: 175-177.

Khristov, A. (1943). Fungi causing spot on the balls and moulding the seed of opium poppy. Bulgaria Agricultural Experiment Station Journal, 13, 13–19.

Klebahn, H . 1925. ZP fiKrankh. PfiPath. PflSchutg . 35: 12- 22

Koch, L. W. and A. A. Hildebrand. 1946. Soybean diseases in southwestern Ontario in 1946. Ann. Rept. Can. Pl. Dis. Surv. 26s 27-28.

Koch de Brotos, L. and Boasso, C. 1955. Publ. Minist. Gonad. Agric, Montevideo,106: 65.

Kochman,J., and T, Majewski. 1970. Grzyby(Mycota). (Flora Polska) Tom IV. Glonowce(Phycomycetes) Wroslikowe(Peronosporales), Panstwowe Wydawnictwo Naukowe Warszawa.

Koike,S., Smith,R., and K, Schulbach.1992. Resistant cultivars, fungicides combat downy mildew of spinach. Calif. Agric. 46:29-30.

Kolte, S.J. 1985. Diseases of annual edible oilseed crops, Vol. 3, sunflower, safflower & nigerseed diseases. CRC Press, Inc., Boca Raton, USA.

Komnenic, M., Obserdovic, A and L.Stankovic. 1995. Resistance of different cucumber genotypes to Pseudoperonospora cubensis, causal agent of downy mildew. Zastita Bilja. 46(1): 69-74.

Kontaxis, D. G. and Guerrero,P. 1978. Plant Dis. Reptr.62: 170-171.

Kontaxis, D.G., Mayberry, KS. and Rubatzky, V.E. 1979. Reaction of cauliflower cultivars to downy mildew in Imperial Valley. California Agric. 33: 19.

Kothari, K. L. and Prasad, N.1971.IndianPhytopathology 23: 674-688

Kothari, K. L., and Verma, A. C. 1967.Control of downy mildew of opium poppy by root application of fungicides. Plant Dis.'Reptr. 51(8):686-687.

Kumudini, B. S. and H.S. Shetty. 2002. Association of lignification and callose deposition with host cultivar resistance and induced systemic resistance in pearl millet to Sclerospora graminicola. Aust. Plant pathol. 31(2):157

Kunene, I.S., Odvody, G.N., and Frederiksen, R.A. 1990.Gaertennomyces sp. as a potential biological control agent of systemic downy mildew infection of sorghum. Sorghum Newsletter 31:82.

Kushalappa, A.C., 2001. BREMCAST: Development of a system to forecast risk levels of downy mildew on lettuce (Bremialactucae). International journal of pest management, 47(1), pp.1-5.

Laemmlen, F.F. and Mayberry, K.S. 1984. Broccoli resistance to downy mildew. California Agric.38: 17.

Lai, L.Y.,and S.D.Zhang. 1992. Disease resistance identification of cucumber varietal resources in Shandong. J. Shundong Agril.Sci. 1: 20-25

Lakra B S. 2002. Role of temperature and humidity in oospore formation of Perenospora trigonella causing downy mildew of fenugreek. Pl. Dis. Res. 17: 339-340

Lal, S. S; C. Saxena and R. N. Upadhyay. 1980. Control of brown stripe downy mildew of maize by metalaxyl. Plant Disease64:874-876.

Lange, L.W. 1989. Zoosporogenesis in Pseudoperonospora cubensis, the causal agent of cucurbit downy mildew. Nordic J. Bot. 8(5): 497-504.

Leach, L. D. (1931). Downy mildew of the Beet, caused by Perono-sporaschachtii Fuckel. Hilgardia, 6(7).

Leach, L. (1945). Effect of downy mildew on productivity of sugar beets, and selection for resistance. Hilgardia, 16(7), 317-334.

Leather, R. L . 1967. A catalogue of some plant diseases and fungi in Jamaica. Bull. Minist. Agric.lands Jamaica .61 : 92 pp

Leppik, E.E. 1966. Origin and specialization of Plasmopara halstedii complex on Compositae. FAO Plant Protection Bulletin 14: 72-76.

Leu,L.S., and B.T.Egan. 1989.Downy mildew. In: Diseases of sugarcane, Major diseases. Eds. C.Ricaud., B.T.Egan.,A.G.Gillaspie. Jr. and C.G. Hughes. Pp.107-133, Elsevier, Amsterdam.

Lewis, Esther A.1937.Phytopathology 27: 951-953,

Liese, A.R.; Gotlieb, A.R.; Sackston, W.E. 1982. Use of enzyme-linked immunosorbent assay (ELISA) for the detection of downy mildew (Plasmopara halstedii) in sunflower. Proceedings of the 10th International Sunflower Conference, Surfers Paradise, pp. 173-175.

162　The Plant Mildews

Lin, C.Y. 1981. Studies on downy mildew of Chinese cabbage caused by Peronospora parasitica. In: Chinese cabbage Proc. 1st Int. Symp. AVRDC, Shanhua, Taiwan. Pp. 105-112.

Lindquist, J. C 1939. Physis. B. AiresI5: 13-20.

Lisitisin, V.N and L.E.Pluzhnikova. 1990. Results of evaluating cucumber varieties for the purpose of breeding for resistance to bacteria and downy mildew. Ovoshchvodstvo I Bakhchevodstvo. 35: 82-84.

Ljubich, A.; Gulya, T.J. 1988. Cotyledon-limited systemic downy mildew infection. Proceedings of 1988 Sunflower Research Workshop, Bismarck, USA, National Sunflower Association, p. 9.

Lo, T. T. 1961. Plant Ind.Ser. Chin.-Am. jt Comm.Rur. Reconstr.23: 52 pp.

Lorenzini, G. and Nali, C., 1994. A new race (race 4) of spinach downy mildew in Italy. Plant Disease, 78(2).

Lorbeer J., Andaloro J. Diseases of Onions. Downy Mildew. - Nyaes, Geneva, NY, 1984. http://www.nysaes.cornell.edu/ent/hortcrops/english/dmildew.html.

Lu, S.Z., Ma, D.H., Huo, Z.R.,Shen,W.Y., Li, S.J and Z.W.Chen. 1994. New released cucumber cultivar Jinchun 4- high-quality, disease resistant and high yielding. China Veg. 2: 1-3.

MacLean, N. A. and Baker, L. F. 1951. Bull. Roses Inc. no. 160, 5- 6.

Mahajan, Vijay; Gill, H.S., Moore, T.A. and Mahajan, V. 1995. Inheritance of downy mildew resistance in Indian cauliflowers (group-Ill). Euphytica 86 :1-3.

Mahrishi, R.P. and Siradhana, B.S. 1984. On the occurrence of oospores of Pseudoperonospora cubensis in Rajasthan. Indian Phytopath. 37(2): 323-325.

Mahrishi, R.P. and Siradhana, B.S. 1988a. Studies on downy mildew of cucurbits in Rajasthan: incidence, distribution, host range and yield in muskmelon. Ann. Arid. Zone Res. 27(1): 67-70.

Mahrishi, R.P. and Siradhana, B.S. 1988b. Effect of nutrition on downy mildew disease caused by Pseudoperonospora cubensis (Berk. & Curt.) Rostow. on muskmelon. Ann. Arid. Zone Res. 29(2): 153-155.

Mahrishi, R.P. and Siradhana, B.S. 1988c. Epidemiology of downy mildew of muskmelon (Cucumis melo L.) caused by Pseudoperonospora cubensis . J. Tar. Phytopath. 17(2): 67-73.

Malein, P.J.1993. Fungicide control of Peronosclerospora sacchari in Papua New Guinea. International Journal of Pest Management 9(3):325-327.

Malhotra S K and Vashishtha B B. 2008. Package of Practices for Production of Seed Spices. National Research Centre on Seed Spices, (ICAR) Ajmer.p.93-98

Marziano, F., Calarese, S.and Stefanis, D.1973.Annali Fac.sci.Agr Univ.Napoli7,

Mayor, 1963. Phytopalh. Z.48:322-328.

McCain,A.H and C, Noviello. 1985. Biological control of Cannabis sativa, pages 635-642 in: Delfosse, E.S(Ed), Agricultural Canada. Proceeding, VI International Symposium on Biological Control of weeds, 19-25 August. 1984, Vancouver, Canada.

McKay,R. 1949. Gardeners'Chronicle126: 28.

McGregor, R.C. 1978. People placed pathogens: the emigrant pests. In: Horsfall JG, Cowling EB ed. Plant disease.An advanced treatise. Volume II. How disease develops in populations. Academic Press. Inc. New York,USA & London UK, 383-306.

McGuire, J. U.and Crandall, B. S. 1967. USD A Int. Agric. Development Service,157 pp.

McKay, A.G., Floyd, R.M. and Boyd, C.J. 1992. Phosphonic acid controls downy mildew (Peronospora parasitica) in cauliflower curds. Aust. J. Exp. Agric. 32: 127-129.

McKee, R. k. 1971. Rep. Res. Tech. Work. Minist. Agric. N. Ireland 1970:105-113.

Mclntosh, A. E. S.1951. Annual Report of the Department of Agriculture, Malaya for the year 1949, 87 pp.

Megenberg,konrad,ca 1388, Buck der Natur. Codex 2669, 305. Osterreichische National bibJiothek,Wien.

Mehra R; Pratap, P. S and Dhawan P. 2002. Evaluation of fenugreek genotypes for resistance against downy mildew disease. Pl Dis Res. 17: 192. 30.

Mehta, N. and Saharan, G.S. 1994. Morphological and pathological variations in Peronospora parasitica infecting Brassica species. Indian Phytopath. 47: 153-158.

Mence,J.M.1971. J.R.Hort.Soc.96: 393-394.

Meyer, E., and R. Ziesenis. 1982. Downy mildew on outdoor cucumbers. Gemuse. 18: 3.

Mikuskovic,M. 1968. Zast.Bilja19: 197-200.

Milani, S., Kamachi, K., Sugimoto, K., Araki, 8. and Yamaguchi, T. 2003. Control of cucumber downy mildew by cyazofamid. J. Pesticides Sci. 28:64-68.

Millardate, A. 1883. Sur le role des spores d'hiver du mildiou(P. Viticola) dans la reinvasion par ce parasite. Mem. Soc.Sci. 5: 24-27

Millardet, P. M. A.1885. Traitement du mildiou et du rot. J. d'Agr. pra£., 2:513-516 .

Minuto A, Pensa P, Garibaldi A, 1999. Peronosporalamii, nuovoparassitafogliaradella salvia. ColtureProtette 28: 63–64.

Mir, N. M., Dhar, A. K., Khan, M. A., Dar, G. H. and Zarger, M. Y. (1987) Downy mildew of Allium cepa and its control with fungicides in Kashmir Valley. Indian Phytopathology 30, 576-577.

Mirakhur, R. K., Dhar, A. K. and Kaw, M. R. (1978) Downy mildew of Allium cepa and its control with fungicides in Kashmir Valley. Indian Phytopathology. 30. 576-577.

Möllerström, G. (1955). The influence of the weather on the development of downy mildew in Sugar Beets, with some observations on control measures. Socker, 11(2).

Moore, W. C.1949.Trans. Br. Mycol. Soc. 32: 95-97.

Moore,W.C. and Moore, F.J.1952. PI.Path. 135-136.

Moore, W. C.1959. "British Parasitic Fungi". Cambridge University Press.

Mujica. F. and Vergara, C. 1960. Boln tec. No. 6 Dep. Invest, agric. Chili6, 60 pp.

Muller, A. S. 1950. A preliminary survey of plant diseases in south Guatemala.Plant. Dis. Reptr.34: 161-164.

Murphy, P. A.1937. Int. Bull. PI. Prot. 11: 176.

My, H. T. 1966. A preliminary list of plant diseases in South Vietnam. 142 pp. Saigon, Directorate of Research. Nakov, B. 1968.

Natti, J.J., Dickson, M.H. and Atkin, J.D. 1967. Resistance of Brassica oleracea varieties to downy mildew. Phytopathology 57: 144-147.

Naqvi,S.A.M.H. 2004. Diseases of Fruits and Vegetables Vol.I. Diagonosis and managements. Kluwer Academic Publisher.

Neergaard, P.1943. Gartnertidende 8: 95-98.

Neumann, P. 1955. Diseases of seedlings and young plants of our Brassicae. Pflanzensckutz.7: 39-44.

Nilsson, L. 1949. Vaxtskyddsnotiser, Vaxtskyddsanst. Stockh.6: 1-3.

Nikoliæ, V., 1952. Plamenjaèa na suncokretu (in Serbian). Zaštita bilja 9: 42-54.

Novotel'nova, N.S. 1966. Downy mildew of sunflower, 150 pp. Nauka, Moscow, Russia.

Ocfemia, G. O. 1925. The occurrence of the white rust of crucifers and its associated downy mildew in the Phillipines. Philipp. Agric.14: 289-296.

Olofsson, J. 1966. Downy mildew of peas in western Europe. Plant Disease Reporter. 50:257-261.

Om, Y.H., Hong,K.H., Oh, D.G. and S.N.Kwack. 1992. Breeding of gynoecious cucumber inbreds' Wonye 502'. Res.Rep. Rural Dev.Admn.Hortic. 34(1):1-4.

Orian, G. 1951. Rep. Dep. Agric. Mauritius. 1949, 66-72; for 1950,80-85.

Orjuela, N. J. 1965. Boln Tec. Inst. Colomb. Agrop.11: 66 pp.

164   The Plant Mildews

Paaske, K., 2000. Chemical control of lettuce downy mildew. Växtskyddsnotiser, 64(3/4): 37-43.

Pai, C. K. 1957. Notes on the Peronosporaceae in northeastern china. Acta Phytopathol. Sinica.3: 137-154.

Palti, J. 1971. Biological characteristics, distribution and control of Leveillula taurica (Lev.) Ann. Phytopath.Mediter. 10: 139-153.

Palti, J., Cohen, Y. 1980. Downy mildew of cucrbits (Pseudoperonospora cubensis)-The fungus and its hosts, distribution, epidemiology and control. Phytoparasitica 8: 109-147.

Pandey, KK., Pandey, P.K., Singh, B., Kalloo, G. and Kapoor, KS. 2001. Sources of resistance to downy mildew (Peronospora parasitica) disease in Asiatic group of cauliflower. Veg. Sci.28: 55-57.

Pape, H.1934.Gartenwelt 37:289-290.

Pape, H. 1955. "Krankheiten und Schadlinge der Zierpflanzen und ihre Bekampfaag' f

Park, M. 1932. Ceylon Adm. Rep. Agric.1931, D103-D111.

Patel JG (1984) Downy mildew of isabgol (PlantagoovataForsk).PhD Thesis, Gujarat Agricultural University, India.

Paulus, A.O. and Nelson, J. 1977. Systemic fungicides for the control of Phycomycetes on vegetable crops applied as seed treatments granular or foliar spray. Proc.9th Brit. Insecticide Fungicide Conf., Brighton, England, 21-24 Nov., Vol. 3, London, UK

Payak,M.M and B.L.Renfro. 1967. A new downy mildew disease of maize. Phytopathology. 57: 394-397.

Payak, M.M; Lal,S and B.L, Renfro.1970. Downy mildew diseases incited by Sclerophthora. Indian Phytopathology. 23: 183-193.

Payak, M.M. 1975. Epidemiology of maize mildew with special reference to those occurring in Asia. Tropical Agricultural Research Series. 8:81-91.

Payak,M.M and R.C.Sharma. 1985. Maize diseases and approaches to their management in India. Tropical Pest Management. 31(4): 302-310.

Pegg,G.F and M.J.Mence. 1972. The biology of Peronospora viciae on pea. The development of local systemic infection and their effect on vining yield. Annals of Applied Biology.71:19-31.

Peleg, J.1953. Outbreaks and new records in Isreal. FAO Plant Prot. Bull.1-4: 60-61.

Peregrine, W. T. H. and Siddiqi, 1972. Phytopathological Papers,No. 16,51 pp. Comm. Mycol. Inst., Kew, UK.

Perišiæ, M., 1949. Prilog poznavanju parazitne mikroflore na gajenim biljkama u FNRJ (in Serbian). Arch. Biol. Sci., Beograd 1(2): 181-184.

Perwaiz, M. S., Moghal, S. M. and Kamal, M. 1969. Studies on the chemical control of white rust and downy mildew of rape. W. Pak. J. Agric. Res.7: 71-75.

Pettersson, S.1954.Vaxtskyddsnotiser ,27-29.

Phillips, D.H.1958.Rep.Sts.Exp.Stn.Jersey1957,24-33pp.

Pickel, B.1939.Bioldgico 5:192-194

Pirone, P. P.1978.Diseases and pests of ornamental plants. Wiley. NewYork.

Pivovarov, V.F., Ushakov,P.P. and A.O. Simanca. 1977. Result of evoluation and selection of cucumber cultivars for improvement of their resistance to downy mildew. Informe Cientifico Tecnico, Instituto de Investigaciones Fundamentals en Agricultura Tropical. 39: 20-25.

Pivovarov, V.F. 1984. Screening cucumber for disease resistance using different ecological conditions. Selektsiya I Semenovodstvo. USSR. 10: 20-22.

Porter. R. H. 1926. Preliminary report of surveys for plant diseases in East China. Plant. Dis. Reptr. Supplement 46: 153-166.

Pound, G. S 1946. Bull. Wash. Agric. Exp. Sin.475, 27 pp.

Prakash S and Saharan G S .2001. Factors affecting downy mildew infection of fenugreek. Indian Phytopath. 54: 193-196

Pratt R.K,and G.D.Janke. 1978. Oospores of Sclerospora sorghi in soils of south Texas and their relationships to the incidence of downy mildew in grain sorghum. Phytopathology, 68: 1600–1605.

Raj, D., Gupta, S.K, Shyam, K.R., Sharma, S.K and Sharma, H.R. 2003. Evaluation of germplasm against downy mildew (Pseudoperonospora cubensis). Plant Dis Res. 18(1): 67.

Ramsay, G. B. 1935. Peronospora in storage cabbage. Phytopathology.25: 955-957.

Ramsay, G.B., Smith,M.A., and W.R.Wright. 1954. Peronospora in radigh roots. Phytopathology. 44: 384-385.

Ramsfjell, T.1960. Mytt Mag. Bot.8: 147-178.

Rao B.M, Shetty S.H, Safeeulla K.M. 1984. Production of Peronosclerospora sorghi oospores in maize seeds and further studies on the seed-borne nature of the fungus. Indian Phytopathology, 37: 278– 283.

Rauka, G.R., Suma,S.,Magarey, R.C and L.S.Kuniata. 2005b. The effect of downy mildew on sugarcane yield in the varietry B72177 at Ramu Sugar, Gusap, Papua New Guinea. Proceedings of the Australian Society of Sugar Cane Technologists. 27: 353-357.

Reuveni, M., Eyal, H. and Cohen, Y. 1980. Development of resistance to metalaxyl in Pseudoperonospora cubensis. Plant Dis. 64:1108-1109.

Reuvoni, R., Shimoni, M., Karchi, Z. and Kuc, J. 1992. Peroxidase activity as a biochemical marker for resistance of muskmelon (Cucumis meld) to Pseudoperonospora cubensis. Phytopathology 82: 749-753.

Richards MC, 1939. Downy mildew of spinach and its control. Cornell University Agricultural Experimental Station Bulletin 718: 1–29.

Robak, J. 1995. Epidemiology and control of cucumber downy mildew Pseudoperonospora cubensis. Biuletyn Warzywniczy 2:145-149.

Rodriguez, & H. 1972. Folleto MiscelaneoINI ANo. 23, 58pp.

Rohner, E., Carabet, A. and Buchenauer, H. 2004. Effectiveness of plant extracts of Paeonia suffruticosa and Hedera helix against diseases caused by Phytophthora infestans in tomato and Pseudoperonospora cubensis in cucumber. Z.. Pflanzenkr. Pflanzenschutz 111: 83-95.

Rondomanski, W. 1967.Studies on the epidemiology of onion downy mildew, Peronospora destructor (Berk.)Fries. Technical Report for 1962-67. Research Institute for Vegetable Crops, Skierniewice, Poland.

Rumberg, V.1974.In "Boleznevstoichivosf Rastenii' (A. Semenova, Ed.)68-330. Tallin Botanical Gardens, USSR.

Russell, G. E. 1969. Recent work on breeding for resistance to downy mildew in sugar beet. Revue de l' Institut International de Recherches Betteravieres. 5:1-5.

Russell, G.E and G.M.Evans. 1972. Some effects of darkness and partial defoliation on the resistance of sugar beet to downy mildew. Annals of Applied Biology. 70(1): 99-103.

Ryan, E.W. 1977. Control of cauliflower downy mildew (Peronospora parasitica) with systemic fungicides. Proc. Ninth British Insecticide Fungicide Conf, Brighton, England, Vol 1 & 2, Res, Rep., London, U.K Session 6B. Pests and Diseases of vegetables 297-300p.

Rydl, R.1968. Uroda 16: 186-187.

Sackston, W.E. 1981. Downy mildew of sunflower. In: The downy mildews (Ed. by Spencer, D.M.), pp. 545-575. Academic Press, London, UK.

Sadravi, M., Etebarian, H. R. and Torabi, M. Downy mildew of spinach and possible mechanisms of resistance in some spinach cultivars. Seed Plant, 2000; 15 (4): 403-412.

Saharan,,G.S;Verma,P.R and N.I.Nishaat. 1997. Monograph on downy mildew mildew of crucifer. Technical bulletin 1997-01, Saskatoon Research Centre, Canada, 197pp.

166　The Plant Mildews

Samual, G. 1925. Rep. Dep. Agric. S. Australia. for 1924, 76-78.

Savulescu O. 1960. Communicarile de botanica, Bucuresti, 1957-1959, 263-267.

Schettini, T.M., E.J. Legg, R.W. Michelmore, 1991. Insensitivity to metalaxyl in California populations of Bremialactucae and resistance of Californian lettuce cultivars to downy mildew. Phytopathology 81: 64–70.

Scherm, H.A and H.C.Van Bruggen. 1995. Comparative study of microclimate and downy mildew development in subsurface drip and furrow irrigated lettuce fields in California. Plant Disease. 79(6):620-625.

Schroeter, J.1874.Hedwigia 13:183.

Schwartz H. F., Mohan S. K. 1995. Compendium of Onion and Garlic Diseases. American Phytopathological Society, St. Paul, MN. – 1995,70 p.

Semeniuk,G and c..J.Mankin. 1964. Occurrence and development of Sclerophthora macrospora on cereals and grasses in south Dakota. Phytopathology.54:409

Serafim, F. J. D. and Serafim, M. C. 1968. Lista des doensas de culturas de Angola. 22 pp. Nova Lisboa, Institute Investigate agronomica de Angola.

Shao, X.H., Chen, Q.Q. and Zhang, F.Q. 1990. Research on relation between some ecological factors and Peronospora parasitica (Pers.) Fr. on Chinese cabbage. Acta .Agric, Shanghai 6: 78-81.

Sharma, S.C., A.S. Khera, S.S.Bains and N.S.Malhi. 1981. Efficacy of fungitoxicant sprays and seed treatment against Phillippine downy mildew of maize. Indian Phytopath.34:498.

Sharma, S.R. and Sohi, H.S. 1982. Effect of fungicides on the development of downy mildew and white rust of radish. Indian J. Agric. Sci. 58 : 521-524.

Sharma, S.R. 1983. Effect of fungicidal spray on white rust and downy mildew diseases and seed yield in radish. Gartenbauwassenschaft. 48 :108-112.

Sharma, B.R., Dhiman, J.S., Thakur, J.C., Singh, A. and Bqjqj, KL. 1991. Multiple disease resistance in cauliflower. Adv. Hortic. Sci. 5: 30-34.

Sharma, R.C; Carlos De Leon and M.M,Payak. 1993. Diseases of maize in South and South-East Asia: Problems and Progress. Crop Protection. 12(6): 414-422.

Sharma, S.R., Kapoor, KS. and Gill, H.S. 1995. Screening against Sclerotinia rot (Sclerotinia sclerotiorum), downy mildew (Peronpspora parasitica) and black rot (Xanthomonas campestris) in cauliflower (Brassica oleracea var. botrytis sub var. cauliflora) .Indian J. Agric. Sci. 65(12): 916-918.

Sharma, D.R., S.K. Gupta and K.R. Shyam. 2003. Studies on downy mildew of cucumber caused by Pseudoperonospora cubensis and its management. J.Mycol. Pl. Pathol. 33 (2): 246-251.

Shaw, C. G.1951.Mycologia 43:448-449.

Shaw, C G. and Yerkes, W. D. 1951. NW Sci. 25:76-82.

Sherf, A.F. and Macnab, A.A. 1986. Diseases of crucifers. In. Vegetable diseases and their control (2 ed.). The Ronald Press. New York, pp. 251-306.

Sherf, A.F. and Macnab, A.A. 1986. Vegetable .John Wiley and Sons, New York, USA, 728pp.

Shetty, H.S. and Safeeulla, K.M. 1981. Effect of some environmental factors on the asexual phase of Peronosclerospora sorghi. Proceedings of the Indian Academy of Science (Plant Sciences) 90: 45-51.

Shimazaki, Y. and Uchiyama, F., 1985.Resistance of spinach cultivars to downy mildew.In Proceedings of the Kanto-Tosan Plant Protection Society (No. 32).

Silvae, D., Nashaat, N.I. and Tidily, Y. 1996. Differential responses of Brassica oleracea and B. rapa accessions to seven isolates of Peronospora parasitica at the cotyledon stage. Plant Dis. 80: 142-144.

Simmonds, J. H.1951. Rep. Dep. Agric. Stk. Qd. 1950-51 p.48.

Singalovsky, Z. (1937). Etude morphologique, cytologiqueetbiologique du mildou de la betterave. (PeronosporaschachtiiFuckel). Ann. Epiph. 3, 551-618.

Singh SD, Singh G (1987) Resistance to downy mildew in pearl millet hybrid NHB-3. Indian Phytopathol 40:178–180.

Singh, S.D., Singh, P., Rai, K.N., Andrews, D.J., 1990.Registration of ICMA 841 and ICMB 841 pearl millet parental lines with A1 cytoplasmic-genic male sterility system.Crop Sci.30, 1378.

Singh, U.P., Srivastava, B.P., Singh, K.P., Mishra, G.D., 1991. Control of pea powdery mildew with ginger extract.Indian Phytopathol. 44, 55–59.

Singh, H and C.H.Dickinson. 1980. Metalaxyl for control of downy mildew of pea caused by Peronospora viciae. Plant Disease. 64: 1090-1092.

Singh,S.D; R.Gopinath and M,N, Pawar. 1987. Effect of environmental factors in asexual sporulation of Sclerospora graminicola. Indian Phytopath.40:186-193.

Singh, P.P. and Sokhi, S.S. 1989. First report of occurrence of oospores of Pseudoperonospora cubensis on two cucurbitaceous hosts. Curr. Sci. 58: 1330-1331.

Singh, P.P., Thind, T.S., Sokhi, S.S. and Grewal, R.K. 1990. Studies on development of downy mildew of muskmelon under hot weather conditions. Plant. Dis. Res. 5(Special): 104-107.

Singh, P.P. and Singh, R. 1994. Epidemiological investigations of downy mildew of muskmelon (Cucumis melo L.). Plant. Dis. Res. 9: 226.

Singh, P.P., Jhorar, O.P., Singh, R. and Sokhi, S.S. 1996. Predictive model for downy mildew of muskmelon. Veg. Sci. 23:186-194.

Singh, P.P., Thind, T.S. and Lai, Tarsem. 1996. Reaction of some muskmelon genotypes against Pseudoperonospora cubensis under field and artificial inoculation conditions. Indian Phytopath. 49:188-190.

Sivanesan, A and J.M.Waller. 1986. Peronosclerospora. In: Sugarcane diseases. pp 58-61. CAB International.

Skalicky, V.1953. Ceskd Mykol.7:133-136.

Skalick, V. 1966. Taxonomie der Gattungen der Familie Peronosporaceae. Preslia(Praha).38:117-129.

Smith, C.O. 1904. Mildew of lima bean. Del.Agric.Exp.Stn.Bull.63:23-24.

Smith , E.F., and R.E.B, McKenney. 1921. The present status of the tobacco blue mold (Peronospora) disease in the Georgia, Florida district. Circ. U.S. Dept. Agric. No.181, pp 1-4.

Soonthronpoct, P. 1969. Tech. Docum. F.A.O. PL Prot. Comm. S.E. Asia70, 23 pp.

Stahl, M.1973.NachBl. dt. PfiSchutzdienst., Stuttg. 25: 161-162.

Stahl, M. and Umgelter, H. 1976. "Pflanzenschutz im Zierpflanzenbau" Eugen 12Stuttgart.

Stegmark,R. 1988. Downy mildew resistance of various pea genotypes. Acta Agric. Scand. 38: 373-379.

Stell, F. 1922. Some common diseases of kitchen garden crops. Proc. Agric. Soc. Trinidad and Tobago. 12:779-785.

Stenina, N.V. 1976. The dependence of the development of downy mildew of onion on climatic conditions. Trudy Kuban S-kh.Inst.24-27.

Stevenson, J.A., and W.A.Archer. 1940. A contribution to the fungus flora of Nevada. Plant Dis. Reptr. 24: 93-103.

Stratton,J.M. 1969. Agricultural Records Ad 220-1968. London:Baker.204pp.

Stryapkova, L.V., Korganova, N.N and M.M.Treeva.1992. Losses caused by Peronosporosis can be avoided. Zashchita Rastenii(Moskova). 12: 10-11.

Sturgis, W.C. 1897. The mildew of lima bean. Connecticut Agric. Exp. Stn. Rep.21:159-166.

168   The Plant Mildews

Sugha, S.K and B.M.Singh. 1991. Status of downy mildew of onion(Peronospora destructor) in Kangra district of Himachal Pradesh and its effect on crop yield.. Plant. Dis. Res. 6:35-38

Suma,S and E.Pais. 1996. Major diseases affecting sugarcane production on Ramu sugar estate, Papua New Guinea. In: Sugarcane Germplasm conservation and exchange.report of an international workshop held in Brisbane, Queensland, Australia, 28-30 June 1995.ACIAR proceeding No.67, 30-33pp.

Suma,S., and R.C.Magarey. 2000. Downy mildew. In: A guide to sugarcane diseases. Eds: Rott,P., Bailey, R.A., Comstock,J.C., Croft, B.J and

Saumtally, A.S. CIRAD & ISSCT.

Sun, M.H., Chang,S.S.,and C.M.Taang. 1976. Research advances in sugarcane downy mildew of corn in Taiwan. Kasetsart Journal. 10: 89-93.

Sutic, C. and KJijajic, R.1954. Zast. A contribution to the knowledge of the parasitic flora of Deliblatska Pescara. Zasht. Bilja.(Plant Prot. Beograd). 24:104-108.

Taylor, J. C. 1963.Sci. Hort. 16:31-34.

Thakur,D.P. 1986. Management of downy mildew disease of pearlmillet in india. Advt. Biol. Res. 4: 17-28.

Thaxter, R. 1889. A new American Phytophthora. Bot.Gaz. 14: 273-274.

Thind, K. S. 1942. J. Indian Bot. Soc. 21:197-215

Thind, T.S. and Mohan, C. 2001. Disease-weather relationship and relative activity of some new fungicides in different application schedules against muskmelon downy mildew. J. Mycol. Pl. Pathol. 31: 174-179.

Thind, T.S., Singh, P.P., Sokhi, S.S. and Grewal, R.K. 1991. Application timing and choice of fungicides for the control of downy mildew of muskmelon. Pl. Dis. Res. 6(1): 49-53.

Thind, T. S., & Mohan, C. (2001). Diseases-weather relationship and relative activity of some new fungicides in different applications schedules against muskmelon downy mildew. J. Mycol. Pl. Pathol, 31, 174-179.

Thinggaard, K., 1985. Investigation of the Danish Bremialactucae (lettuce downy mildew) population in the period 1979-1984 [virulence phenotype, physiological race]. Tidsskrift for Planteavl (Denmark).

Thomas, C.E. 1978. Reaction of susceptible and resistant contaloupes to Pseudoperonospora cubensis. Pl. Dis. Reptr. 62: 221-222.

Thomas, C.E., Inaba, T. and Cohen, Y. 1987a. Physiological specialization in Pseudoperonospora cubensis. Phytopathology 77:1621-1624.

Thomas, C.E., Cohen, Y. Jourdain, E.L. and Eyal, H. 1987b. Use of reaction types to identify downy mildew resistance in muskmelon. Hortic. Sci. 22: 638-640.

Thomas, C.E. and Jour dam, E.L. 1990. Evaluation of broccoli and cauliflower germplasm for resistance to Race 2 of Peronospora parasitica. Hortic. Sci. 25: 1429-1431.

Thung T. H. 1926b. Peronospora parasitica attacking cabbage heads.Phytopathology.16: 365-366.

Tikhonov, O.I. 1975. Diseases of sunflower. In: Pustovoit V.S., editor.Sunflower .Moskow; Kolos. 401- 409 pp (in Russian).

Todd, F.A.1981. The blue mold story. P. 9-25. In: Report: 29th Tobacco Workers Conference. Lexington, KY.

Tramier, R.1960. C.r. Hebd. Seanc. Acad. Agric. Fr. 46:622-624.

Tramier, R.1963. Annls Epiphyt. 14:311-323.

Tramier, R.1965. Phytiat. Phytopharm. 14:49-56.

Tuleen D.M, Frederiksen R.A, Vudhivanich P. 1980. Cultural practices and the incidence of sorghum downy mildew in grain sorghum. Phytopathology, 70: 905–908.

Uppal, B. N., Patel, M. K. &Kamat, M. N. (1935) Pea powdery mildew in Bombay.Bombay Dept. Agric. Bul. 177:12 p.

Ullasa, B.A. and Amin, K.S. 1988. Ridge gourd downy mildew epidemics in relation to fungicidal sprays and yield loss. Mysore J. Agril. Sci.22(1): 62-67.

Ullrich, J. and Schdber, B.1968.Jber. Biol. BundAnst. Land-u. Forts Bramrsrirm.32.

USDA, Crop Profile for Onion. 2003. http://pestdaa.ncsu.edu/cropprofiles/docs/ txonions.html.

Van der Gagg, D.J and H.D.Frinking. 1996a. Homothallism in Peronospora viciae f.sp.pisi and the effect of temperature on oospore production. Plant Pathology. 45:990-996.

Van der Gagg, D.J and H.D. Frinking. 1996b. Extraction from plant tissue and germination of oospore of Peronospora viciae f.sp.pisi. Journal of Phytopathology.144:57-62.

Van der Gagg, D.J and H.D.Frinking. 1997a. The infection court of faba bean seedlings for oospore of peronospora viciae f.sp.fabae in soil. Journal of Phytopathology. 145(5-6),DOI:10.1111/J.1439- 0434.1997.tboo396.x.

Van der Gagg, D.J and H.D.Frinking.1997b. Factors affecting germination of oospore of Peronospora viciae f.sp.pisi in vitro. European Journal of Plant Pathology. 103(6):573-580.

Vasileva, S.D. 1976. Damage by Peronospora disease to cabbage and the effectiveness of polycarbacin in its control. Referativnyi Zhurnal . 155:1299.

Verma, T.S.,and P.C.Thakur. 1989. Comparative field resistance of cabbage collection to downy mildew at seedling stage. Indian Journal of Plant Protection. 7: 79-80.

Viala, P. 1893. Les Maladies de la Vigne. Montpellier, France.

Viennot-Bourgin, G.1954. Revue Path. Veg. Em.Agrk. Fr. 33:31.

Virányi, F. 1978. Harmful incidence of Plasmopara halstedii in downy mildew "resistant" sunflowers. Phytopathologische Zeitschrift .91: 362-364.

Viranyl,F. 1981. Downy mildew of Onion. In: Spencer, D.M.(Ed). Downy mildews. Academic Press,London.pp.461-472.

Virányi, F.1984. Recent research on the downy mildew of sunflower in Hungary. Helia 7:35-38.

Virányi, F.1988a. Plasmopara halstedii. In: European handbook of plant diseases (Ed. by Smith, I.M.; Dunez, J.; Phillips, D.H.; Lelliot, R.A.; Archer, S.A.), pp. 228-230. Blackwell Scientific Publications, Oxford, UK.

Virányi, F. 1988b. Factors affecting oospore formation in Plasmopara halstedii. Proceedingsof the 12th International Sunflower Conference, Novi Sad 2: 32-37.

Virányi, F.; Oros, G. 1990. Developmental stage response to fungicides of Plasmopara halstedii (sunflower downy mildew). Mycological Research 95: 199-205.

Vitanov, M. and E, Elenkov. 1971.Growth of onion and infection by Peronospora destructor. Gradinarska I Lozarska Nauka.8:81-88(Bulgarian with English summary).

Von Heydendorff, R.C. 1977. De Untersuchungen zur Biologie von Peronospora pisi, dem Erreger des falschen Mehltaus fur die uchtung resistenter Sorten. Ph.D.Thesis.Technischen universitat Hannover.

Wadhwa, N.S. Jain, H.K.L. Chawla and M.S. Panwar.2001. Changes in activities of elicitor-releasing enzymes in pearl millet leaves infected with Sclerospora graminicola, Indian Phytopath. 54(1):414.

Wager, V. A.1970.Flower Garden Diseases and Pests. Puraeil, CapeTown.

Wallace, J. G.1948.Gardeners' Chronicle 124:21.

Walker, J. C. 1950. Plant pathology, McGraw Hill Book Company, New York, 1950. pp. 222-228.

Walker, J.C. 1952. Downy mildew of lima bean. In: Diseases of Vegetable Crops. McGraw-Hill, New York.

Waterston, J. M. 1940. Rep. Dep. Agric. Bermuda 1939,13 pp.

Weber, G. F. 1932. Bull. Fla. Agric. Exp. SwNo. 256.

170   The Plant Mildews

Weille, G.A.de. 1975. An approach to the possibilities of forecasting downy mildew infection in onion crops. Meded.Verhandel.Kon.Nederlands Meteorol.Inst.97.

Weston, W.H. 1920. Philippine downy mildew of maize. Journal Agricultural Research.19:97-122

White, J.G.,Crute,I.R.,and E.C.Wynn. 1984. A seed treatment for the control of Pythium damping off and Peronospora parasitica on brassica. Ann.Appl.Biol.104:241-247.

Whitwell J. D. and Griffin, G. W. 1967. Proc. 4th Br Insect. Fung. Conf.,239-242.

Wicks, T.J., Hall,B and P. Pezzaniti. 1994. Fungicidal control of metalaxyl-insensitive strains of Bremia lactucae on lettuce . Crop Prot.13:617-623.

Wicks, T.J., McLachlan, D., Campbell, K., Biggins, L. Magarey, P. and Emmett, R. (1995) Oils – new soft fungicides for grape powdery mildew (Uncinulanecator). 10th Biennial Australasian Plant Pathology Society Conference, Christchurch, New Zealand, p. 98.

Wilkinson,H.T and D.Pedersen. 1993. Yellow tuft of Kentucky bluegrass sod identified in Illinois. Plant Disease Note.77: 647.

Wilson, G.W. 1907. Studies in North American Peronosporales-II. Phytophthoreae and Rhysotheceae. Bulletin of the Torrey Botanical Club. 34(8):387-416.

Wilson, G. W. (1907). Studies in North American Peronosporales. I. The genus Albugo. Torrey Botanical Club Bulletin, 3461–3485.

Wilson, G.W. 1914. Studies in North American Peronosporales- VI. Notes on miscellaneous species. Mycologia. 6:192-210.

Wolf, F. A. and S. G. Lehman. 1924. Soybean diseases. N. C. Agr. Expt. Sta. Ann. Rept. 47: 82-83.

Wolf, F.A. 1947. Tobacco downy mildew, endemic to Texas and Mexico. Phytopathology.37: 721-729.

Wright, C.M. & Yerkes, W.D., 1950.Observations on the overwintering of the pathogen causing downy mildew of spinach in the Walla-Walla area. Pl. Dis. Reptr 34: 28.

Wright, P. J., Chynoweth, R. W., Beresford, R. M., and Henshall, W. R. 2002.Comparison of strategies for timing protective and curative fungicides for control of onion downy mildew (Peronospora destructor) in New Zealand. Proc. Br. Crop Prot. Council Conf., Pests Dis. 2002:207-212.

Wu, Q.X. 1991. New cucumber varieties. Bull.Agril.Sci Techn. 3: 34-35.

Yakimenko, L.N. 1983. Promosing cucumber varieties for summer sowing in the Krasnodar area. Trudy po Prilladnoi Botanike Genetike I Selektsii. 77: 22-25.

Yarwood, C E. (1952)• Moving particles of vacules of Eryslphaceae. Bio. Abs. 65 (7) : 3656/383 43.

Yarwood, C. (1947). Snapdragon downy mildew. Hilgardia, 17(6), 239-250.

Yarwood, C. E. (1957) • Powdery mildews. Bot Rev 23: 235 -300.

Yarwood, C. E. 1943. Onion downy mildew.Hilgardia 14:595-691.

Yerkes WD, Shaw CG, 1959. Taxonomy of the Peronospora species on Cruciferae and Chenopodiaceae. Phytopathology 49: 499–507.

Yossifovitch, M. 1929.Peronospora arborescens, parasite tres important de papaver somniferum en Yougoslavie. Revue de Pathologie vegetale et d'entomologie agricole de France.16:235-270.

Zarzycka H. 1970. Acta Mycol. 6:7-19.

Zimmer, D.E. and Hoes, J.A. 1978. Diseases. In: Sunflower science and technology (Ed. by Carter, J.F.), pp. 225- 262. American Society of Agronomy, Madison, US.

Zuo, Q.Y., Huang, T., Lin, Y., Chen, Q.H., Xie, S.D.A and T.S.Zhu. 1995. Creation and utilization of breeding parents of cucumber. Acta Hort. 402: 334-339.

Section II: Powdery Mildew

12

An Introduction to Powdery Mildews

The term "powdery mildew" is generally used for such disease where mildew fungus forms a powdery growth on the infected surfaces. Such powdery growth can be easily removed from the infected surfaces with a pressure of air or water. The term "Surface mildew" is also used for this disease present on many hosts.

Powdery mildew fungi are obligate plant pathogens that attack approximately 10,000 species of plants belonging to more than 1600 genera. As obligate biotrophs, powdery mildew fungi obtain their nutrients from living cells of their host plants through specialized feeding organs, known as haustoria. Powdery mildews evolved effective secretive ways of feeding and pathogenesis, effective counter defense mechanisms that neutralize the host's defenses, or effective pathways for scrambling defense signaling. Numerous pathogen and host genes become involved in each of the steps in a successful infection, including recognition of host and pathogen, adhesion of fungal spores to host surfaces, spore germination, appressorial initiation and development, penetration peg development, peg penetration into host cell, haustorial initiation and development, neutralization of host defenses, removal of nutrients from host cell, hyphal growth, and sporulation

Powdery mildews are probably the most common, conspicuous, widespread, and easily recognizable plant diseases. They affect all kinds of plants except gymnosperms.

12.1. History of Powdery mildews

Homer (1000 BC) mentioned mildew and its control with blasting of sulphur. The oldest text on Indian agriculture "Krishi Parashar" written by Parashar (400BC) mention plant protection in one verse with respect to powdery mildew as enemies of crop and invokes the wind god to move them away from his field. The mention of powdery mildew is in Vedic period also.

The powdery mildew on vine was first observed in European glass house at Margate, England in 1845 by Edward Tucker, a gardener (Berkeley, 1847). The fungus was first described by Berkeley and Curtis in 1848, followed by De Bary in 1863, Farlow in 1876 and Berlese and De Toni in 1888 when it was assigned to present genera. The disease spread to France sometimes previous to 1874 from eastern United States. Subsequent appearances of powdery mildews on other crop plants were noticed and reported in several parts of the world.

Gontier, a French gardener obtained excellent results in controlling the mildew in glass house by applying sulfur dust to moistened vines by means of blow duster in 1850 (Mares, 1856). John Roberton (1924) carried out a careful study of peach mildew in 1921 in Ireland and reported its control by repeated application of sulfur and soap water. This treatment or modification of it came into use by gardeners and recommendation of preparation of lime and sulfur soon followed (Lodeman, 1896).

12.2. Losses due to Powdery mildew

Powdery mildew causes serious yield losses. Rawal (2013) reported powdery mildew losses upto 25.31 percent in crop yield in grapes. In Japan, powdery mildew losses in wheat and barley are reported upto 2 percent.

12.3 Symptoms

Powdery mildew appear as spots or patches of a white to grayish powdery growth on young plant tissues or on entire leaves and other plant parts (Fig.12.1). Tiny, pinhead-sized, spherical, at first white, later yellow-brown, and finally black cleistothecia as sexual fruiting bodies of the fungus may be present singly or in groups on the white to grayish mildew in the older areas of infection (Fig.12.2). Powdery mildew is most common on the upper side of leaves, but it also affects the underside of leaves, young shoots and stems, buds, flowers, and young fruit.

Fig. 12.1: Growth of powdery mildew fungus on infected leaf surface

Fig. 12.2: Formation of brown black sexual fruiting bodies known as Cleistothecia on leaf surface

Powdery mildews are so common, widespread, and ever present among crop plants and ornamentals that the total losses, in plant growth and crop yield, they cause each year on all crops probably surpass the losses caused by any other single type of plant disease. Powdery mildews seldom kill their hosts but utilize their nutrients, reduce photosynthesis, increase respiration and transpiration, impair growth, and reduce yields, sometimes by as much as 20 to 40%.

Among the plants affected most severely by powdery mildew are the various cereals, such as wheat and barley, primarily because the chemical control of plant diseases in these crops is difficult, impractical, or not cost effective. Other crops that suffer common and severe losses from powdery mildew are the cucurbits, especially cantaloupe, squash, and cucumber; sugar beets; strawberries; clovers; many ornamentals, such as rose, begonia, dephinium, azalea, and lilac; grape; and many trees, particularly apple, catalpa, and oak.

12.4. The Pathogen

The powdery mildew fungus belongs to the Ascomycetes group of fungi, in the order Erysiphales. They are obligate parasite of Angiosperm extending from tropics to Artics and from below sea level to 40000 M height (Hirata, 1966). Erysiphaceae are parasitic on some 7187 host species in 1289 genera, 149 families and 44 orders of angiosperms (Hirata, 1966).

Fungi causing powdery mildews are obligate parasites: they cannot be cultured on artificial nutrient media, but recently the powdery mildew fungus of barley, *Blumeria graminis* f. sp. *hordei*, was grown in culture. They produce mycelium that grows only on the surface of plant tissues but does not invade the tissues themselves. They obtain nutrients from the plant by sending haustoria into the epidermal cells of the plant organs.

The mycelium produces short conidiophores on the plant surface; each conidiophore produces chains of rectangular, ovoid, or round conidia that are carried by air currents (Fig 12.3). When environmental or nutritional conditions become unfavorable, the fungus may produce cleistothecia containing one or a few asci (Fig.12.4). Powdery mildew fungi although are common and cause serious diseases in cool or warm humid areas, are even more common and severe in warm dry climates. This happens because their spores can be released, germinate, and cause infection even when there is no film of water on the plant surface as long as the relative humidity in the air is fairly high. Once infection has begun, the mycelium continues to spread on the plant surface regardless of the moisture conditions in the atmosphere.

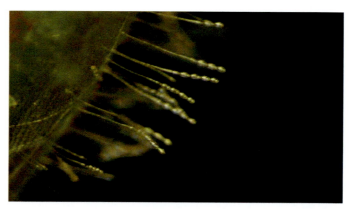

Fig. 12.3: Formation of asexual fungal spore conidia in pseudochains for spread of the infection

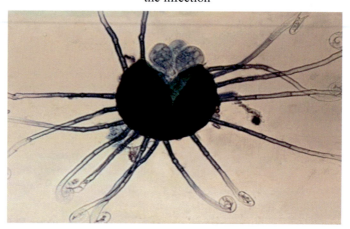

Fig. 12.4: Repture Cleistothecia showing asci containing sexual ascospores

An Introduction to Powdery Mildews   175

Powdery Mildew causing fungi belonging to:

Division	Ascomycota
Class	Leotiomycetes
Order	Erysiphales
Family	Erysiphaceae

Erysiphales is the exceptional order as it produces cleistothecium instead of perithecium. The reason is that the asci are grouped in fascicles or form a basal layer (hymenium) at maturity and ascospores are released violently withforce. Cleistothecia are formed on superficial mycelium without formation of stroma.Members cause a disease called powdery mildew because they produce enormous number of conidia on the surface of infected host plants which appear to the naked eye as a white powdery coating.Mycelium is hyaline and mostly ectophytic. Members are obligate parasites of plants and nourishment through haustoria. Asci are persistant, globose to pyriform and explode at the time of release of ascospores.

12.4.1 Somatic characteristics

Mycelium is well developed, septate, uninucleate, profusely branched, entirely superficial (ectophytic) except *Leveillula* (endophytic) and *Phyllactinia* (semi endophytic), produce haustoria into epidermal cells to absorb nourishment. Conidiophores and conidia are produced on the mycelium, which appear white on the plant surface. In the advance stages of fungal growth or in unfavourable environmental conditions the fungus produces fruiting bodies called cleistothecia which are spherical and dark with or without appendages.

12.4.2. Asexual reproduction

Asexual reproduction is through conidia produced on conidiophores. Conidiophores are long, erect and hyaline.

Three types of conidial stages are recognised in powdery mildews:

1. Oidium *(Acrosporium)*

Mycelium is ectophytic, hyaline. Conidia are developed from a flask shaped mother cell(spore mother cell) formed on a short conidiophore. Conidia are barrel shaped with flat ends and are produced in chains.The conidia are also referred to as meristem arthrospores as these are formed by fragmentation of hyphae. The powdery mildew genus viz., *Erysiphe, Podosphaera,Uncinula, Sphaerotheca and Microsphaera* produce *Oidium* as conidial stage.

2. Oidiopsis

Mycelium is endophytic. Conidiophores may be branched or unbranched, erect, septate, hyaline and emerge through stomata. Conidia are produced singly and cylindrical in shape. Conidia are of two types i.e. with blunt tip or with pointed tip. The powdery mildew genus *.Leveillula* sp.produce Oidiopsis as conidial stage .

3. Ovulariopsis

Mycelium is partly ectophytic and partly endophytic.The conidiophores are hyaline, septate, unbranched, and bear a single conidium.Conidia are rhomboid in shape. In some species, the conidiophores are spiral in shape. e.g. *Phyllactinia subspiralis. Phyllactinia sp* produce ovulariopsis as conidial stage

Powdery mildew conidia do not require free water for germination and are able to germinate at very low humidity levels.

12.4.3. Sexual reproduction

Some species are homothallic and some are heterothallic. Antheridia and ascogonia are sex organs. Both gametangia are uninucleate. Fruiting body is cleistothecium which is produced on superficial mycelium as a result of gametangial contact. The cleistothecia are first white and finally black in color when mature. The wall is made up of pseudoparenchymatous tissue of several layers called peridium. Over wintering of powdery mildews takes place in clestothecial stage which are resistant to winter conditions. In perennials, the mycelium may over winter in the dormant buds of host. In warm weather, many species never form cleistothecia and perpetuate by means of conidia. The cleistothecia are provided with characteristic appendages which vary considerably in length and character.

12.4.4. Types of cleistothecial appendages

Different types of cleistothecial appendages are observed (Fig.12.6).

A. **Mycelioid appendages:** These are flexible, flaccid and resemble somatic hyphae. e.g. in *Erysiphe, Sphaerotheca, Leveillula.*

B. **Circinoid / hooked / coiled appendages:** These are rigid with curled or coiled tips. e.g. in *Uncinula .*

C. **Bulbous base with pointed tip:** These are rigid, spear like with bulbous base and pointed tip. e.g. in *Phyllactinia.*

D. **Dichotomously branched tips:** These are rigid, flattened with dichotomously branched tips. e.g. in *Podosphaera, Microsphaera.*

An Introduction to Powdery Mildews 177

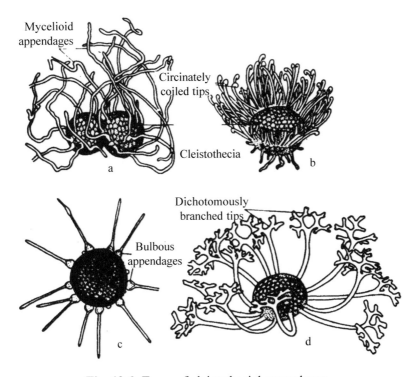

Fig. 12.6: Types of cleistothecial appendages

12.4.5. Key used for the identification of powdery mildew genera

1. Types of cleistothecial appendages

   a. mycelioid

   b. dichotomously branched

   c. circinoid

   d. bulbous base with pointed tip

2. Number of asci in cleistothecium

   a. one

   b. many

3. Type of conidial stage

   a. Oidium

   b. Oidiopsis

   c. *Ovulariopsis*

4. Nature of mycelium

a. ectophytic

b. endophytic

c. semi- endophytic

Identification of Powdery mildew genera on the basis of given keys:

1. **Mycelium superfacial**........Cleistothecia with appendages

   1.1 Cleistothecia with single ascus

      1.1.1 Cleistothecial appendages simple......*Sphaerotheca*

      1.1.2 Cleistothecial appendages dichotomously branched..................*Podosphaera*

   1.2 Cleistothecia with several asci

      1.2.1 Cleistothecial appendages simple.......***Erysiphe***

      1.2.2 Cleistothecial appendages dichotomously Branched.......... ***Microsphaera***

      1.2.3 Cleistothecial appendages coil at tip......***Uncinula***

2. **Mycelium partly internal**...........Cleistothecia with appendages

   2.1. Cleistothecia appendages simple.........***Leveillula***

   2.2. Cleistothecial appendages with basal swelling..........***Phyllactina***

3. **Cleistothecia without appendages**........***Braziliomyces*, *Astomella***

4. Cleistothecia lacking .... *Oidium*

Besides the above generas, the identification characteristic of some other powdery mildew generas is as follows.

Brasiliomyces: have single layer of cleistothecium peridium cells (Zneng 1984) not differentiated into inner and outer layer.

Cystotheca: The peridium cell layer is differentiated into inner and outer layer. Appendages present, poorly developed, myceloid.

Sawadaea: Cleistothecia with several asci, cleistothecial appendages stiff, radiating, with dichotomously or trichotomously branched, curved tips. Conidia with fibrosin bodies . The genus is almost confined to maples.

Blumeria: Differ from Erysiphe in its digitate haustoria and in details of conidial wall. Blumeria is considered to be phylogenetically distinct from Erisiphe as it

An Introduction to Powdery Mildews   179

hosts solely on the true grasses of poaceae. Ascocarp is dark brown, globose with filamentous appendages, asci oblong, ascospore hyaline, ellipsoid, 20-30x 10-13 μm in size. Anamorph produces on hyaline conidiophores; catenate conidia of oblong to cylindrical shape, not including fibrosin bodies, 32-44 x 12-15 μm in size. Haustoria are palmate.

Parauncinula: This particular fungi is found exclusively on Quercus (oak tree) in Japan and china, has relatively large chasmothecia, containing many eight spored asci with many uncinate circinate apiced appendages. Asexual stage produced conidia in chain (euoidium type) and is ectoparasitic.

Golovinomyces: Hyphae substraight to waxy, occasionally geniculte 40-80 μm long, 4-7 μm vide, mostly branching at right angle, with a septum near the branching point. Appressoria poorly developed, nipple shaped and single. Conidiophores single on hyphal cell, arising from the upper part of mother cell, producing 2-4 swollen cells in chain with a sinuate outline, 120-200 μm long, and foot cell straight, 65-120 x 9-12 μm, simple, with a basal septum at the branching point or slightly displaced from the mycelium. Conidia oval to ellipsoid, 30-42 x 20-26 μm (length /breadth ration 1.3-1.9), Lacking distinct fibrosin bodies, producing germtubes on the perihilar position, with reticulate wrinkling of the outer walls. First formed conidia apically conical, basally round to more or less subtruncate and generally shorter (30-36 μm long) than the secondary conidia.

Neoerysiphe : Conidia 29-43 x 15-19μm, conidiophore or its foot cell 114-200 x 9 x 11 μm, chasmothecia 104-148 μm, Asci 47-63 x 19-29 μm. Ascospore 18-27 x 12-16 μm. Number of ascospores 5-6 and upto 8. Mycelium amphigenous i.e. it occurs on the upper and lower leaf surfaces. Superficial hyphae branched, septate, hyaline, 4-7 μm wide. Hyphal appressoria lobed to multilobed, single or in opposite pair. Conidiophores erect, simple, hyaline, with straight foot-cells, 20-45 x 8-12 μm followed by 1-3 shorter cells and 2- 4 (-6) conidia in chains with sinuate edge line. Conidia hyaline, unicellular with faintly striated surface without fibrosin bodies, 23-42 x 12-22 μm. First conidium ovoid with rounded apex and a flattened base, following (secondary) conidia mostly doliform and rarely subcylindric, producing germ tube below the shoulder. The anamorph belongs to oidium subgenus striatoidium. Chasmothecia gregarious to subscattered, glosose, blackish brown, 105-210 μm in diam. Appendages numerous (>. 25 per chasmothecium), myceloid, septate, hyaline , brown at base, generally simple, rarely irregularly branched, arising from the lower half of chasmothecium, 0.5-2 times as long as the chasmothecia diam, interlaced with each other. Asci 6-12 per chasmothecia, oblong- elliptic, stalked 45-60 x 20-30 μm.

Powdery mildew genera are now grouped into five tribes,i.e. Blumeriae, Erysipheae, Cystotheceae, Golovinomyceteae and Phyllactineae(Takamatsu,2013) and some genera have been added or merged.The chart below shows the tribes and some representative genera of each; the previous teleomorphic names (and less commonly used anamorphic names) are given to aid in reference to the older literature(Heffer et.al, 2006).

Tribe Genus	New holomorphic Genus	Anamorphic Genus	Former Teleomorphic Genus	Common Host
l. Phyllactineae	Phyllactinia	Ovulariopsis	Phyllactinia	Trees and Shrubs
	Leveillula	Oidiopsis	Leveillula	Solanaceae
2. Erysipheae	Erysiphe section Erysiphe	Oidium	Erysiphe	Legumes
	Erysiphe section Microsphaera	Oidium	Microsphaera	Trees and shrubs
	Erysiphe section Uncinula	Oidium	Uncinula	Trees and shrub
3. Blumeriae	Blumeria	Oidium	Blumeria/Erysiphe	Grasses
4. Golovinomyceteae	Golovinomyces	Oidium	Golovinomyces	Cucurbits & Composites
5. Cystotheceae	Podosphaera section Podosphaera	Oidium	Podosphaera	Rosaceae
	Podosphaera section Sphaerotheca	Oidium	Sphaerotheca	Rosacea

Previously, identification was based largely on the teleomorph (sexual stage) and the morphology of the cleistothecium and its appendages, but the morphology of structure is not as conserved as originally assumed. With the new taxonomy, identification of powdery mildews now also requires attributes of the anamorph (asexual stage), so that it incorporates characteristics of the whole fungus (anamorph plus teleomorph, i.e., the holomorph).

The taxonomy of powdery mildew fungi (Fig 12.5) (order Erysiphales) recently underwent extensive revision based on rDNA sequence data.

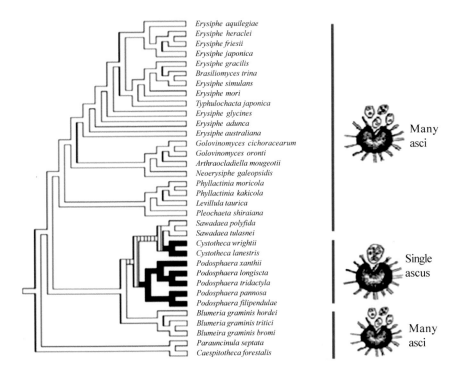

Fig. 12.5: Taxonomy of powdery mildew fungi

12.5. Disease Cycle

Powdery mildew fungi have a polycyclic life cycle typical of its phylum, Ascomycota. Powdery mildew fungi overwinters in the form of cleistothecia as dormant sexual fruiting body of the fungus in plant debris. Under warmer conditions, however, the fungus can overwinter as asexual conidia or mycelium on living host plants. It can persist between seasons most likely as ascospores in plant debris left in the field. Ascospores are sexual spores produced in the ascus in the cleistothecia. These spores, as well as conidia, serve as the primary inoculum and are dispersed by wind. Neither spore requires free water to germinate; only high relative humidity is required for their germination. Powdery mildew fungus thrives in cool humid conditions and cloudy weather increases chances of disease. When conidia land on a host leaf's hydrophobic surface cuticle, they release proteins which facilitate active transport of lightweight anions between leaf and fungus even before germination. This process helps the powdery mildew fungus to recognize it's correct host and directs growth of the germ tube. Both ascospores and conidia germinate directly with a germ tube. Conidia can recognize the host plant and within one minute of initial contact, the direction of germ tube growth is determined. The development of appressoria then begins infection following the growth of a germ tube. After initial infection,

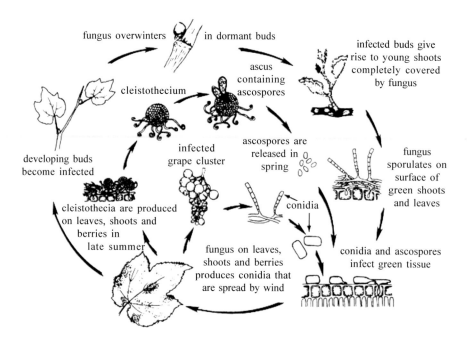

Fig. 12.7: Representatative disease cycle of powdery mildew.

the fungus produces haustoria inside of the host cells and mycelium grows on the plant's outer surface. Powdery mildew fungus produces conidia during the growing season as often as every 7 to 10 days. These conidia function as secondary inoculum as growth and reproduction repeat throughout the growing season to increase the incidence and intensity of the disease.

12.6. Disease Control

The control of powdery mildews particularly in grapes and some other crops depends on dusting the plants with sulfur. In cereals and several other annual crops, powdery mildew control is primarily through the use of resistant varieties. More recently, powdery mildew control has been obtained with systemic fungicides used as seed treatments or as foliar sprays. The same chemicals are used as sprays for the control of powdery mildews in other crops and in ornamentals. Several powdery mildew fungi, however, have developed resistance and are no longer controlled by some systemic fungicides. Powdery mildew on trees, such as apple, is controlled effectively with sprays of any of several sterol-inhibiting systemic fungicides. Powdery mildews have also been controlled experimentally with sprays of phosphate salt solutions and detergents or ultrafine oils and, in the greenhouse, by using blue photosensitive polyethylene sheeting. Experimentally, powdery mildew control has also been obtained through sprays with the biocontrol fungus *Ampelomyces quisqualis* and with plant activator compounds.

13

Powdery Mildew of Cereals

The disease is caused by Erysiphe graminis. Specialised forms of the species attack different cereals like *Agropyron cristatum*, A. *dasystachyum, A. inerme, A. repens, A. riparium, A. sibiricum, A. smithii, A. spicatum, A. striatum, A. subsecundum, A. trachycavlum, A. trichophorum, Agrostis alba, A. diegoensis, A. exarata, A. verticillata, Avena byzantina, A. fatua, A. sativa, Beckmannia syzigachne, Bromus breviaristatus, B. carinatus, B. catharticus, B. erectus, B. inermis, B. mollis, B. racemosus, B. rigidus, B. sterilis, B. suksdorfii, B. tectorum, B. vulgaris, Buchloe dactyloides, Calamagrostis canadensis, C. rubescens, Catabrosa aquatic, Cinna arundinacea, Cynodon dactylon, Dactylis glomerata, Digitaria sarujuinalis, Elymus canadensis, E. condensatus, E. dahuricus, E. glaucus, E. junceus, E. sibiricus, E. triticoides, E. villosus, E. virginicus, Festuca idahoensis, F. ovina, F. rubra , F. rubra* var. *commutata, Glyceria striata, Hordeum brevisubulatum, H. distichon*, *H. jubatum, H. nodosum, H. pusillum, H. vulgare, Hystrix patida, Koeleria cristata, Melica calijornica, Milium cffvsurii, Phalaris arundinacea, Phleum pratense, Poa alpina, P. ampld, P.arachnifera, P. arida, P. canbyi, P. compressa, P. cusicki, P. epilis, P. glaucifolia, P. gracillima, P. interior, P. junci-folia, P. leptocoma, P. longifolia, P. nemoralis, P. nervosa, P. nevadensis, P. palustris, P. pratensis, P. scabrella, P. secunda, P. silvestris, P. vaseyochloa, Polypogon monspeliensis, Puccinellia distans, Secale cereale* L., *Sitanionhansenii , S. hystrix, S. jubatum, Sphenopholis obtusata, Sporobolus giganteus, Stipa calijornica, Triticum aestivum.*

The host range of the species is large, as indicated by the above list. Among these host, Powdery mildew of wheat (*Triticum aestivum* L.), barley (*Hordeum distichon* L.), oat (*Avena sativa* L.), and rye (*Secale cereale* L.) is caused by *Erysiphe graminis tritici, E. graminis hordei, E graminis avenae* and *E graminis secalis* respectively. In each formae specialis, distinct physiologic races, specialising on a particular cultivar of the cereal are found.

The pathogen has the superficial sparingly branched mycelium with flat appressoria and elliptic, finger-like branched haustoria. The pathogen forms scattered, elongate to ellipsoid effuse patches of mycelia growth on the affected parts. Oidiophores arising from the hemispherical swellings on the mycelia web, at right angles to the leaf surface, are rather short, each with a terminal generative cell and a swollen basal cell. Each oidiophore produces a long chain of 10-20 oidia. The oidia are hyaline, elliptic or barrel-shaped, 25-35 x 8-10 μm in size. Oidia are produced in abundance and are wind disseminated. Under favourable conditions, cleistothecia may also be produced. They are immersed in the mycelial weft, globose to strongly depressed, dark brown to black, 135-250 μm in diameter, bearing rudimentary, myceloid, subhyaline to pale brown appendages. Each cleistothecium contains 8-25 asci. Asci are ovate to cylindric, more or less sessile, 70-108 x 25-40 μm in size, each with four to eight ascospores. Ascospores are elliptic, subhyaline to pale brown, 20-25 x 10-15 μm in size.

In areas where conditions for development of cleistothecia are favourable, the disease is soil borne. Oidia do not retain viability at high summer temperatures. During growing season, the disease spreads in the field through oidia. They are produced abundantly during cool and moist conditions, and are easily disseminated by wind. Unlike most powdery mildews, which are favoured by dry weather, this powdery mildew thrives well at low as well as high relative humidities. About 100% R.H. and 15-20⁰C temperature are optimum for oidial germination.

In most areas, where cereals are cultivated, conditions for cleistothecia development are rarely favourable. Even if asci and ascospores are formed, there is no possibility of their germination to cause infection. Some workers, however, could later show that asci remain immature in cleistothecia on fallen leaves. Ascospores in the asci, developing in the cleistothecia of such leaves, may form after 9-10 months-if such leaves are subjected to alternating dry and wet soil conditions. Such conditions exist in low lying fields where this disease is common and soil borne.

In India the disease is common and severe in the cool, lower hills of northern India and in low lying areas and in certain parts of Rajasthan. It is recently reported that cleistothecia are source of primary inoculum in temperate regions like Lahul and Spiti in Himachal Pradesh whereas in other parts of north India oidia are the main source of primary infection. In most areas where cleistothecia fail to develop, the disease appears to be introduced in the plains in every season through oidia blown down by wind from the hilly areas. The pathogen in cooler areas is available as oidia throughout the year.

13.1. Powdery Mildew of Wheat (*Triticum aestivum* L.)

Powdery mildew of wheat is generally distributed wherever wheat is grown. It is most prevalent in cool and lower hills of North India. Occurrence of this disease along with rust is also reported. Mehta (1930) have reported powdery mildew in plains of Rajasthan. Powdery mildew is more severe in fields, which have had heavy applications of nitrogen fertilizer. Cool, humid, and cloudy weather are conductive to disease severity.

Pathogen: *Erysiphe* (*Blumeria*) *graminis* f. sp. *tritici*

Erysiphe (*Blumeria*) *graminis tritici* is an obligate parasite.The mycelium is superficial, branched with small appresoria, and haustoria. Hyphae are septate and uninucleate of which some undergo swelling to act as conidiophore.

Conidiophore produces elliptical, hyaline uninucleate conidia in chains. Conidia are produced, and cause the powdery appearance;damp or muggy weather favours the production of conidia and their germination after dissemination. Conidia are disseminated by other agencies such as wind, water and insects. After reaching to suitable host conidia germinate to produce one or more germ tubes. These germ tubes develop into appresorium. From appresorium, penetration peg develop, which penetrate the epidermal cells and after entering to host cell it swells up into vesicle and form branched haustoria, which is the only structure which develop inside the host. After haustorium formation in the host epidermal cells, elongating secondary hyphae develops from appresorium, grows rapidly over the surface of the affected parts (leaf or stem) and branches rapidly to form mycelium. In summer, from mycelium sex organs develops, and after sexual reproduction black cleistothecium develops on the leaf surface, which are globose without an opening, and contains asci and ascospores.

Ascospores are shot out and dispersed by wind; liberated ascospores germinate immediately on falling on a suitable host to start new loci of infection.

Geographical Distribution

Powdery mildew can be found in all wheat growing areas of the United States but usually will be most severe in the east and southeast.It is more common in areas with a humid or semi-arid environment where wheat is grown. In India, the disease is mostly confined to northern and southern hills but sporadically appear in plains and foothills of the country.

Symptoms

Triticum sp. (wheat) is the only host of *Blumeria graminis* f. sp. *tritici*.

The symptoms start as superficial greyish small patches of white cottony growth on leaves, stem, sheath and even on ear (Fig.13.1). The patches may appear on both surfaces of the leaves. The white cottony growth can cover the whole leaf and other aerial parts of the plant. The white patches turn into brownish to dull tan colour late in the season due to formation of cleistothecia. The mildew colonies are surrounded with chlorotic or yellow areas resulting decease in photosynthetic rate. The affected leaves die off prematurely. The cleistothecia remain on wheat debris. At times, development of the ears is checked partially or wholly. Symptoms progress from lower to upper leaves. Lower leaves are commonly the most infected because of higher humidity around them.

Fig. 13.1: Powdery mildew (*Erysiphe(Blumeria) graminis* f.sp. *tritici*) on wheat.

Epidemic and Losses

Powdery mildew has become a more important disease in some areas because of increased application of nitrogen fertilizer, which favors the development of the fungus. Severe symptoms of powdery mildew can cause stunting of wheat. If unmanaged, this disease can reduce yields significantly by reducing photosynthetic areas and causes non-seed producing tillers. Powdery mildew causes reduced kernel size and lower yields. It depends on the intensity of the mildew, area of plant canopy covered and the progress of the disease in the field due to environmental conditions. Yield Losses up to 45 percent have been shown in Ohio on susceptible varieties when plants are infected early and weather favors the disease. In Canada the infection of the disease was observed up to 80 percent during years of heavy infections.

Disease cycle

Blumeria graminis f. sp. *tritici* has a polycyclic life cycle (Fig.13.2) typical of its phylum, Ascomycota. Powdery mildew of wheat overwinters as cleistothecia dormant in plant debris. Under warmer conditions, however, the fungus can overwinter as asexual conidia or mycelium on living host plants. It can persist between seasons most likely as ascospores in wheat debris left in the field. Ascospores are sexual spores produced in the ascus in the cleistothecia. These spores, as well as conidia, serve as the primary inoculum and are dispersed by wind. Neither spore requires free water to germinate; only high relative humidity is required for their germination.

Wheat powdery mildew thrives in cool humid conditions and cloudy weather increases chances of disease. When conidia land on a wheat leaf's hydrophobic surface cuticle, they release proteins which facilitate active transport of lightweight anions between leaf and fungus even before germination. This process helps Blumeria recognize that it is on the correct host and directs growth of the germ tube. Both ascospores and conidia germinate directly with a germ tube. Conidia can recognize the host plant and within one minute of initial contact, the direction of germ tube growth is determined. The development of appressoria then begins infection following the growth of a germ tube. After initial infection, the fungus produces haustoria inside of the wheat cells and mycelium grows on the plant's outer surface. Powdery mildew of wheat produces conidia during the growing season as often as every 7 to 10 days. These conidia function as secondary inoculum as growth and reproduction repeat throughout the growing season.

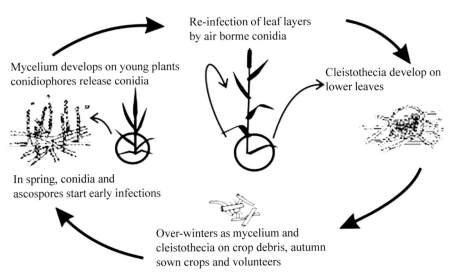

Fig. 13.2: Life cycle of powdery mildew of wheat

Environment

Powdery mildew of wheat thrives in cool, humid climates and proliferates in cloudy weather conditions. The pathogen can also be an issue in drier climates if wheat fields are irrigated. Ideal temperatures for growth and reproduction of the pathogen are between 16 °C and 21 °C with growth ceasing above 25 °C. Dense, genetically similar plantings provide opportune conditions for growth of powdery mildew.

Management

Controlling the disease involves eliminating conducive conditions as much as possible by altering planting density and carefully timing applications and rates of nitrogen. Since nitrogen fertilizers encourage dense leafy growth, nitrogen should be applied at precise rates, less than 31.75 kg per acre, to control disease severity. Crop rotation with non-host plants is another way to keep mildew infection to a minimum; however the aerial nature of conidia and ascospore dispersal makes it of limited use. Wheat powdery mildew can also be controlled by eliminating the presence of volunteer wheat in agricultural fields as well as tilling under crop residues.

Chemical control is possible with fungicides such as triadimefon and propiconazole. Another chemical treatment involves treating wheat with a silicon solution or calcium silicate slag. Silicon helps the plant cells defend against fungal attack by degrading haustoria and by producing callose and papilla. With silicon treatment, epidermal cells are less susceptible to powdery mildew of wheat.

Powdery mildew fungi are unique among the plant pathogens showing high response to sulphur and related fungicides. Several systemic fungicides such as benlate (0.1 %), karathane (2 lb /ha) etc. can control powdery mildew through foliar sprays. Seed dressing and soil drenching with 0.01 % calyxin is also found effective. One spray of propiconazol (Tilt 25EC@0.1%) on disease appearance is highly effective.

Systemic fungicides like Benlate (0.1%) even upto30 days, completely control the disease.Triazole fungicides, Bayton, Bayleton and Vigil gives 100% control for 42 days. HD 2204, HB208, CPAN 1676, VL 401 are highly resistant. CPAN 1922 was found resistant to all races of *E. graminis tritici.*

Milk has long been popular with home gardeners and smallscale organic growers as a treatment for powdery mildew. Milk is diluted with water (typically 1:10) and sprayed on susceptible plants at the first sign of infection, or as a preventative measure, with repeated weekly application often controlling or eliminating the disease. Studies have shown milk's effectiveness as comparable to some conventional fungicides, and better than benomyl and fenarimol at higher concentrations. Milk has proven effective in treating powdery mildew of summer squash, pumpkins, grapes, and roses. The exact mechanism of action is unknown, but one known effect is that ferroglobulin, a protein in whey, produces oxygen radicals when exposed to sunlight, and contact with these radicals is damaging to the fungus.

Another way to control wheat powdery mildew is breeding in genetic resistance, using "R genes" (resistance genes) to prevent infection. There are at least 25 loci on the wheat genome that encode resistance to powdery mildew. If the particular variety of wheat has only one loci for resistance, the pathogen may be controlled only for a couple of years. If, however, the variety of wheat has multiple loci for resistance, the crop may be protected for around 15 years. Because finding these loci can be difficult and time consuming, molecular markers are used to facilitate combining resistant genomes. One organization working towards identifying these molecular markers is the Coordinated Agricultural Project for Wheat. With these markers established, researchers will then be able to determine the most effective combination of resistance genes.

13.2. Powdery Mildew of Barley (*Hordeum vulgare* L.)

Pathogen: *Erysiphe* (*Blumeria) graminis* f. sp. *hordei*

Symptoms

The symptoms of this disease are generally limited to the upper surface of the leaves. Under favourable conditions, the sheaths, stems and glumes may also be affected. In the beginning the fungus appears as superficial flocculent growth on the upper surface of the leaves, which later on spreads to sheath and floral bracts. The pathogens may be observed in isolated white patches (Fig 13.3). Sometimes whole of the leaf is found to be affected. Ultimately the fungus changes to grey or reddish-brown colour, and the cleistothecia may be observed as black dots. As a result of powdery mildew, the leaves transpire more profusely, and the photosynthesis is reduced. Infected plants become stunted in size. The leaves that are not shed become wrinkled, spirally twisted and deformed.

The disease is commonly found in the hilly regions of northern India. It has been noticed inUttar Pradesh and Bihar.

Fig. 13.3: Powdery mildew of Barley (*Erysiphe graminis* f.sp. *hordei*)

Pathogen

Blumeria graminis f. sp. *hordei* is an obligate parasite. The mycelium is ectophytic, superficial, profusely branched, consisting of uninucleate cells. It forms a white web-like coating over the leaf and sends haustoria into the epidermal cells of the host.

It has branched haustorium and forms finger-like structures. It is frequently provided with an external disc or appresorium, from which the haustorium arises and pushes into the epidermal cell. Superficial mycelium develops conidiophores (swollen at the base), which arise at right angles on the leaf surface

Conidiophores elongate and divide into two cells. The upper cell acts as the mother cell of the conidia. The mother cell cuts off conidia in acropetal succession from its distal end and gives rise to a basipetal chain of conidia.

Each conidial chain contains 10-12 conidia. The conidium is oval or barrel-like, hyaline, single-celled, uninucleate. Each conidium is capable to germinate in short time and produce several germ tubes. The conidia are disseminated by wind. Conidial germination is maximum at 12 °C, while the best growth of germ tubes is at 21 °C. Late in the season cleistothecium may be formed on leaves.The cleistothecia are globose, black, partly immersed in the mycelium, measuring upto 200 µm in diameter. Each cleistothecium is furnished with simple or slightly branched appendages. Each cleistothecium contains as many as 25 asci, each with 4-8 elliptical ascospores. The cleistothecia are rarely formed in India.

Disease Cycle

The disease is soil-borne through cleistothecia. It has been observed that cleistothecia can survive upto thirteen years. While other powdery mildews are favoured by dry weather, the powdery mildew of cereals is an exception and survice and develops well in low and high relative humidities.

Disease Management

In general, barley is more heavily affected than wheat, and losses can be very severe if the disease is not controlled. Barley is infected only by forma specialis *hordei* of *E. graminis*, so powdery mildew from wheat or rye cannot infect barley and vice versa. Winter barley should, however, not be grown next to spring barley. A range of cultural practices exists that may somewhat reduce the infection by *E. graminis*. The growing of resistant cultivars is recommended, particularly with the durable Mlo resistance gene. The use of mixtures of resistant cultivars considerably reduces the risk of powdery mildew. If possible, winter barley should be sown late in regions favourable to powdery mildew. An open stand of barley reduces the incidence of powdery mildew as compared with a dense stand, heavily fertilized with nitrogen. If powdery mildew infection becomes serious, one or more fungicide sprays may be necessary; these should not be applied after growth stage 55. In most cases, it is not useful to treat winter barley in the autumn. However, occasionally, a treatment in autumn may be recommended especially on light soils, where mildew infection may exacerbate winter crop losses. Treatment in spring barley may commence as soon as mildew is seen, and in any case if the third-youngest leaf, after elongation has begun, is affected. In areas where powdery mildew infection can regularly be expected, fungicides may be applied as a seed treatment, but this may lead to problems of resistance. Chemical treatment particularly sulphur dusting is helpful in controlling the disease. Spraying wattable Sulphur, Karathane E.C. is also effective in disease management.

Problems with resistance

E. graminis has been reported to show reduced sensitivity to fungicides of the sterol-biosynthesis inhibitor group, but this has not led to loss of control.

Main fungicides

Sprays: bromuconazole, cyproconazole, cyprodinil, epoxiconazole, fenpropidin, fenpropimorph, flusilazole, flutriafol, kresoxim-methyl, prochloraz, propiconazole, tebuconazole, triadimefon, triadimenol, tridemorph, triforine. Seed treatments: flutriafol, triadimenol.

13.3 Powdery Mildew of Rye (*Secale cereale* L.)

Pathogen: *Erysiphe (Blumeria) graminis*f. sp.*secalis*

The specialized rye form of the fungus B. graminis f. sp. secalis is adapted to Secale species.

Light infections of powdery mildew are common on rye although of no economic importance.

Distribution

Disease is distributed throughout all continents, where the host-plant is present. In Russia it occurs in all regions of rye cultivation.

Economic significance

Disease causes reduction of spikelets and grain number in head, and weight of the grain. At high disease development (to 60%), under artificial inoculation, the yield losses may reach 30-50%. In natural conditions the average long-term loss is estimated as 5%.

Symptoms

The fungus develops usually on the upper surface of leaves, and under favorable conditions it may attack all above-ground parts of the plant. Surface mycelium producing conidia gives the leaves a powdery appearance (Fig.13.4). These conidia are a source for other plant infections. The symptoms and etiology are similar to those on barley.

Fig. 13.4: Symptoms of powdery mildew on rye leaves

Pathogen

Erysiphe (Blumeria) graminis f. sp. *secalis* conidia develop on terminal generative cells of short and simple conidiophores. Conidia are hyaline, unicellular, and ellipsoidal to ovate (8-10 x 25-30 μ). They are produced basipetally and diurnally in long chains. The conidial stage is named Oidium monilioides (Nees) Link. Germ tubes penetrate leaves, establish haustoria in epidermal cells, and give rise to surface sporulating colonies. Haustoria (5-10 x 10-30 μ) bear distinct fingerlike appendages up to 20 μ long at each apex. Later the surface mycelium condenses and forms small pads of dirty-gray, brown, rust-brown color, on which cleistothecia are formed, being dark, 135-280 μ in diameter. Each cleistothecium's asci contain 6-30 ascus. Mature asci are cylindrical to ovate (25-40 x 70-110 μ). There are 8 hyaline ascospores (10-13 x 20-23 μ) in each ascus. The fungus is heterothallic, and two mating types occur in about an equal ratio in natural populations.

Initial infection of winter rye may already be carried out in autumn by ascospores or conidia from non-harvested or wild-growing plants. The fungus over-winters as mycelium on infected leaves or in cleistothecia, from where it infects plants in the spring. Conidia and ascospores may be carried a great distance by wind

Weather

The best condition for disease development is damp cool weather. Optimum temperature for the Powdery Mildew is 15-18^{0}C. In these conditions the disease incubation period lasts 3 to 7 days. Hot and dry weather stops disease development. Conidia are able to germinate in a wide range of temperatures (1-30^{0}C), without presence of water drops, at a relative humidity near 100%.

Control measures

Control measures include: resistant varieties, crop rotation, deep ploughing, treatment of seeds and plants by fungicides used for wheat and barley powdery mildew. According to Germar (1934) and Mains (1926), mildew-resistant rye selections are not difficult to secure.

13.4 Powdery Mildew of Oat (*Avena sativa* L.)

Pathogen: *Erysiphe (Blumeria) graminis* f. sp. *avenae*

The powdery mildew symptoms are not common on most of the cultivated varieties of oats. Reed (1920) demonstrated that many of the older varieties of oats are susceptible to certain specialized races of the fungus with typical powdery mildew symptoms (Fig.13.5).

This variety of the fungus is specialized on species of *Avena* and *Arrhenatherum*. See Powdery Mildew of barley for other details of the disease like disease cycle and control measures.

Fig. 13.5: Powdery mildew on leaves of a maturing oat crop

14

Powdery Mildew of Pulse Crops

14.1. Powdery Mildew of Pea

This disease is much more destructive than the downy mildew and occurs worldwide. The disease usually develops late in the season reaching its maximum intensity at the time of pod formation (January-end or February). Unlike the downy mildew which is favoured by moist weather, this mildew is worst during dry weather. Besides pea some other leguminous crops like bean, urdbean and lucerne and non leguminous ones like coriander, turnip, cabbage etc and many others are also attacked by different physiological races of the fungus.

Pathogen: *Erysiphe pisi* Syd. (syn. *E. polygoni* DC.)

Symptoms

The disease appears first on leaves and then on other green parts of the plant. On leaves, there appear small, irregular or circular, powdery spots on the upper surfaces. These spots enlarge rapidly, to cover the entire upper surface, and later on the lower surface may also become infected. The fungus grows rapidly over the leaf surfaces and in advanced stage it appears if the leaves are covered with a talcum-like powder (Fig 14.1). The leaves become reduced in size, turn yellow and are finally shed.At the time of flowering and pod formation (January-February), the disease assumes a severe state. The whitish powdery spots completely cover the leaves, petioles, tendril, stem and even pods.The attacked plant assumes a greyish white appearance

Fig. 14.1: Symptoms of Powdery Mildew on Pea leaves and pods

Pathogen

The pathogen is an obligate parasite found on *Pisum, Medicago, Vicia, Lupinus* and *Lens* spp belongs to the Class: Ascomycetes; Order: Erysiphales; and Family:Erysiphaceae. The mycelium is superficial, and forms a white web-like coating over the leaf, consists of delicate and persistent hyphae, attached to leaf by means of appressoria.Infection threads developing from appressoria penetrate the epidermal cells and swell into lobed and round sac-like haustoria. From the superficial mycelium on the leaf, several oidiophores arise vertically. Each oidiophore bears a chain of unicellular oidia. Oidiophore is septate and its cells not much different from the oidia. The oidia are elliptical, barrel-shaped, or even cylindric, hyaline and 35-45 x 15-20 μm in size.Late in the season, but not always on the pea leaves in the field, sharp, black specks appear scattered in the mycelial web. These are cleistothecia (ascocarps) of the pathogen. The cleistothecia are globose, 85-125 μm in diameter,covered by few to many, free or interwoven appendages resembling the vegetative hyphae.Each cleistothecium contains 2-8 asci. Each ascusis almost sessile,sub-globose to ovate, 50-60 x30-40 μm in size containing 3-8 ascospores, which are unicellular, elliptical, hyaline and 19-25 x9-15 μm in size.

Cleistothecia are not common on intact leaves in the field. They may develop on infected plant debris in the soil, influenced by soil moisture and temperature. They persist in soil until the following season.

Disease cycle

The cleistothecia persisting in the soil or in plant debris are the source of primary infection during next season. Ascospores released from them infect first the lower most leaves near the soil. Secondary spread in the field occurs by conidia formed in abundance on the infected leaves. Mundkur (1946) suggest seed borne nature of the disease where dormant mycelium in the pea seeds may also initiate primary infection during next season.However, recent work of some

Indian workers could rule out the seed-borne nature of the disease. They suggested that this disease in India has a cycle more or less similar to that of wheat rusts. They could not detect the mycelium of the fungus on or in the seed. Cleistothecia were also not reported in the plains of north India as well as in hills. Possibility of any alternate hosts has also been ruled out. According to them, pea is grown throughout the year in one or the other regions of the plains (Punjab, U.P.) and hills (H.P.), and the pathogen is present in oidial stage throughout the year in nature. It survives in mycelial and conidial stages on the main pea host in plains during winter and in cooler regions or higher altitudes of H.P. during summer season. The primary inoculum comes from these hills by wind to plains to infect main crop.

Disease management

Since cleistothecia surviving in diseased plant debris lying in soil, are source of primary inoculum, field sanitation is useful in reducing the incidence of disease. Diseased plant debris should be collected and burnt in the field. Early varieties for vegetable crops should be preferred.Several fungicides have been recommended for use as foliar sprays. These include sulphur dust (200 mesh; @ 25-30 kg/ha); Elosal (0.5%); Karathane W.D. (0.2% @ 2 kg/ha); Morocide (0.1%); Sulfex(2.5 kg/ha); Bavistin (0.1%, 2-3 sprays).

Resistance to powdery mildew is readily available in commercial cultivars and this resistance has been stable for at least 30 years. Currently, there is some controversy as to whether there are one or two genes which govern resistance to powdery mildew. It is thought that the single recessive gene, er1, will not suffice underconditions of extreme pathogen pressure (Hagedorn, 1984). In that case, resistance conferred by a second gene. er2, is needed in combination with er 1. Some resistant varieties of pea developed in India are P 185, P 388, P 6583 and P 6588. In areas where the disease frequently occurs and resistant cultivars are not readily available, early maturing cultivars plus chemical sprays are the most viable control option. Chemical control includes spraying powdery mildew infested fields with flowable, elemental sulphur at a rate of 3-4 kg ha^{-1}. Other control measures include crop rotation and immediate ploughing of the infested crop debris.

14.2. Powdery Mildew of Pigeon pea

Pathogen: *Leveillula taurica* Arnaud (anamorph Oidiopis taurica (Lev.) Salmon)

Powdery mildew is a widespread disease affecting pigeon pea (Nene *et al.,* 1996), and many other economic plants (Holliday, 1980). Probably the first report of its occurrence on pigeonpea was from Tanzania (Wallace, 1930). It is particularly common in southern and eastern Africa (Reddy *et al.*, 1990b; Nene *et al.*, 1996). Powdery mildew usually assumes importance during their productive stage in long-duration types but can severely infect short-duration types as well. Long-duration land races often escape from disease if they flower and pod when the season is dry and warm.Short-duration pigeonpea is vulnerable as it flowers when the season is cool and humid, especially in some areas of the southern Indian states of Karnataka and Tamil Nadu. Although the disease has potential to be serious, limited research has been done (Raju, 1988).

Symptoms

Infected plants show white powdery fungal growth on all aerial parts, especially leaves (Fig, 14.2), flowers and pods. Severe infection results in heavy defoliation. The disease causes stunting of young plants, followed by the visible symptoms of white powdery growth that appear gradually before flowering. The initial symptoms develop as small chlorotic spots on the upper surface of individual leaves and subsequently with corresponding lower surfaces. When the fungus sporulates, this white powdery growth covers the entire lower leaf surface. In severe infection, leaves turn yellow, twist and crinkle, then fall (Narayanaswamy and Jaganathan, 1975).

Epidemiology

Although the pathogen is generally considered to be monocyclic, it is reported (Raju,1988) that initial infection with powdery mildew in pigeonpea was followed by secondary spread. Infection is directly proportional to the quantity of conidial inoculum available and disease progress is exponential (Raju, 1988). Indian cultivars with thin, succulent leaves that are easily colonized by the fungus are more susceptible than those from Kenya that have thicker leaves. The disease develops at temperatures ranging from 20 to 35°C, but 25°C is optimal. A cool, humid climate favours fungal infection and colonization while a warm humid climate is suitable for sporulation and spore dispersal. Sporulation is more frequent on young leaves than on older ones. Plants attacked by sterility mosaic or phyllody support abundant sporulation and since such plants remain green in the field for long periods, they provide a continuous source of inoculum

Fig. 14.2: Powdery mildew on pigeon pea leaves

(Reddy *et al.*, 1984;Raju, 1988; Prameela *et al.*, 1989).The fungus survives on perennial pigeonpea and volunteer plants growing in the shade, and on the ratoon growth of harvested stubbles (Raju, 1988). It also survives as dormant mycelium on infected plant parts such as the axillary buds. In India, early sowing, shade and irrigation encourage disease establishment (Raju, 1988).

Effect on Yield and Quality

Although limited information is available on the effect of powdery mildew on yield of pigeonpea, the disease caused 100% defoliation and yield loss when newly developed short-duration pigeonpea lines were tested in eastern and southern Africa (Raju, 1988).

Disease Control

Cultural:

(i)   Select fields distant from perennial pigeonpea affected with powdery mildew and

(ii) Sow late (after July) in India, to reduce disease incidence.

Chemical:

Control measures include spraying with wettable sulphur at 1 g l⁻¹ or triadimefon at 0.03% (Raju, 1988).

Biological:

Cladosporium sp.was identified as a hyperparasite of the powdery mildew pathogen and has potential to control the disease (Raju, 1988).

Resiatant varieties:

Table 14.1: Multiple sources of resistance to pigeonpea diseases including powdery mildew.

Diseases	Resistant varieties/lines
Wilt+Sterility mosaic+ Powdery mildew	ICP 7867, ICP 8861 ()ICP 7035, ICP 8862 (Hy 3C)
Sterility mosaic+ Powdery mildew+ Halo blight	Hy 3C, ICPM7035

ICP 9177, a germplasm accession from Kenya, was immune to the disease. Some of the resistant lines, such as ICP 8862 and ICP 7035, also have resistance to wilt and sterility mosaic. In general germplasm from Kenya was found to be highly resistant (Raju, 1988).

14.3 Powdery mildew of green gram (Mungbean)

The disease occurs throughout Indian subcontinent and south-east Asian countries. In India it causes serious damage in the states like Maharashtra, Andhra Pradesh, Tamil Nadu, MP, Rajasthan, Himachal Pradesh and UP.

Pathogen: *Erysiphe polygoni* DC

Economic importance

Yield losses up to 21% have been reported in severely infected crop of mungbean. In India, the losses due to powdery mildew in winter sown mung bean are more as compared to rainy season crop.

Symptoms

Powdery mildew is one of the widespread diseases of green gram. White powdery patches appear on leaves and other green parts which later become dull colored. These patches gradually increase in size and become circular covering the lower surface also. When the infection is severe, both the surfaces of the leaves are completely covered by whitish powdery growth (Fig 14.3). Severely affected parts get shriveled and distorted. In severe infections, foliage becomes yellow causing premature defoliation. The disease also creates forced maturity of the infected plants which results in heavy yield losses.

Fig. 14.3: Symptoms of Powdery mildew on green gram leaves

Pathogen

The fungus is ectophytic, spreading on the surface of the leaf, sending haustoria into the epidermal cells. Conidiophores arise vertically from the leaf surface, bearing conidia in short chains. Conidia are hyaline, thin walled, elliptical or barrel shaped or cylindrical and single celled. Later in the season, cleistothecia appear as minute, black, globose structures with myceloid appendages. Each cleistothecium contains 4-8 asci and each ascus contains 3-8 ascospores which are elliptical, hyaline and single celled.

The pathogen has a wide host range and survives in oidial form on various hosts in off season.secondary spread is through air-borne oidia produced in the season

Disease cycle

The Pathogen is an obligate parasite and survives as cleistothecia in the infected plant debris. Primary infection is usually from ascospores from perennating cleistothecia. The secondary spread is carried out by the air-borne conidia. Rain splash also helps in the spread of the disease.

Control measures

1. Use disease resistant varieties HUM 1, ML 131, ML 395, OUM 15-3, Pant M1, Pusa 105, Pusa 9072, TARM 1, TARM 2, TARM 18, BPMR 145, LBG-17, PDU 10 etc.

2. Delayed sowing and wider spacing considerably reduces the disease intensity.

3. Field sanitation and clean cultivation reduces the disease intensity.

4. On initiation of symptoms two spray of 0.1% Calyxin or Karathane at 10 days interval.

5. Uproot and destroy diseased plants

6. Spray Carbendazim 500g or Wettable sulphur 1.5 kg orTridemorph 500 ml/ha at the initiation of disease and repeat 15 days later.

7. The seeds must be sown early in the month of June to avoid early incidence of the disease on the crop.

14.4. Powdery mildew of black gram (Urdbean)

The disease occurs throughout Indian subcontinent and south-east Asian countries. In India it causes serious damage in the states like Maharashtra, Andhra Pradesh, Tamil Nadu, MP, Rajasthan, Himachal Pradesh and UP.

Pathogen: *Erysiphe polygoni* DC

Economic importance

The average yield loss due to this disease in urdbean has been estimated to be 29.1%. In India, the losses due to powdery mildew in winter sown urd bean are more as compared to rainy season crop.

Symptoms

Small, irregular powdery spots appear on the upper surface of the leaves (Fig 14.4) and sometimes on both the surfaces. The disease becomes severe during flowering and pod development stage. The white powdery spots completely cover the leaves, petioles, stem and even the pods. The plant assumes grayish white appearance; leaves turn yellow and finally shed. Often pods are malformed and small with few ill-filled seeds.

Fig. 14.4: Symptoms of powdery mildew on black gram leaves

Pathogen

The fungus is ectophytic, spreading on the surface of the leaf, sending haustoria into the epidermal cells. Conidiophores arise vertically from the leaf surface, bearing conidia in short chains. Conidia are hyaline, thin walled, elliptical or barrel shaped or cylindrical and single celled. Later in the season, cleistothecia appear as minute, black, globose structures with myceloid appendages. Each cleistothecium contains 4-8 asci and each ascus contains 3-8 ascospores which are elliptical, hyaline and single celled.

Weather

Infection is favoured by warm dry weather with 22-26^{0}C and 80-88% relative humidity. The disease is severe generally during late *kharif* and *rabi* seasons.

Disease cycle

The fungus survives as cleistothecia in the infected plant debris. Primary infection is usually from ascospores from perennating cleistothecia. The secondary spread is carried out by the air-borne conidia. Rain splash also helps in the spread of the disease.

Management

1. Adjusting the sowing dates and planting of crop with wider specing.
2. Remove and destroy infected plant debris.
3. Foliar spraying of wettable sulphur 0.3% or carbendazin 0.05, 0.1% or benlate 0.05% and topsin M 0.15%.
4. Grow resistant varieties like Urd bean lines OBG 16.

14.5. Powdery Mildew of Lathyrus *(Lathyrus sativum* L.)
Pathogen: *Erysiphe polygoni*

Symptoms
Affected plants show white powdery mass on leaves (Fig 14.5) in the initial stage and later on stem and pods.

Fig. 14.5: Symptoms of powdery mildew on Lathyrus leaves

Pathogen, disease cycle and environment is the same as described in other pulse diseases for E.polygoni.

Control measures
1. Field sanitation reduces the disease intensity.
2. On initiation of symptoms, two spray of 0.1% Calyxin or Karathane at an interval of 15 days.
3. Up root and destroy diseased plants.

14.6. Powdery Mildew of Lentil
Pathogen: *Erysiphe polygoni* f. sp. *lentis*

Symptoms
Affected plants show white powdery mass on leaves (Fig 14.6) in the initial stage and later on stem and Pods.

Pathogen, disease cycle and environment is the same as described in other pulse diseases for E.polygoni.

Fig. 14.6: Symptoms of powdery mildew on lentil leaves

Control measures

1. Field sanitation reduces the disease intensity
2. Proper spacing also reduces the disease up to some extent.
3. On initiation of symptoms two spray of 0.1% Calyxin or Karathane at an interval of 15 days.

14.7. Powdery Mildew of Cowpea *(Vigna unguiculata)*, Mothbean *(Vigna aconitififolin)* and Horsegram *(Macrotyloma uniflorum)*

Pathogen: *Erysiphe polygoni* DC

Symptoms

Powdery mildew is one of the wide spread diseases in cowpea. White powdery patches appear on leaves and other green parts which later become dull colored (Fig 14.7). These patches gradually increase in size and become circular covering the lower surface. When the infection is severe, both the surfaces of the leaves are completely covered by whitish powdery growth. Severely affected parts get shriveled and distorted. In severe infections, foliage becomes yellow causing premature defoliation. The disease also creates forced maturity of the infected plants which results in heavy yield losses.

Pathogen, disease cycle and environment is the same as described in other pulse diseases for E.polygoni.

Fig. 14.7: Symptoms of powdery mildew on cowpea leaves

Control measures

1. Use disease resistant varieties.
2. Delayed sowing and wider spacing considerably reduces the disease intensity.
3. Field sanitation and clean cultivation reduces the disease intensity.
4. On initiation of symptoms two spray of 0.1% Calyxin or Karathane at 10 days interval.
5. Uproot and destroy diseased plants

14.8. Powdery mildew of minor legumes

The powdery mildews on minor legumes with the host infected, pathogen responsible and their distribution are given in tabular form.

Disease	Pathogen	Distribution/ Importance	References
Bambarra groundnut : *Vigna subterranean*			
Powdery Mildew	*Erysiphe pisi* DC ex. St- Am.	Zimbabwe, Tanzania, Madagascar, Zambia, Malawi	Rothwell (1983)
	E. poligoni DC		Riley (1960)
	Sphaerotheca voandzeiae Bour		Bouriquet(1946)
			Angus (1962-1966)
	Oidium sp.		Peregrine and Siddiqi (1972)
Chickpea			
Powdery Mildew	*Leveillula taurica* (Lev.) Arn.	India, Ethiopia, Sudan/ Minor	Allen (1983)
			Nene *et al.* (1996)
Common bean : *Phaseolus vulgaris*			
Powdery Mildew	*Erysiphe poligoni* DC	Worldwide, exacerbated by shade	Allen (1983)
Tropical Pasture Legumes : *Desmodium* spp.			
Powdery Mildew	*Oidium* sp.	South America, Caribbean, India/Minor	Lenne (1990a)
Feba bean:			
Powdery Mildew	*Erysiphe pisi* DC (*E. poligoni* sensu lato)	Widespread, Middle East, Canada	Iqbal et al. (1988)
	E. cichoracearum DC		Al-Hassan (1973).
	Leveillula taurica (Lev.) Arn.		Tarr (1955)
	Microsphaera penicillata (Wall ex Fr.) Lev.var. ludens (Salmon) Cooke		Morrall and McKenzie (1977)
Cowpea :			
Powdery Mildew	*E. poligoni* DC	Widespread/Minor	Allen (1983)
	Sphaerotheca fuliginea (Schlecht.ex.Fr.) Poll.		
Ground nut:			
Powdery Mildew	*Oidium arachidis Chorin*	Widespread in subtropics and temperate climates/ Minor	Chorin (1961)
			Hirata (1966)

Contd.

Crop / Disease	Pathogen	Distribution	References
Hyacinth bean :*Lablab purpureus*			
Powdery Mildew	*E. poligoni* DC *Leveillula taurica* (Lev.) Arn.	Venezuela, India, Ethiopia, Kenya, Nicaragua	Ondieki (1973) Lenne (1990)
Lentil:			
Powdery Mildew	*E. poligoni* DC *Leveillula taurica* (Lev.) Arn.	Bangladesh, Chile, Cyprus, Ethiopia, India, Jordan, Spain, Sudan, Syria, Former USSR	Golovin (1956) Sankhla *et al.* (1967) Mengistu (1979) Photiades and Alexandrou (1979) Fakir (1983) Sarrag and Nourai (1983) Mamlouk et al.(1984) Sepulveda (1987) Bayaa (1989) Diaz Morall (1993)
Lima bean : *Phaseolus lunatus*			
Powdery Mildew	*Leveillula taurica* (Lev.) Arn.	Tanzania	Riley (1960)
Lupins:			
Powdery Mildew	*E. poligoni* DC *E. cichoracearum* DC	Denmark, France, New Zealand, Peru, South Africa, Spain/ Generally minor but can be a major problem of *L. angustifolius* in spain	Lewartowska and Frencel (1994)
Macroptilium atropurpureum:			
Powdery Mildew	*Oidium sp.*	South America/Minor	Lenne (1990a)
Soyabean:			
Powdery Mildew	*Microsphaera diffusa Cooke & Peck*	Worldwide	Lohnes and Bernard (1992) Lohnes and Nickell (1994)
Asiatic Vigna :			
Powdery Mildew	*E. poligoni* DC	Worldwide and important	Bose (1932) Lawn and Ahn (1985)
Winged bean : *Psophocarpus tetragonolobus*			
Powdery Mildew	*E. cichoracearum* DC *Oidium* sp.	Papua New Guinea	Price (1977)

15

Powdery Mildew of Oil Seed Crops

15.1. Powdery mildew of Rapeseed and Mustard

Pathogen: *Erysiphe cruciferarum*

Losses

Damage to the crop may be very much if the disease appears during early stages of plant growth. Heavily infected pods remain empty or produce a few seeds at base with sterile twisted tips. Number of pods per plant, pod length, number of seeds per pod and 1000 grain weight, total yield and per cent oil content of plants are reduced due to infection.

Symptoms

All the above ground parts of plant are attacked. The disease manifests itself in the form of dirtywhite, circular, floury patches on leaf, stem and pod of the plant (Fig 15.1). As the temperature rises, these patches increase in size and coalesce with each other covering the entire surface of the infected plant part. On severely infected parts it appears as a white granular chalk-like powdery mass dusted on its surface. Overall growth of the plant is reduced and it produces lesser pods. Stems covered with powdery mildew appear purplish or bleached brown. On young green pods, white to dirty white circular patches of mildew growth appear in the beginning. Later the pods become covered completely with white powdery mass composed of mycelium, oidiophores and oidia of the fungus. Severely infected pods remain smaller in size and produce few small-sized seeds or remain empty. Later in the season, minute, spherical, dark brown or black, scattered or gregarious structures i.e. cleistothecia appear on leaves, stem and pods.

Fig. 15.1: Symptoms of powdery mildew on mustard leaves

212 The Plant Mildews

In India, the disease was first reported in 1963.It was not considered much harmful earlier to the yield of rapeseed and mustard. However, during last few years the disease could assume an epidemic form in Brassica growing areas of the country.

Pathogen

The disease is caused by *Erysiphe cruciferarum*, also described as *E. cichoracearum* by some workers. The mycelium is superficial, amphigenous, with white, dense, branched, septate and spreading hyphae. Oidia are hyaline, singly or in short chains on each oidiophore, cylindrical 25-45 x 12-16 μm in size. Cleistothecia are scattered, globose, yellow orange in the beginning, becoming brown to dark brown or black at maturity, and 90-130 μm in diameter. Appendages are many, myceloid, all over the surface, narrow,hyaline to faintly coloured, seldom branched,unequal in length. Asci are 3-12, but usually 6-8 in each cleistothecium, each with a distinct stipe,oval to pyriform, 50-70 x 30-45 μm in size. Ascospores are ovoid, 2-7 in each ascus, and 16-22 x 11-14 μm in size.

Oidia germinate only at tip end; the germ tube is branched, with up to three branches. Cleistothecia of the pathogen develop rarely in field. However, they have been frequently observed in abundance in climatic conditions of Haryana State.

An isolate of *E. cruciferarum* from the cultivar Prakash of mustard (*Brassica juncea*, also called Indian mustard or rai or raya) could also infect rapeseed (*Brassica campestris* var. brown sarson, *B. campestris* var. yellow sarson and*B. campestris* sp. *oleifera*), taramira (*Eruca sativa*), *B. chinensis* and *B. tournefortii* under artificial inoculation.

Disease Cycle

The pathogen survives in the form of mycelium and oidia on off season host plants (Brassica spp) and other weeds. Oidia are the source of primary inoculum. Since cleistothecia are reduced in abundance on diseased plant parts at the maturity of the crop at least in some areas of the country, like Haryana, it appears that they may survive on diseased plant debris lying in the field. However, it is yet to be confirmed. The disease spreads in the field through air disseminated oidia.

Weather

Moderate temperature (16-28°C), low humidity (below 60%), minimum rainfall or dry season during February-March favour the outbreak of the disease in the field. Production of cleistothecia in abudance is favoured by alternating low and moderate temperatures, whereas heavy oidia formation occurs at low nutrition of host, low relative humidity, dry soil and ageing of the host.

Disease management

The best method of control of the disease is the use of resistant cultivars of rapeseed and mustard. *Brassica alba*, *B. carinata* and *B. rapus* are reported to be immune to the fungus. Such sources of resistance should be exploited for the purpose of development of resistant varieties. Some cultural practices have been recommended for reducing the incidence of the disease. Crops sown during the end of October are very often attacked more severely and the time of sowing must, therefore, be adjusted accordingly so that crop escapes infection. The disease has been effectively controlled by using timely sprays of foliar fungicides, soon after the appearance of the disease. Sprays of Karathane (0.1%), Calyxin (0.1%), and Sulfex (0.2%) at 10 days interval immediately after the appearance of the disease proved effective to minimise the disease incidence. There is 95.7% reduction in disease incidence when Karathane (0.1%) was sprayed twice at an interval of 10 days soon after the appearance of the disease.

15.2. Powdery mildew of sesame

Pathogen: *Erysiphe cichoracearum*

Symptoms

Initially grayish-white powdery growth appears on the upper surface of leaves. When several spots coalesce, the entire leaf surface may be covered with powdery coating (Fig 15.2). In severe cases, the infection may be seen on the flowers and young capsules, leading to premature shedding. The severely affected leaves may be twisted and malformed. In the advanced stages of infection, the mycelial growth changes to dark or black because of development of cleistothecia.

Fig. 15.2: Symptoms of Powdery mildew on sesamum leaves

Pathogen

The fungus produces hyaline, septate myceliums which are exophytic and send haustoria into the host epidermis. Conidiophores arise from the primary mycelium and are short and non septate bearing conidia in long chains. The conidia are ellipsoid or barrel-shaped, single celled and hyaline. The cleistothecia are dark, globose with the hyaline or pale brown myceloid appendages. The asci are ovate and each ascus produces 2-3 ascospores, which are thin walled, elliptical and pale brown in colour.

Disease cycle

The fungus is an obligate parasite and disease perenneates through cleistothecia in the infected plant debris in soil. The ascospores from the cleistothecia cause primary infection. The secondary spread is through wind-borne conidia.

Weather

Dry humid weather and low relative humidity favours the disease.

Management

1. Remove the infected plant debris and destroy.

2. Spray Wettable sulphur@0.2% or dust Sulphur at 25 kg/ha and repeat after 15 days.

3. Grow resistant varieties like Rajeshwari, SI-1926, KRR-2, etc.

15.3. Powdery mildew of sunflower

Pathogen: *Erysiphe cichoracearum/ Leveillula compositarum*

These pathogens affect leaves and stalks of sunflower. *E. cichoracearum* infection appears as a continuous white felt fur mainly on the upper surface of leaves. *L. compositarum* forms separate angular yellow spots with brown bordering, with mycelium that are partly drowned into vegetative tissue on the lower side of leaves or, sometimes, on their upper surface.

Symptoms

The disease produces white powdery growth on the leaves. White to grey mildew appears on the upper surface of older leaves (Fig 15.3). As plant matures black pin head sized fruiting bodies of the pathogen are visible in white mildew areas. The affected leaves loose luster, curl, become chlorotic and die.

Fig. 15.3: Symptoms of Powdery mildew on sunflower leaves

Pathogen

Conidia on mycelium are formed as chains (*E. cichoracearum*) or individual formations *(L. compositarum)*. At the end of sunflower vegetation, the sexual stage (ascus) appears on affected parts as closed fruiting bodies or cleistothecia (perithecia). The fungus *E. cichoracearum* has round cleistothecia; their size is 120-122 µm; asci are 58-65 x 30-32 µm, ascospores are 21-24 x 11-13 µm and conidia are 24-32 x 17-19 µm in size. *L. compositarum* has flat cleistothecia; their size is 160-250 µm, asci are 72-120 x 27-35 µm, ascospores are 26-38 x 14-24 µm, and conidia are 48-63 x 15-24 µm in size. Cleistothecia and over-wintered mycelium in affected sunflower leaves and stalks are the source of initial infection by the disease pathogen.

Disease cycle

The fungus is an obligate parasite and disease perenneates through cleistothecia in the infected plant debris in soil. The ascospores from the cleistothecia cause primary infection. The secondary spread is through wind-borne conidia. During the vegetation of affected sunflower the pathogen form several generations of asexual spores or conidia, which cause new infections on plants.

Favourable conditions

The disease is more severe under dry condition towards the end of the winter months.

Management

1. Complete field and crop sanitation.
2. Early varieties should be preferred.
3. Removal of infected plant debris.
4. Application of karathane or calyxin 1L/ha or wettable sulphur 2 kg/ha is found effective in reducing the disease incidences.

15.4. Powdery Mildew of Castor

Pathogen: *Leveillula taurica*

Symptoms

It is characterized by typical mildew growth which is generally confined to the undersurface of the leaf. When the infection is severe the upper-surface is also covered by the whitish growth of the fungus (Fig.15.4). Light green patches, corresponding to the diseased areas on the under surface, are visible on the upper side especially when the leaves are held against light.

Fig. 15.4: Symptoms of powdery mildew on castor leaf.

Pathogen

Conidiophores emerged through leaf stomata, singly or branched, usually in groups of two to three, and formed dimorphic conidia. Primary conidia lanceolate with distinct apical points, 12.5-19×37.5-70 µm; secondary conidia cylindrical,12.5-20×37.5-77.5 µm; both conidia hyaline with angular/reticulated wrinkling of the outer walls. These morphological features are typical of the anamorphic stage description of Leveillula *taurica*.

Management

1. When weather is comparatively dry, spray twice with wettable Sulphur 2g/lit at 15 days interval, starting from 3 months after sowing.

2. Spray 1ml hexaconazole or 2ml dinocap / litre of water at fortnight intervals.
3. The variety Jawala is resistant to this disease.

15.5. Powdery mildew of olive

Pathogen: *Leveillula taurica* (Lev.) *Arnaud*

Leveillula taurica is a powdery mildew pathogen with endoparasitic habit, entering the host tree through stomata.

Host range: Woody perennials such as olive, green pepper, aubergine, potato, tomato, cucurbitaceae, cotton, okra and artichoke.

Occurrence

This disease is widely distributed, but is more frequently found in dry areas of Europe, Asia and around the countries of the Mediterranean basin. The importance of the disease is that it causes the fall of the leaves especially in young saplings and tender shoots.

Symptoms

The most conspicuous symptom of the disease appears on the upper leaf surface of the mature leaves as scattered chlorotic spots among the veins, while the white powdery mass of the pathogen occupies the lower leaf surface. Chlorotic spots, followed by necrotic ones, sometimes lead to leaf fall (Fig 15.5).

Fig. 15.5: Symptoms of Powdery mildew on olive leaves

The pathogen is wind-borne and many outbreaks of the disease are related to dry conditions. Yields production may be reduced.

Disease cycle

The causal fungus persists on volunteer olive. Spores are spread to olive tree anytime during the production season by wind. Infection occurs over a very wide temperature range with 15° to 22°C being the optimum. During the period of dry and warm days with cool nights, the conditions are ideal for disease development.

Control

Foliar spays with the appropriate products control the disease.

The useful fungicides are presented in table 15.1..

Table 15.1: Useful fungicides against powdery mildew of olive

Name of the active ingredient	Type of action	Name of the product	Used rates	Application	Last application before harvest
Benomyl	Systemic	Benazol 50 WP, Benlate 50 WP, Benomyl 50 WP, Benor 50 WP, Fundazol 50 WP	200-250 gr /100 lt	Foliar	14 days
Thiophanate-methyl	Systemic	Cequiphanate 70 WP, Thiophanate Methyl - Inagro 70 WP, Thiophanic - M 70 WP, Neotopcin 70 WG/WP	200-250 gr /100 lt	Foliar	14 days

15.6. Powdery mildew of safflower

Pathogen: *Erysiphe cichoracearum*

Geographic distribution

Reported from Afghanistan (Gattani, 1962), France (Darpoux, 1946), India (Saluja and Bhide,1962), Israel (Ashri, 1971), USA (Zimmer, 1961), Ukraine (Milovtzova, 1937).

Symptoms

Presence of white growth of the fungus on the safflower leaves characterizes the disease symptoms (Fig 15.6).

Spineless varieties are more susceptible to this disease.

Pathogen

The mycelium of the fungal pathogen is grayish-white, septate and profusely branched.

Fig. 15.6: Symptoms of powdery mildew on safflower leaves and flower capsules.

Cleistothecia are black with numerous hyphae like appendages, spherical, 103-154 μm in diameter, each having 8-20 stalked, hyaline, ovate ,47-70 x 23.5 μm asci. Each ascus contains two hyaline to light yellow, 14-32.5x 11-24 μm size ascospore.

Control measures

Foliar spray with 0.2 per cent wettable sulphur (Wetsulf/Sulfex), Karathane or Miltox or dusting with sulphur at flowering and capsule formation stages help to control the disease (Reddy, 1982).

15.7. Powdery mildew of Linseed (Flax)

Pathogen: *Oidium lini*

Geographical distribution: Worldwide

Epidemics and losses

Epidemic of the disease in the early stage of crop causes heavy reduction in yields and deterioration in the quality of seed and fibre.

Symptoms

The symptoms are characterized by a white powdery mass of mycelia that starts as small spots and rapidly spreads to cover the entire leaf surface (Fig 16.2). Heavily infected leaves dry up, wither and die. Early infections may defoliate the flax plant and reduce the yield and quality of seed.

Fig. 16.2: Symptoms of powdery mildew on flax leaves

Disease cycle

The life cycle of the fungus is initiated by airborne conidia which are the asexual reproductive spores of the fungus. When the conditions are favourable, the conidia start to germinate after four hours and produce a single thick germtube from its one corner. Germinating conidia produces primary hyphae at about 30 h after incubation, secondary hyphae after 36 h and tertiary hyphae after 48 hours of incubation. After about 60 hours the other hyphae elongate. The white patches of infection became visible to the naked eye only after 120 hours.

The mature conidia are released about seven days after inoculation. The conidia are released when the host leaf and the atmosphere are relatively dry during day time. Detached conidia are carried away by the wind and are deposited over the fresh host leaves on which they get germinated again.

Weather

When conidia or ascospores fall on a plant surface, they start to germinate in 2 to 4 hours, reaching a maximum number in about 25 hours. The optimum temperature for germination is about 22°C; the minimum about 5°C and the maximum is close to 35°C.

Spore germination occurs on the surface of a host over a range of relative humidity from 23 to 99 percent. Free moisture is detrimental to spore germination of the powdery mildew fungus. Once released from the conidiophores, the thin-walled conidia do not live long. At 32°C, and a relative humidity of 70 percent or less, germination reaches 95 to 100 percent in 2 hours and drops to 8 to 20 percent after 5 hours. At 21°C, and a relative humidity of 70 percent or less, germination is only 20 to 40 percent after 5 hours. Although conidia remain viable longer at a relative humidity of 80 to 90 percent, essentially all conidia are dead after 48 hours at 21°C and after 24 hours at 32°C.

The environment most favorable for conidial production, maturation, release and spread, germination, and infection include repeated day-night cycles where the nights are cool (about 16°C) and damp with a relative humidity of 90 to 99 percent, and the days are warm (about 27°C) and dry with a relative humidity of 40 to 70 percent.

When spring and summer rainfall is high, epidemics of powdery mildew are most common during the late summer or fall. The disease cycle i.e.production of conidia, release, germination, infection, and production of conidia can be as short as 72 to 96 hours. If left uncontrolled, powdery mildew can quickly become epidemic when cool, damp nights are followed by warm, dry days.

Disease management

Crop resistance

- Use resistant varieties like

For seed type: L 34, L 1066, L 1453, L 1743 &L 2502 Aoyogi, Flak I

For fibre flax: Tua, Diana, EC 278967, EC 278980, EC 278988A, Ottowa 770, Band L 1720

Cultural control

- Space planting for good air circulation.
- Remove and destroy infected plants.
- Plant in sunny open locations rather than shady.

Chemical control

- Spraying Sulfex (3g/litre of water) can effectively control the disease.
- Karathane (40 E.C) @ 0.5% or dusting sulphur powder at 15 days interval.

16
Powdery Mildews of Fibre Crops

16.1. Powdery mildew of cotton

Pathogen: *Leveillula taurica*

Geographical distribution: Worldwide

Epidemics and losses

Cotton suffers significant losses due to *Leveillula* powdery mildew when attack starts at younger growth stages and develops fully at further growth stages. The epidemics develop well in dry and warm regions. The potential for losses is greater in crops that are irrigated.

Symptoms

The host plants parts most affected by *Leveillula* is the leaf blade. Petioles, stalks and flowers are rarely affected and fruits are occasionally infected. *L. taurica* penetrates the interior leaf tissues and symptoms are usually apparent on the ventral side as powdery, whitish spots that gradually expand (Fig 16.1). On the dorsal side, yellow spots of varying intensity develop opposite the spotted ventral side.

On the dorsal side, powdery spots may also develop. Spots may later become necrotic. Leaf shedding occurs in cotton. Once leaves are shed the fruits may be exposed to direct sun irradiation and damaged.

Fig. 16.1: Symptoms of Powdery mildew on cotton leaves

Pathogen

The *Leveillula taurica* mycelium is septate, endophytic and produced dimorphic conidia (pyriform and cylindrical) on long and oftenly branched conidiophores, which emerged through stomata. Both the types of conidia are single celled, hyaline, pyriform or cylindrical and borne either singly or in short chains which are cylindrical. The average size of pyriform conidia is found to be 70.20 x 19.80 µm, while that of cylindrical is 60.50 x 15.80 µm.

Disease cycle

The polycyclic disease cycle of *L. taurica* is similar to that of other powdery mildew species. It overwinters (as chasmothecia) in crop residues above the soil surface. Under favorable climatic conditions, the chasmothecia open and release ascospores which are wind dispersed.

The ascospores germinate on leaf, enter the host through its stomata, and colonize the host's tissues with its mycelia. The pathogen then begins to produce its asexual conidia either on individual or on branched conidiophores which emerge through the host's stomata. These conidia serve as a secondary inoculum to spread disease after initial infection. In the fall, the pathogen undergoes sexual reproduction and again produces chasmothecia, which are dormant, over wintering structure.

Weather Parameters

In general, high humidity favors germination of conidia. Infection of plants can occur over a wide temperature range of 18° to 33°C under both high and low humidity.

Under favorable conditions, secondary infections occur every 7 to 10 days, and disease can spread rapidly. Temperatures over 35°C temporarily suppress the development of the disease.

Disease management

Crop resistance

Resistance programmes are carried out in some countries; however, the number of tolerant crops is very limited. Some cultivars or varieties like AKA-7 in India show resistant reaction against powdery mildew pathogen.

Cultural control

- Proper irrigation according to the needs of the crop is essential in order to prevent drought stress of ageing plants.

Powdery Mildews of Fibre Crops 225

- Overhead irrigation was found, in some cases, to be related to lower disease severity

- In perennial crops where foliage is cut and subsequently regrowth occurs, the intervals between cuttings should be reduced in order to minimize the existence of old senescing leaf tissues.

- Proximity to other crop or weed hosts and sequential cropping of susceptible hosts should be avoided or minimized.

- Crops that are grown as annual or bi-annual should be held as annual, thus avoiding the major increase in disease at the later stage of the growth.

Chemical control

- Chemical control is possible with fungicides such as triadimefon, dinocap, sulfur and propiconazole.

- Another chemical treatment involves treating with a silicon solution or calcium silicate slag and Potassium bicarbonate.

- Milk has long been popular with home gardeners and small-scale organic growers as a treatment for powdery mildew. Milk is diluted with water (typically 1:10) and sprayed on susceptible plants at the first sign of infection, or as a preventative measure.

Biological control

- Use *Bacillus subtilis* or *Trichiderma harzianum.*

16.2. Powdery mildew of Jute

Pathogen: *Oidium* spp.

Geographical distribution: Worldwide

Losses

Infection of the disease in the early stage of crop causes heavy reduction in yields and deterioration in the quality of seed and fibre.

Symptoms

The symptoms are characterized by a white powdery mass of mycelia that starts as small spots and rapidly spreads to cover the entire leaf surface (Fig 16.3). Heavily infected leaves dry up, wither and die.

Fig. 16.3: Symptoms of Powdery mildew on jute leaves

Disease cycle

The life cycle of the fungus is initiated by airborne conidia which are the asexual reproductive spores of the fungus. Under favourable environmental condition, the conidia start to germinate after four hours and produce a single thick germtube from its one corner. Germinating conidia produces primary hyphae at about 30 hr after incubation, secondary hyphae after 36 hours and tertiary hyphae after 48 hours of incubation. After about 60 hours the other hyphae elongate. The white patches of infection became visible to the naked eye only after 120 hours.

The mature conidia get released about seven days after inoculation. The conidia get released when the host leaf and the atmosphere are relatively dry during day time. Detached conidia get carried away by the wind and get deposited over the fresh host leaves on which they get germinated again.

Weather parameters

When conidia or ascospores fall on a plant surface, they start to germinate in 2 to 4 hours, reaching a maximum number in about 25 hours. The optimum temperature for germination is about 22°C; the minimum, about 5°C and the maximum is close to 35°C.

Spore germination occurs on the surface of a host over a range of relative humidity from 23 to 99 percent. Free moisture is detrimental to spore germination of the powdery mildew fungus. Once released from the conidiophores, the thin-walled conidia do not live long. At 32°C, and a relative humidity of 70 percent or less, germination reaches 95 to 100 percent in 2 hours and drops to 8 to 20 percent after 5 hours. At 21°C, and a relative humidity of 70 percent or less, germination is only 20 to 40 percent after 5 hours. Although conidia remain viable longer at a relative humidity of 80 to 90 percent, essentially all conidia are dead after 48 hours at 21°C and after 24 hours at 32°C.

The environment most favorable for conidial production, maturation, release and spread, germination, and infection include repeated day-night cycles where the nights are cool (about 16°C) and damp with a relative humidity of 90 to 99 percent, and the days are warm (about 27°C) and dry with a relative humidity of 40 to 70 percent.

When spring and summer rainfall is high, epidemics of powdery mildew are most common during the late summer or fall. The disease cycle i.e production of conidia, release, germination, infection, and production of conidia can be as short as 72 to 96 hours. If left uncontrolled, powdery mildew can quickly become epidemic when cool, damp nights are followed by warm, dry days.

Disease management

Crop resistance
- Use of resistant varieties.

Cultural control
- Space planting for good air circulation.
- Remove and destroy infected plants.
- Plant in sunny open locations rather than shady.

Chemical control
- Spraying Sulfex (3g/litre of water) can effectively control the disease.
- Karathane (40 E.C) @ 0.5% or dusting sulphur powder at 15 days interval.

16.3. Powdery mildew of Sunhemp

Pathogen: *Leveillula taurica*

Geographical distribution: Worldwide

Epidemics and losses
Crop suffers significant losses due to *Leveillula* powdery mildew when attack starts at younger growth stages and develops fully at further growth stages. The epidemics develop well in dry and warm regions. The potential for losses is greater in crops that are irrigated.

Symptoms

The host plant parts most affected by *Leveillula* are the leaf blade. Petioles, stalks and flowers are rarely affected and fruits are occasionally infected. *L. taurica* penetrates the interior leaf tissues and symptoms are usually apparent on the ventral side as powdery, whitish spots that gradually expand (Fig.16.4). On the dorsal side, yellow spots of varying intensity develop opposite the spotted ventral side. On the dorsal side, powdery spots may also develop.

Fig.16.4: Symptoms of powdery mildew on hemp

Spots may later become necrotic. Leaf shedding occurs. Once leaves are shed the fruits may be exposed to direct sun irradiation and damaged.

Pathogen

The *Leveillula taurica* mycelium is septate, endophytic and produced dimorphic conidia (pyriform and cylindrical) on long and oftenly branched conidiophores, which emerged through stomata. Both the types of conidia are single celled, hyaline, pyriform and cylindrical and borne either singly or in short chains. The average size of pyriform conidia is found to be 70.20 x 19.80 µm, while that of cylindrical is 60.50 x 15.80 µm.

Disease cycle

The polycyclic disease cycle of *L. taurica* is similar to that of other powdery mildew species. It overwinters (as chasmothecia) in crop residues above the soil surface. Under favorable climatic conditions, the chasmothecia open and release ascospores, which is wind dispersed. The ascospores enter the host through its stomata, germinate, and colonize the host's tissues with its mycelia. The pathogen then begins to produce its asexual conidia, either on individual or on branched conidiophores which exit through the host's stomata and serve as a secondary inoculum to spread disease after initial infection. In the fall, the pathogen under goes sexual reproduction to produce chasmothecia, which is a dormant and overwintering structure.

Weather

In general, high humidity favors germination of conidia. Infection of plants can occur over a wide temperature range (18° to 33°C) under both high and low humidity. Under favorable conditions, secondary infections occur every 7 to 10 days, and disease can spread rapidly. Temperatures over 35°C can temporarily suppress development.

Disease management

Crop resistance

Resistance programmes are carried out in some countries; however, the number of tolerant crops is very limited. Some cultivaras or varieties e.g. Tropic Sun has resistant reaction against powdery mildew pathogen.

Cultural control

- Proper irrigation according to the needs of the crop is essential in order to prevent drought stress of ageing plants.

- Proximity to other crop or weed hosts and sequential cropping of susceptible hosts should be avoided or minimized.

- Crops that are grown as annual or bi-annual should be held as annual, thus avoiding the major increase in disease at the later stage of the growth.

Chemical control

- Chemical control is possible with fungicides such as triadimefon, dinocap, sulfur and propiconazole.

- Another chemical treatment involves treating with a silicon solution or calcium silicate slag and Potassium bicarbonate.

- Milk has long been popular as a treatment for powdery mildew. Milk is diluted with water (typically 1:10) and sprayed on susceptible plants at the first sign of infection, or as a preventative measure.

Biological control

Use *Bacillus subtilis* or *Trichiderma harzianum*.

17

Powdery Mildew of Cash Crops

17.1 Powdery mildew of Tobacco

Pathogen: *Golovinomyces cichoracearum* var. *cichoracearum*

Geographical distribution

Powdery mildew, caused by *Golovinomyces cichoracearum* var. *cichoracearum*, is a relatively important disease of tobacco in several Asian countries, Oceania, the Mediterranean and Africa where it occasionally causes significant losses. On the American continent, it is mainly present in Canada.

It can also be present in several European countries. In recent years, it seems that its incidence has increased in some of them (particulary Italy, Spain, and Greece). For a long time, in France, powdery mildew of tobacco was consideredas a minor disease that caused severe losses only occasionally (in Lot and Garonne regions). However, since the 1990s the situation has become rather alarming especially on Virginia type tobacco where it usually appears in early July and develops during the summer.

Symptoms

The plants affected by *Golovinomyces cichoracearum* var. *Cichoracearum* show highly characteristic symptoms, regardless of the plant size and the affected plant parts. Greyish, white, felt-like powdery patches first appear on the upper leaf surface (Fig 17.1). They gradually spread and cover eventually the entire lamina. Afterwards the affected tissues show irregular brown lesions resulting from cell death. The leaves infected with oidium often have thinner tissues and lose their commercial value. The disease initially appears on the lower leaves and as disease advances,

Fig. 17.1: Symptoms of powdery mildew on tobacco leaves

the rest of the leaves are also infected and sometimes powdery growth can also be seen on the stem. The affected leaves turn to brown, wither, and show scorched appearance.

Pathogen

Golovinomyces cichoracearum var. *cichoracearum* (DC.) V.P. Heluta (1988) synonym : *Erysiphe cichoracearum* var. *nicotianae.*

Golovinomyces cichoracearum var. *cichoracearum* may survive from one year to another in many ways. The cleistothecia that ensure its perpetuation may be form on the oidium infected leaves late in the season. They do not occur very often, but they form as an overwintering storage. They can be maintained for several years on plant debris of tobacco. Generally, the asci in cleistothecia are formed in autumn, but the ascospores are not ejected until the following spring. The fungus can also overwinter through alternative hosts.

Erysiphe cichoracearum var. *nicotianae* is ecotophytic and produces hyaline, septate and highly branched mycelium. Short, stout and hyaline conidiophores arise from the mycelium and bear conidia in chains. The conidia are barrel shaped or cylindrical, hyaline and thin walled. Cleistothecia are black, spherical with no ostiole, with numerous densely woven septate, brown-coloured appendages. They contain 10-15 asci, which are ovate with a short stalk. Each ascus contains two ascospores, which are oval to elliptical, thin walled, hyaline and single celled.

Disease Cycle

The infection of the host is cause from a primary inoculum consisting of either ascospores or conidia. These, once in contact with the host, germinate quickly, likely in two hours if the environmental conditions are appropriate. They form an appressorium and penetrate directly into epidermal cells, where they form haustoria, which act as sucking organs by extracting nutrient elements necessary for the mycelium growth of the fungus. Within 4-6 days after infection, short conidiophores born on secondary hyphae formed on the surface of the lamina and produce conidia. Sporulation is abundant. Conidia are quite fragile and live only a few hours or days, when conditions are favorable. They are very light and can be easily carried away and spread by wind and incidentally by rainfall or water splashes during sprinkler irrigation.

Weather

Golovinomyces cichoracearum var. *cichoracearum* seems to be able to multiply at temperatures between 4 and 32° C. Optimum temperature is around 20-25° C. It dislikes the presence of free water on the leaves as well as very high humidity (close to 100 %). It prefers humidity ranging from 60 to 86 %. The light also influences its development. It appreciates the diffused light, while direct sunlight disturbs its growth. This is certainly the reason why it can be found more on lower leaves, inside the plant canopy, where the surrounding microclimate is particularly mild.

The age of the plants significantly alters the development of the disease. Seedlings are generally less susceptible. In addition, climatic conditions in the nurseries are not always favourable to the fungus. One may consider that in France, in principle, there is no fear of attacks in nurseries. After planting and in the following weeks, the plants are still not very susceptible to powdery mildew, but it is no longer the case when the first formed leaves reach their final size, about 6 weeks after transplanting.

Disease Management

Cultural
Carefully select the location of the plot so that it is in a rather airy and sunny area.

Avoid excessive watering and avoid stagnation of water in the plot.

Ensure a balanced fertigation,

Destroy weeds that act as relays to the pathogen in the plot and in its surroundings. Avoid the vicinity of crops that host the pathogen.

Harvest the lower leaves as soon as possible.

Chemical
Apply sulphur powder (200 mesh) @ 40 kg/ha to soil in between plant row, 6-8 weeks after planting. Mix ash or sand to sulphur for easy application. See that sulphur does not fall on tobacco leaves. This is recommended for black soils only. Spray 0.2% Karathane or Thiovit or 0.05% Carbendazim just before the disease sets in. Repeat at 10-12 days interval if necessary.

Varietal Resistance
Use resistant varieties like Swarna or Line 2359 developed at Central Tobacco Research Institute, in disease endemic area. In France, several varieties of Virginia and Burley tobacco types, available to producers, are resistant to powdery mildew. Rsistant varieties are also available in the countries like Zimbabwe, Japan and South Africa where disease is severe.

Several species of *Nicotiana* are resistant to *Golovinomyces cichoracearum* var. *cichoracearum*, but only 3 inter-specific hybrids were created between *Nicotiana tabacum* and *N. debneyi, N. glutinosa* and *N. tomentosiformis*. The resistance conferred by these three species seems to be monogenic and only partial. The most used source of resistance until now comes from a Japanese cultivar Kokobu (Kuo-fan). Two recessive genes control it. It provides a very high level of resistance, although some spots are some times observed on tobacco seedlings in nurseries.

17.2 Powdery mildew of Rubber

Pathogen: *Oidium heveae* Steinm.

Symptoms

The fungus forms white powdery growth on tender leaflets (Fig 17.2). The leaflets of seedlings and mature trees show such symptoms. Diseased leaflets curl, crinkle and their edges roll inwards and such leaves fall off. The petioles attached to the twigs give a broomstick appearance. After a few days the petioles also fall. On the older leaves white patches of the fungus appears which later causes necrotic spots and reduce the photosynthetic efficiency. The infected tender shoots dry off and dieback symptoms appear. Infection on the flowers and tender fruits cause them to shed. The disease reduces the seed production also.

Fig. 17.2: Symptoms of powdery mildew on rubber leaves

Pathogen: *Oidium heveae* Steinm.

The fungus is known to have only the conidial stage. Mycelium is hyaline and septate. The fungus sends pyriform or round haustoria into the epidermal cells. The conidiophores bear the conidia in chain. Conidia are 2 to 7 in each chain, barrel-shaped or cylindrical and measure 25 to 45 x 12 to 25 µm. The youngest conidium is at the base of the chain and the mature one at the tip.

Mode of spread and survival

The fungus survives as dormant mycelium inside the bark. It becomes active during cloudy days with warm weather and low humidity. The fungus also survives on *Euphorbia hirta.* In the plantation, it spreads through wind-borne conidia.

Disease Management

Sulphur dusting have been found to be most effective. Carbendazim 0.05 per cent a.i. (1 gm in 1 litre of water) spray is more effective.

17.3. Powdery mildew of Betelvine

Pathogen: *Oidium piperis.*

Geographical distribution

India, Denmark, Burma and Brazil.

Symptoms

The disease appears on the under surface of the leaves as white to light-brown powdery patches (Fig 17.3). These patches gradually increase in size and often coalesce with each other. The infected portion, later on .turn grey. The infected leaves has no market value and there is total loss to the betelvine grower.

Fig. 17.3: Symptoms of powdery mildew on Betelvine

Pathogen: *Oidium piperis.*

The vegetative mycelium of the fungus is superficial and consist of delicate, white, septate hyphae, frequently branched and more or less densely interwoven. The hyphae are 5 to 8.2µ wide, form haustoria inside the cuticle cells to absorb nutrients. Conidiophore emerge from the stomata are erect,simple and usually 2 to 3 septate.These measures 66 to 132μ long and bear conidia in chain of 3 to10 in basipetal succession. The conidia are unicellular, colourless and elliptical or barrelshaped, and measure 20.4 – 74.7 x 6.8 – 23.8μ.

Management

On newly planted vines, single dusting of sulphur @ 80-90kg/ha is useful. Aureofungin,elosol, cosan, and thiovit is also effective against this pathogen.

18

Powdery Mildew of Vegetables

Powdery mildew is a common disease on many types of vegetables plants. There are many different species of powdery mildew fungi (such as *Erysiphe* species, *Sphaerotheca* species) and each species only attacks specific plants. A wide variety of vegetable crops are affected by powdery mildews, including artichoke, beans, beets, carrot, cucumber, eggplant, lettuce, melons, parsnips, peas, peppers, pumpkins, radicchio,radishes, squash, tomatoes, and turnips.

Powdery mildews generally do not require moist conditions to establish and grow, and normally do well under warm conditions; thus, they are more prevalent under dry summer conditions than many other leaf-infecting diseases.

Symptoms

Powdery mildew first appears as white, powdery spots that may form on both surfaces of leaves, on shoots, and sometimes on flowers and fruits. These spots gradually spread over a large area of the leaves and stems. An exception is one of the powdery mildews that affect artichokes, onions, peppers, and tomatoes where it produces yellow patches on leaves but little powdery growth.

Leaves infected with powdery mildew may gradually turn completely yellow, die, and fall off, which may expose fruit to sunburn. On some plants, powdery mildew may cause the leaves to twist, buckle, or otherwise distort. Powdery mildew fungal growth does not usually grow on vegetable fruits, although pea pods may get brownish spots. Severely infected plants may have reduced yields, shortened production times, and fruit that has little flavor.

Life Cycle

All powdery mildew fungi require living plant tissue to grow. Year-round availability of crop or weed hosts is important for the survival of some powdery mildew fungi. Special resting spores are produced, allowing overwinter survival of the species that causes the disease in cucurbits, lettuce, peas, and certain other crops.

Most powdery mildew fungi grow as thin layers of mycelium (fungal tissue) on the surface of the affected plant part. Spores, which are the primary means of dispersal, make up the bulk of the white, powdery growth visible on the plant's surface and are produced in chains that can be seen with a hand lens.

Powdery mildew spores are carried by wind to new hosts. Although humidity requirements for germination vary, all powdery mildew species can germinate and infect in the absence of free water. In fact, spores of some powdery mildew fungi are killed and germination is inhibited by water on plant surfaces for extended periods. Moderate temperatures (15.5° to 26.6°C) and shady conditions generally are the most favorable for powdery mildew development. Spores and fungal growth are sensitive to extreme heat (above 32.2°C) and direct sunlight.

Disease Management Strategies

The best method of control is preven-tion. Planting resistant vegetable varieties when available, or avoiding the most susceptible varieties, planting in the full sun, and following good cultural practices will adequately control powdery mildew in many cases . However, very susceptible vegetables such as cucurbits (cucumber, melons, squash, and pumpkins) may require fungicide spray treatment. Several leasttoxic fungicides are available but must be applied no later than the first sign of disease.

Resistant Varieties

In some cases, varieties resistant to powdery mildew may be available. If available, plant resistant varieties of cantaloupe, cole crops, cucumber, melons, peas, pumpkins, and squash. If you plant more susceptible varieties, you may need to take control measures.

Cultural Practices

Plant in sunny areas as much as possible, provide good air circulation, and avoid applying excess fertilizer. A good alternative is to use a slow-release fertilizer. Overhead sprinkling may help reduce powdery mildew because spores are washed off the plant. However, overhead sprinklers are not usually recommended as a control method in vegetables because their use may contribute to other pest problems.

Fungicide Application

In some situations, especially in the production of susceptible cucurbits, fungicides may be needed. Fungicides function as protectants, eradicants, or both. Apply protectant fungicides to highly susceptible plants before the disease appears. Use eradicants at the earliest signs of the disease. Once mildew growth is extensive, control with any fungicide becomes more difficult.

Several least-toxic fungicides are available, including horticultural oils, neem oil, jojoba oil, sulfur, and the biological fungicide Serenade. With the exception of the oils, these materials are primarily preventive. Oils work best as eradicants. *Be careful, however, to never apply an oil spray within 2 weeks of a sulfur spray otherwise the plants may be injured. Also, oils should never be applied when temperatures are above 32°C or to drought-stressed plants.* Some plants may be more sensitive than others, and the interval required between sulfur and oil sprays may be even longer. Always consult the fungicide label for any special precautions.

Sulfur products have been used to manage powdery mildew for centuries but are only effective when applied before disease symptoms appear. The best sulfur products to use for powdery mildew control in gardens are wettable sulfurs that are specially formulated with surfactants similar to those in dishwashing detergent (such as Safer Garden Fungicide) However, sulfur can be damaging to some squash and melon varieties. *To avoid injuring, do not apply sulfur when air temperature is near or over32°C and do not apply it within 2 weeks of an oil spray.* Other sulfur products, such as sulfur dust, are much more difficult to use, irritating to skin and eyes, and limited in terms of the plants they can safely be used on.

Biological Fungicides

Biological fungicides (such as Serenade) are commercially available beneficial microorganisms formulated into a product that, when sprayed on the plant, destroys fungal pathogens. The active ingredient in Serenade is a bacterium, *Bacillus subtilis,* that helps prevent the powdery mildew from infecting the plant. While this product functions to kill the powdery mildew organism and is nontoxic to people, pets, and beneficial insects, it has not proven to be as effective as the oils or sulfur in controlling this disease.

Table18.1: Powdery mildew on Vegetable Host Plants with inciting pathogen and Control Measures.

Hosts	Fungus species	Controls
Cucumbers, endive, lettuce, melons, potato, pumpkin, squash	*Erysiphe cichoracearum*	Resistant varieties of lettuce, cucumber; water sprays; fungicides if necessary on squash and pumpkin
Broccoli, Brussels sprouts, cauliflower, and other cole crops; radicchio, radishes, turnips	*Erysiphe cruciferarum*	Not usually required
Tomatoes	*Erysiphe lycopersici*	Fungicides if necessary
Peas	*Erysiphe pisi*	Resistant varieties; sprinkler

Contd.

		irrigation
Carrots, parsley, parsnips	*Erysiphe heraclei*	Tolerant varieties
Beets	*Erysiphe polygoni*	Tolerant varieties
Artichoke, eggplant, peppers, tomatillo, tomatoes	*Leveillula taurica*	Rarely required; fungicides if necessary
Beans, black-eyed peas, cucurbits, okra	*Sphaerotheca fuliginea*	Resistant varieties for some; fungicides if necessary

18.1. Powdery Mildew of Cucurbits

This is common and sometimes a destructive disease of cucurbits, especially bottle gourd and pumpkins. In India, the disease occurs almost every year in all cucurbit-growing areas. Besides cultivated species, the same pathogen also attacks wild plants of Cucurbitaceae and other families.

Pathogen: Erysiphe cichoracearum and *Sphaerotheca fuliginea*

Geographical distribution

The disease is worldwide in distribution. In Asia, the disease is reported to occur in Iran, Iraq, Saudi Arabia, Israel, Malaysia, Singapore, China, Japan, Taiwan and India. From India Butler in 1918 reported the disease for the first time from U. P. and Bihar.

Symptoms

The disease appears first in the form of minute, white to dirty grey spots on leaves and stems. The mildew develops most extensively on the lower surface of leaves. At later stage, the spots increase in size and give a powdery appearance. The superficial powdery mass finally covers the whole surface of affected parts (Fig 18.1). Rarely in the late winter season, minute dark-brown, pinhead bodies appear intermixed with the white powdery mass. These are the

Fig. 18.1: Symptoms of powdery mildew on pumkin leaves and pumpkin

Powdery Mildew of Vegetables 241

perfect state fruiting bodies i.e. cleistothecia of the fungus. These have been reported in India only during winter months. In severe attack, leaves may fall off permanently and the fruits remain undersized and deformed.

Pathogen

Cleistothecia of six species of *Erysiphaceae* have been recorded on cucurbits in various parts of the world. Of these, the two most common and most widely distributed species throughout the world are *Erysiphe cichoracearum* and *Sphaerotheca fuliginea*. Situations on identity of the causal organisms of this disease differ in different countries. There are two situations *viz.*two-pathogen disease situation and one pathogen disease situation.

Two pathogen disease situation means that the same disease is caused by two pathogens separately. This situation should not be confused with disease caused by two pathogens in combination like, disease complexes, which involves the intimate association of two or more pathogens. Two-pathogen disease situation has been established in U.S.A, Germany, U.K., Italy, Bulgaria, and Hungary, exUSSR, Israel, Japan and India. *E. cichoracearum* dominates over *S. fuliginea* in U.S.A., Germany, ex-U.K. and Bulgaria, whereas *Sphaerotheca fuliginea* is predominant in Italy, Israel, Japan and India. In ex-U.S.S.R. and Hungary, both are more or less equally important. One pathogen disease situation caused by *Erysiphe* is found in Canada, France, Norway, Sweden, Austria, Switzerland, Mozambique, Egypt, Malta, Fizi, Kenya, Bolivia, Brazil, Peru, Nicaragua, West Indies, Iraq, Saudi Arabia, Malaysia and Singapore. One pathogen disease situation caused by *Sphaerotheca* is recognised in Netherlands, Greece, Turkey, Czeckoslovakia, Rumania, Australia, New Zealand, South Africa, Sudan, Malawi, Iran, China and Taiwan. In India, the disease is caused by *Erysiphe cichoracearum* as well as *Sphaerotheca fuliginea*.

Biology of Erysiphe cichoracearum DC ex. Merfat

Mycelium developes evanascent but sometimes persistent effused, superficial growth with well-developed haustorias in host cells. Oidiophores arise on mycelia and are unbranched, erect, producing oidial chains at their apex. Oidia ellipsoidal barrel-shaped, 25-45 x 14-26 µm, produced in abundance and disseminated by wind. Cleistothecia gregarious or scattered, globose becoming depressed or irregular, 90-135 µm in diameter, cell wall usually indistinct 10-20 µm wide. Appendages numerous, myceloid, basally inserted hyaline to dark, interwoven with mycelium, 1-4 times as long as the diameter of cleistothecium, rarely branched. Asci 10-25 per cleistothecium, ovate to broadly ovate, rarely subglobose, more or less stalked, 60-90 x 25-50 µm. Ascospores two in each ascus, very rarely three, 20-30 x12-18 µm.

Biology of *Sphaerotheca fuliginea* (Schlecht ex Fr.) Poll

Mycelium hyaline, occasionally brown when old, usually evanascent but sometimes persistent forming white circular to irregular patches on the host surface, intercellular with haustoria. Oidiophores simple, short, producing long chain of oidia at tips. Oidia often with distinct fibrosin bodies, ellipsoid to barrel shaped, 25-37 x14-25 µm. Cleistothecia scattered to densely gregarious, 66-98 µm in diameter, usually 85 µm, wall cells usually over, 25 µm wide. Appendages variable in number, usually, as long as the diameter of cleistothecium, myceloid, brown, tortuous, interwoven with mycelium. Single ascus in each cleistothecium, broadly elliptic, to subglobose, 50-80 x 30-60 µm. Ascospores eight in each ascus, ellipsoid to nearly spherical, 17-22 x 12-20 µm. Three distinct pathological forms of *S. fuliginea* have been reported on different cucurbits in India in 1985 on the basis of differential reactions of cucurbit hosts. Race 3 is predominant in Punjab. In most of the agroecological zones of India, these three races are present. Races 3 is most widespread and infect most of the cucurbits in different states.Race 2is restricted to green melon (*Citrulus vulgaris* var. *fistulosus)*and race 1to spongegourd *(Luffa cylindrical).* Race 1 does not infect *C. vulgaris* var. *fistulosus* and *Cucumis sativus.*Race 2 does *not* infect *L cylindrical* and *C. vulgaris* var *fistulosus.* The fungus is heterothallic in nature.

Eight race of *S. fuliginea* have been identified in the USA, Africa, and Europe and around the Mediterranean Sea. Four new races were reported from greenhouse melons in Japan where prevalence of races varied with season. Race 5 was most common in early season and race 1 in late season.

Disease cycle

There are many possible means of survival of these fungi between two crop seasons. Where cleistothecia are formed, they can explain the mode of perennation from one crop season to the next. In India, these sexual fruit bodies develop on leftover cucurbit crops during winter in isolated areas or in the sub-mountainous areas in the north. These may initiate the disease in the local hosts and from there the primary inoculum in the form of conidia might be blown by wind to the main crop in the plains. However, as in the downy mildew, the main source primary inoculum seems to be the existence of wild and cultivated cucurbits in one or the other locality of the country from where the conidia are blown by wind currents to the new crop.

These powdery mildew fungi are greatly influenced by the age of the host plant and air humidity and temperature. 16 to 23 days old, leaves are highly susceptible while very young leaves are almost immune. The fungi can sporulate and cause infection in a very dry as well as wet atmosphere but infection increases as the atmospheric humidity increases, and heavy dew deposits favours the penetration by germ tubes. Penetration is direct and confined to the epidermal cells. When conidia land and establish on the leaf surface of a susceptible host, one or two germ tubes develop and penetrate one, or two epidermal cells, the penetration zones are surrounded with a callose like material. Penetration of the host by germ tubes from ascospores is entirely mechanical which results in dislocation of epidermal cells. The cells considerably increase in size. The infection peg does not form haustoria. Instead, it gives rise to several fine branches, which extend to neighbouring cells and form primary hyphae. The conidial stage develops from these hyphae. In resistant cultivar, the fungus develops a single germ tube. Calloses like deposition also occur along with liginification in the epidermal cell. The rapid collapse of the penetrated cell in the resistant cultivar is accompanied by accumulation of callose like deposits in cell walls and around haustoria, electron-opaque deposits in the plasma membrane and between the cell wall and the plasma membrane.

Weather parameters associated with disease

The minimum and maximum temperature for conidia formation and host penetration are 10° and 32°C, respectively, the optimum being about 26°-28°C. Gupta *et al.* (2001) have reported 25°C as the optimum for conidial germination at 100% relative humidity. Moderate temperature of 25°C with high relative humidity (<95%) and reduced sunshine hours significantly help in disease development by *Sphaerotheca fuliginea.*

Disease Management

Cultural Methods

The cucurbit weed hosts should not be allowed to grow near the cultivated fields. The diseased crop debris should be burnt .

Chemical Method

Fungicides are the only commercially available options for the management of powdery mildew of cucurbits. Sulphur dust (15-30 kg/ha) had been an old recommendation for effective control. Elasol (0.5%) was used as a substitute of elemental sulphur. Other effective fungicides are Sulfex (0.2%), Calixin (0.1%), and Karathane (0.05-0.2%). Bavistin (0.1%), Mildex Ovatram, etc. Use of Sulfex is cheaper. One to two sprays of Calixin or 2-3 sprays of Karathane are required. Since the infection starts on the abaxial surface of leaves which are difficult to reach during the sprays of protectant fungicides, the systemic fungicides such as benomyl, carbendazim (Bavistin) and triazoles were extensively used in many countries. The resistance to triadimefon in the pathogen population spreads rapidly. When less than 50% of the population has developed resistance, application of triadimefon and chlorothalonil is effective but when resistance has developed in 80% of the population triadimefon is ineffective. Similarly, a population with 40% resistance to benomyl shows no effect of this systemic fungicide. Although sensitivity of *S. fuliginea* to triazole fungicides myclobutanil and propiconazole also decreased after they were applied, these fungicides were more effective than triadimefon. To overcome this problem it has been suggested that fungicides like triadimefon should be applied once or twice with a protectant fungicide and then subsequent sprays should be with only protectant fungicides.

Thind *et al.* (2002) have reported a comparative evaluation of three strobilurin fungicides, azoxystrobin (Quadns), kresoxim methyl (Flint) and trifloxistrobin (Stroby), for their efficacy against powdery mildew of grapevines and powdery mildew of cucurbits. All the three are highly effective against Uncinula (Erysiphe) necator and Sphaerotheca fuliginea on the basis of inhibition of conidial germination. Under field conditions, azoxystrobin proved highly effective in checking powdery mildew on summer squash. It was equal to standard fungicides triademefon and mancozeb. Among more recent fungicides against cucurbit powdery mildew is cyzofamid, which has given as good or even better control than fungicides in common use.

Induced Systemic Resistance

Solutions of mono or dipotassium phosphate and potassium nitrate applied on a 7 or 14 days schedule were highly protective against S. fuliginea in cucumber. A single spray of solution of micronutrients (boron, copper, manganese) on upper surface of cucumber leaves provides protection from S. fuliginea. The effect is similar to the systemic protection given by potassium phosphate. A single spray of 0.1 M phosphate solution is reported to induce systemic protection against powdery mildew of cucumber (Sphaerotheca fuliginea). It suppresses the lesions on diseased leaves. Oxalates and phosphates applied to upper leaf surface of cucumber induce systemic resistance against S. fuliginea. Mineral oil (1%), potassium bicarbonate (0.5%), sodium bicarbonate (0.5%) and milk powder (10%), whitewash and clay and antitranspirants reduce mildew on leaves as much as a wettable sulphur fungicide.

Wurms, *et al.*, (1999) studies the effect of two activator compound, Milsana and benzothiadiazole (BTH) on haustoria of the powdery mildew fiungus in cucumber. Localized application of Milsana caused collapse of the haustoria within 4 days. The haustoria were encapsulated by an amorphous material impregnated by electron-opaque substances. Possible role of phenolics but not of chitinolytic activity was suggested. Application of high dose of BTH elicited weak resistance response from the host.

Disruption of pathogen hyphae by silicon is reported for P. aphanidermatum as well as Sphaertotheca. In healthy cucumber leaves, silicon is mainly distributed in cells around the base of trichome hairs. During infection by S. fuliginea areas of host cell wall adjacent to the growing germ tube shows altered surface morphology and high concentrations of silicon. Silicon was found in papillae, in the host cell walls, around the haustorial neck and in between the host cell wall and plasma membrane. These morphological alterations in Pathogen and host and depositions were thought to be responsible for induced resistance. Silicon treatments significantly reduce the time taken for initiation of production and /or accumulation of phenolic substances. The number of haustoria produced per colony of S. fuliginea is also significantly reduced and conidiophores development is delayed. Silicon mediated accumulation of flavonoid phytoalexins in cucumber leaves was reported. Presence of these antifungal compound induced resistance in leaves of *S. fuliginea*

Betiol (1999) has reported efficacy of fresh cow milk in suppressing *S. fuliginea* on squash leaves. Lactic acid bacteria in milk or from other sources like vegetables are known to be antimicrobial. A single spray of 0.5% clay (nonswelling chlorite mica clay) is reported to reduce powdery mildew (*Sphaerotheca fuliginea*) on cucumber without eradicating the fungus. Spray after inoculation was more effective than spray before inoculation.

Biological Methods

The mycoparasite *Ampelomyces quisqualis* is a biocontrol agent against powdery mildews caused by *Erysiphe* and *Sphaerotheca*. The parasite is wholly internal within the mycelium, conidiophores, conidia and ascocarps of the mildew fungus. *Verticillium lecanii (Lecanicillium lecanii)*, as a biocontrol agent, is another valuable alternative to current management strategies for powdery mildew. Its hyphae colonize the structures of *S.fuliginea* by tight binding with the help of mucilage matrix. Within 24 hours of application of the antagonist, increased vacuolation and disorganization of the cytoplasm of the host hyphae occurs. By 36 hours, plasmalemma retraction and local cytoplasmic aggregations are seen. There is no change in cell wall of the host hyphae except at the point of penetration by mechanical pressure. By 72 hours, hypal cells of *S. fuliginea* collapse depleted of their protoplasm. Romero *et al.* (2003) have reported that the mycoparasite fungi *Ampelomyces quisqualis* and *Lecanicillium supp*ress powdery mildew.

Lima *et al.*(2003) tested several yeasts, synthetic, and biofungicides against powdery mildew of cucurbits. Pre-treatment of leaves with the mycoparasite reduced development and *spreading* of mildew colonies. Spray on mildew colonies also reduced percentage of colonized area. *Trichoderma harzianum* is another fungal biocontrol agent against *Sphaerotheca fusca.* Involvement of local and systemic resistance has been demonstrated. Cells of the biocontrol agent applied to roots and dead cells applied to the leaves of cucumber plants induced control of powdery mildew.

The yeasts *Rhodotorula glutinis, Cryptococcus laurentii* and *Aureobasidium pullulans* applied alone or with mineral oil or gum xanthan significantly reduced disease severity on melon leaves. The effect was comparable with a synthetic fungicide Topas and a biofungicide based on *Ampelomyce quisqualis.* The antagonists survived on leaves in the field at high level even in hot dry climate. The yeast-like, blastospore-forming fungi, *Tilletiopsis* spp., also suppress *Sphaerotheca (Podosphaera).*These fungi produce exo- and endo-glucanase and chitinase that inhibit the germ tube growth of the powdery mildew fungus and plasmolyze its conidia at 130µg/ml (Urquhart and Punja, 2002). Spray of the antagonist *Tilletiopsis pallescens* formulated in natural oil reduce powdery mildew severity. The oil formulation improves biocontrol efficiency of the antagonist, when inoculated on healthy leaves, *Tilletiopsis pallescens* grows better at 90% relative humidity than at 70%, It forms colonies adjacent to leaf veins on healthy leaves. In presence of the powdery mildew the growth of *Tilletiopsis* equally good at both relative humidities and extensive colonies are formed near the base of tichomes, Growth and survival of the fungus are enhanced by high relative humidity.

Isolates of several bacterial antagonists from rhizosphere and leaf surface of cucurbits suppress powdery mildew by 80% (Romero *et al,* 2004). These bacterial isolates remained stable on the leaf surface and formed micro-colonies with extracellular matrix. Some strains of the bacterium *Bacillus subtilis* are reported as very strong antagonists of the cucurbit powdery mildew fungus *(Podosphaera fusca)* giving as good control as the mycoparasites. They inhibit germination of conidia, thus reducing the number of colonies on leaves. The lipopeptides produced by the antagonists cause morphological damage to conidia that includes presence of large depressions in conidial cell wall, loss of turgidity, severe modifications in the plasma membrane and disorganization of the mildew cytoplasma (Romero *et al.,* 2007).

Infection of tobacco necrosis virus induces systemic resistance against *S. fuliginea.* Systemic infections of viruses generally elicit defense responses in plants. Prior inoculation of cucumber leaves with a non-pathogenic isolate of *Alternaria cucumerina* or *Cladosporium fulvum* also induces systemic resistance to powdery mildew. Increased inoculum of the resistance inducer has provided up to 71.6-80% reduction in mildew colonies (Reuveni and Reuveni, 2000).

In a study on effects of plant extracts of *Renautria sachalinensis* on *S. fuliginea* and Physiology of cucumber leaves, the extract has no effect on conidial germination but induced rapid accumulation of phenolics phytoalexins in the leaves, which protected the plant against Powdery mildew. In greenhouse grown cucumber, even under high disease pressure, *sachalensis* extract controlled powdery mildew and increased fruit yield by 49%. Ethanolic extracts of leaves of *Aloe vera,* for medicinal value, reduce the number of mildew colonies on treated cucumber leaves by 21%. Extracts of the fruit bodies (basidiocarps) of the higher basidiomycetes fungi *Oudemansiella* and *Ganoderma* reduce number of mildew colonies on treated leaved 79 and 65%, respectively. They also reduce diameter of the colonies by 45 and 70%. The extract of *Oudemansiella* reduces conidial germination of *S. fuliginea* by 71% (Stadnik *et al.,* 2003).

Varietal resistance

The hydrolytic enzyme β-1, 3-glucanase provides defense against powdery mildew in muskmelon. Cultivars Diguria and Haragola are reported to be immune to the disease. In *Cucurbita pepo*(pumpkin, squash) resistance to *Podosphaera xanthii* is conferred by a single incompletely dominant gene (Cohen *et al.,*2003). In resistant melon, development of the fungus is checked at the primary haustorium stage irrespective of temperature. The resistance genes in the host may be temperature sensitive. When the temperature sensitive resistant cultivar is infected the temperature of incubation of the host has a clear effect on the outcome of infection. At 21 °C it takes longer for symptom expression to appear on the resistant cultivar and at 26°C the resistance is complete and no symptoms develop.

Resistance to *E.cichoracearum* and *S. fulginea* occurs in many species of wild cucurbits. African accessions of *Cucurbita ficifolius, C. anguria, C. dinteri* and *C. saggitatus* are reported to possess resistance to both pathogens.

18.2. Powdery mildew of Lettuce

Pathogen: Erysiphe cichoracearum

Distribution

Powdery mildew, is generally considered to be a minor disease of lettuce, however; it may cause significant losses under certain conditions. It is found throughout the world.

Symptoms

The disease is characterised by a white powdery fungal growth on both the upper and lower leaf surface (Fig 18.2). Older, outer leafs are principally affected, becoming chlorotic in due course and becoming a brown, scorched in appearance. Small, black specks (cleistothecia) may appear on such tissues. Early powdery mildew infections may reduce head size and quality.Masses of spores produced in chains are found with the fungal growth on leaf surfaces.

Fig. 18.2: Symptoms of powdery mildew on lettuce leaves

The Disease Cycle and Weather Parameters

E. cichoracearum may survive in the conidial state on old lettuce or closely related weeds. However, initial infections may also arise from ascospores released from cleistothecia persisting from the precious crop (Schnathorst 1959). Ascospore release is triggered by free moisture and a temperatures of 15-22°C. Conidia *of E.cichoracearum* are easily dislodged during periods of low humidity and are wind-disseminated over long distances. Unlike those of most fungi, conidia of E.cichoracearum are inhibited from germinating by the presence of free moisture, with optimal germination occurring between 95 and 98% relative humidity (Schnathorst 1960). Germination and mycelial growth are optimized at 18°C. Thus, powdery mildew thrives in climates that are somewhat cool and dry.

Disease Management

Sulfur applied at the initial appearance of the disease and at subsequent intervals is quite effective in controlling powdery mildew. Many newer compounds are also demonstrating efficacy and are worthy of investigation (Matheron and Porchas 2000).

18.3. Powdery mildew of Carrot

Pathogen: *Erysiphe heraclei* DC. and *Leveillula lanuginose Fuckel*

Distribution

Powdery mildew occurs wherever carrots are grown. Powdery mildew is particularly important in Mediterranean climates. E. heraclei occurs on many other umbelliferous crops, including anise, caraway, chervil, dill, parsnip, parsley. The other powdery mildew that occurs on carrot is Leveillula lanuginosa Fuckel (synonym: Erysiphe lanuginosa (Fuck.) Golovin), which is generally limited to the Middle East, Armenia, India, Kazakhstan and other countries of Central Asia, Pakistan, and the Mediterranean regions of Europe and Africa. In addition to carrots, it infects anise, caraway, celery, coriander, dill, fennel, and parsley. It is sporadic and of minor economic importance.

Symptoms

All aboveground plant parts, including leaves and petioles, as well as flower stalks and bracts, are susceptible and exhibit powdery fungal growth (Fig 18.3). As spots enlarge on leaves, the foliage becomes chlorotic. Leaves can survive heavy infections, although they may senesce prematurely. The disease appears first on the older leaves and then spread to the younger foliage. Depending on the crop, the severity of the disease, and the growth stage of the crop at disease onset, significant yield reductions can occur.

Leveillula lanuginose causes pale yellow areas on the upper leaf surface with associated whitish sporulation on the lower leaf surface. The infected areas may be limited by veins, thus giving the lesions an angular appearance. In advanced stages, sporulation also appears on the upper side of the leaf and the yellow areas turn brown. Severely affected areas eventually dry. Petioles are also infected.

Fig. 18.3: Symptoms of powdery mildew on carrot leaves

Pathogen associated

Two species of powdery mildew attack umbelliferous crops. The most common one on carrots is *Erysiphe heraclei* DC. Synonyms of *E. heraclei* that appear in the literature are *E. polygoni* DC and *E. umbelliferarum* de Bary (Braun, 1995). *The other powdery mildew that occurs on carrot is Leveillula lanuginose*

Fuckel (synonym: Erysiphe lanuginosa (Fuck.) Golovin). The asexual stage is *Oidium*. *E. heraclei* produces white mycelium and sporulation, which are conspicuous and often dense.

E. heraclei is ectophytic, i.e., it grows primarily external to the plant with only hausioria penetrating the host epidermal cells. Sporulation on carrot tissue occurs 7 to 14 days after infection. The mycelium of E. heraclei is highly branched and produces lobed haustoria. Hyphal cells are 55 to 85 µm long and 4 to 5 µm wide. The conidiophores are moderately long (60 to 140 µm) and straight. They possess a cylindrical foot cell that measures 20 to 35 x 8 to 10 µm followed by a longer cell and one or two shorter cells. Cylindrical conidia (25 to 45 x 12 to 21 µm) are form singly. Germ tubes, which are located at the ends of conidia, form lobed or club-shaped appressoria. Cleistothecia, the sexual fruiting structures, are 80 to 120 µm in diameter with few to numerous appendages that are basally inserted, mycelioid, and brown. These appendages are mostly as long as the cleistothecial diameter and are usually irregularly branched, resulting in a coral-like appearance. There are 3 to 6 asci per cleistothecium (rarely as few as two or as many as10) and 3 to 5 ascospores (rarely 2 or 6) per ascus. Ascospores are relatively large (18 to 30 x 10 to 16 µm) and ovate to elliptic.

L. lanuginose produces Oidiopsis type conidia with mycelium that is both endophytic and external (as compared with the Oidium-type mildew, which is only external). Fungal growth is typically persistent, but is not as conspicuous as the Oidium type mildew. The conidia of L lanuginose are cylindrical (around 40 to 80 x 13 to 20 µm) with distinctive rings near the ends. The conidiophores of L. lanuginose are 200 to 250 µm long. Cleistothecia of L. lanuginose are gregarious, subspherical, about 170 to 250 µm in diameter and decorated with a few to numerous appendages on the lower half of the ascocarp. These appendages are typically shorter than the diameter of the cleistothecium, mycelioid. hyaline to yellowish, septate, often irregularly branched, interwoven with each other and with the mycelium, and measure about 4 to 10 µm wide. The asci are numerous (mostly more than 20 per cleistothecium), stalked, slender (75 to 100 x 25 to 35 µm), and two-spored. Ascospores are hyaline, one-celled, ovoid, and measure about 30 to 35 x 15 to 20 µm.

The disease cycle and weather parameters

Conidia of both Erysiphe and Leveillula are light and can be carried long distances in the air. The spores are unique among fungal pathogens in their lack of a requirement for free water for germination. High humidity and moderate temperatures favor infection and disease development. Powdery mildew is more severe under shady conditions, as sunlight damages the spores and mycelium.Crops become more susceptible as they age. In Israel, the earliest age at which carrots were affected was 50 days after sowing (Palti, 1975). Rain or sprinkler irrigations tend to reduce disease severity. In general, powdery mildews tend to be more common and severe in warm, dry climates. This is particularly true of Leveillula. For example, in Israel, Leveillula on carrot occurs only in the driest part of the country.

Cleistothecia, if formed, may survive on debris and have been reported as contaminants in seeds of carrot, fennel, parsley, and parsnip, but transmission via seed has not been documented. In the absence of cleistothecia, infection of new crops probably depends on air-borne conidia from other crops or wild umbelliferous hosts.

Disease Management

Applications of sulfur is the most common chemical control but fungicides are not typically warranted unless the disease appears early in the growing season. Cultural controls include the use of tolerant cultivars, maintenance of good plant vigor while avoiding excess fertigation, and avoiding shady growing conditions and/or water stress. In Israel, mulches applied to carrot crops to reduce drought stress significantly reduced severity of powdery mildew (Palti, 1975).

18.4. Powdery mildew of beans

Pathogen: *Erysiphe polygoni*

Symptoms

Whitish floury patches on the leaves characterize the disease (Fig 18.4). Under favourable conditions, these spots enlarge and cover the whole leaf lamina. Ultimately, leaves turn yellow to brown in colour and then dry up. The infection may spread on pods, get wrinkled and finally dry up.

Fig. 18.4: Symptoms of powdery mildew on bean leaves

Pathogen

The fungus *Erysiphe polygoni* DC produce superficial powdery mass consists of mycelium, conidiophores and spores of the fungus. The mycelium is ectophytic, septate and produces the tubular haustoria. The cleistothecia are black with myceloid appendages over the body. Asci are ovate and nearly sessile. Hyaline ascospores are elliptical and unicellular. The pathogen is an obligate parasite.

The Disease Cycle and Weather Parameter

Primary infection is by means of dormant mycelium in the seed and cleistothecia present in the soil and plant debris. Secondary infection is by means of oidia disseminated by wind.

Disease Management

Dust the crop with 300 mesh sulphur at the rate of 20 kg/ha or spray the crop with Wettable sulphur (0.25%) or dinocap or tridemorph or triadimefon or penconazole, or carbendazim (0.1%).

18.5. Powdery mildew of okra (Leadyfinger)

Pathogen: *Erysiphe cichoracearum* DC.

The disease is of common occurrence , wherever okra is grown throughout the world. In Himachal Pradesh the disease is reported on almost all the commercial cultivars of okra (Raj *et al* 1992).

Losses

Powdery mildew affects plants of all growth stages and may result yield losses to the tune of 17 to 86.6 per cent (Sridhar and Sinha 1989). Crop yield losses are significant under favourable weather conditions if the infection takes place in early stages of plant growth (Gupta and Thind 2006) Among biotic factors powdery mildew disease is the principal one (Franco, 1983) which cause 20-40% yield losses (Agrios, 2005).

Symptoms

The leaves show the presence of white or grayish patches of powdery fungal growth on the upper surface of the lower older leaves and then spreads to younger ones. The fungal growth is diffused without any marked boundary covering the entire leaf surface. The lower surface may also show infection. Grayish white powdery coating is visible on severely affected leaves (Fig 18.5). Leaves finally show necrosis resulting in withering, drying and defoliation.In severe infection, the affected leaves dry up and fall off prematurely. Early infection causes more effect on plant growth and yield.

Fig. 18.5: Symptoms of powdery mildew on okra leaves

Pathogen

Erysiphe cichoracearum DC.conidia are single celled, hyaline, barrel-shaped and in long chains and 30 to 64 x 13 to 32 μm in size. Cleistothecia are globose and dark with hyaline to dark brown myceloid appendages. The asci are pedicellate, ovate or ellipsoid and 30 to 90 x 22.5 to 50 μm. The number of ascospores is usually 2 rarely 3 per ascus. The ascospores are single celled, hyaline, oval to sub cylindrical and 18 to 30 x 12 to 18 μm.

Weather

Dry weather conditions favour powdery mildew. The disease is observed commonly during September to December. Favourable temperature for disease development is 15 to 30°C.

Disease Management

Application of Wettable sulphur 0.2 per cent or sulphur dust at 25 kg/ha thrice at 20 days interval or four times at 15 days interval is effective. First spray should be given immediately after the appearance of the disease. Spraying with carbendazim 0.1 per cent or benomyl 0.1 per cent is also effective.

18.6 Powdery mildew of cruciferous vegetables

Pathogen: *E. polygoni* De Candolle

The disease is distributed throughout the world and causes the powdery mildew on crucifers (Koch and Slusarenko, 1990), including Brassica crops (Karakaya *et al.*, 1993; Koike, 1997; Kumar and Saharan, 2002) and Arabidopsis thaliana (Koch and Slusarenko, 1990; Kunkel, 1996).

Symptoms

Erysiphe spp. can infect any above ground plant part and can cause heavy yield losses in Brassica crops by reducing plant growth and consequently, the quantity and quality of seeds (Kumar and Saharan, 2002). Appears as a white talcum like growth, either in spot or more or less completely covering the upper surface of the leaves and steam (Fig 18.6). Later the leaves become pale green to yellow or tan. In severe cases leaves curl, die and may drop off. Usually plants are stunted and partially defoliated depending on the stage of growth when infected.

Fungal infections can also generally reduce the cold hardiness of plants, increasing the amount of frost damage (Paul and Ayres, 1986). As the infection progresses, spreading chlorosis, dehydration, and necrosis can reduce the plant fitness

Fig. 18.6: Symptoms of Powdery mildew on cabbage

Pathogen

E. polygoni De Candolle occurs on more than 359 plant species from 154 genera. The nomenclature was clarified in 1967 as *Erysiphe cruciferarum* Opiz ex LK. Junell.

Disease Cycle and Weather Parameters

The fungus can survive in the sexual stage (Cleistothesia). Heavy rains wash spores to the ground. Conidia can germinate in 0 to 100% RH. Germ tubes are short and short lived under low relative humidity. At high temperature the *E. polygoni* grows well only when the humidity is high.

Disease Management

Cultural control

Crop rotation, eradication of crucifer weeds, destruction of volunteer crucifers plants and good soil drainage.

Chemicals

Fungicide spray of Wettable Sulphur (0.25%) or Tridemorph or Dinocap or Triadimefon or Benomyl (0.1%). Dusting with 300 mesh sulphur dust is also effective.

Resistant Varieties

Use of resistant varities like Globelle, Hybelle and Sanibel.

18.7. Powdery mildew of tomato

Pathogen: *Oidium neolycopersici*

Geographical distribution

Powdery mildew of tomato occurs through out the world on greenhouse and field-grown tomatoes. In U.S.A, it occurs in California, Nevada, Utah, North Carolina, Ohio, and Connecticut and sporadically on Long Island. Outdoors it tends to be more common in gardens than commercial crops, perhaps reflecting different environmental conditions and crop management practices. It also develops on tomatoes grown in high tunnels where it can be a very important disease.

Losses

Losses in fruit production due to decreased plant vigor can reach up to 50% in commercial production regions where powdery mildew is severe. Although this level of damage has not been observed on tomatoes in fields, plants grown in greenhouses in North Florida reached 50-60% disease incidence.

Symptoms

Symptoms of the disease occur only on the leaves. Symptoms initially appear as light green to yellow blotches or spots that range from 1/8 - ½ inches in diameter on the upper surface of the leaf. The spots eventually turn brown as the leaf tissue dies. The entire leaf eventually turns brown and shrivels, but remains attached to the stem. A white, powdery growth of the fungal mycelium is found on the top of leaves (Fig 18.7).

Fig. 18.7: Symptoms of powdery mildew on tomato plant

Yield and fruit quality can be reduced by powdery mildew because the disease can develop quickly, and severely affected leaves are killed.This results in less fruit being produced, especially in cherry, Heirloom, and other indeterminate tomato plant types. Fruit that forms typically does not taste as good as fruit produced on a plant with a full canopy of photosynthetically active leaves, and it is more likely to develop sunscald damage with less protective leaf cover.

Pathogen

The fungus, *Oidium neolycopersici*, causes the disease. Pathogens causing powdery mildew typically have narrow host ranges. Sometimes weeds are also hosts and thus can function as a potential source of a powdery mildew pathogen. The pathogen causing powdery mildew on Long Island and throughout the eastern U.S. is *Oidium neolycopersicum*.

A different fungus, *Leveillula taurica*, occurs in other areas, including California. Both pathogens produce characteristic white, powdery growth. *Leveillula taurica* only produces this on the underside of leaves. *Leveillula taurica* has been observed twice on Long Island and only in pepper.

Disease cycle and weather parameters

Like other powdery mildews, the white, powdery growth is mostly the asexually produced spores (conidia) of the pathogen plus the structures on which these spores are form i.e conidiophores of the fungus. The spores are easily dispersed by wind. A spore landing on a tomato leaf can infect and in about one week develop a new disease spot with an abundance of spores ready to be dispersed. Powdery mildew fungi do not require leaf wetness or high humidity to infect leaves, as do other fungi causing foliar diseases. Their ability to develop under a range of conditions combined with their ability to quickly produce a lot of spores, means powdery mildew diseases can develop rapidly. While moisture is not required, tomato powdery mildew develops best when the air is somewhat humid, but not above 95% RH.

Disease Management

Resistance to powdery mildew caused by *Leveillula taurica* has been bred into a few varieties that are adapted for being grown under greenhouse conditions because this disease has been more problematic under those conditions than outdoors for growers. Grace is a greenhouse variety that grew well and demonstrated good suppression of *Oidium lycopersici* under field conditions in an experiment conducted in Connecticut. Noticeable differences in susceptibility to powdery mildew caused by this fungus have been observed among other varieties growing outdoors.

Powdery mildews are relatively easy to control with fungicides. There are several conventional and biological products that have proven effective in efficacy experiments with this and other powdery mildews. Conventional fungicides include those containing sulfur, copper, chlorothalonil or mineral oil as the active ingredient. Botanical oil (including sesame, rosemary, and thyme), plant extracts (giant knotweed), biocontrol microorganisms (including species of *Bacillus* and *Streptomyces*), and potassium bicarbonate are some of the active ingredients in biological fungicides, most of which are approved for organic production. Typically, fungicides needs to be applied weekly to maintain control. Removing affected leaves is not considered a viable approach to managing any powdery mildew disease because once spots are seen spore dispersal has likely already occurred from the spots, spores will likely be disrupted in the process of removing leaves, and these likely will continue to be sources of spores from other plants in the area. Rotating where tomatoes are grown is not a viable practice because these are obligate pathogens, thus they need living host plant tissue to survive (they cannot live in diseased crop debris over winter).

18.8. Powdery Mildew of Onion (*Allium cepa*)

Pathogen: *Leveillula allii and Oidium taurica*

Geographic distribution

Powdery mildew caused by *O. taurica* affecting onion and/or garlic has been reported in Sudan (Schwartz & Mohan, 1996), United States (Laemmlen & Endo, 1985), Israel (Palti, 1959) and in Brazil (Café Filho *et al.*, 2001; Maffia *et al.*, 2002; Mendes *et al.*, 1998).

Symptoms

Circular to oblong (0.2 to 0.75 inches), yellow (chlorotic) to white (necrotic) areas appears on the older leaves. Whitish, powdery patches, typical when the fungus is producing asexual spores (conidia), are apparent on infected tissues. Occasionally, lesions coalesce, and cover larger areas of the leaf surface (Fig 18.8). Symptoms can appear on both young and mature leaves. Powdery appearance, Chlorosis and eventually necrosis may develop around areas of sporulation. Lesions may coalesce to cover large areas of the leaf surface. This disease appears to be most

Fig. 18.8: Symptoms of powdery mildew on onion leaves

common on varieties with glossy leaves, which are associated with thin cuticular waxes.

Pathogen Associated

Leveillula allii (formerly *L. taurica*), a fungus overwinters as chasmothecia on diseased tissue or on alternate hosts. The sexual stage has not been reported in the Pacific Northwest.

Leveillula taurica , the first report of *Leveillula taurica as* powdery mildew pathogen of onion (*Allium cepa*) in the Pacific Northwest was given by du Toit, et.al,(2004).

Disease Cycle and Weather Parameters

Conditions favoring the disease are not well known, but infection appears to be favoured by high relative humidity during warm weather. The spores do not require rain or persistent dew to cause infection. The disease apparently causes little damage to most onion varieties.

Disease Management

Cultural control

Plant varieties resistant to powdery mildew.

Chemical control

The disease is not common and generally causes too little damage to warrant fungicide sprays. The Fungicide spray of Wettable Sulphur (0.25%) or Tridemorph or Dinocap or Triadimefon or Benomyl (0.1%). Dusting with 300 mesh sulphur dust is effective to manage the disease.

18.9. Powdery mildew of Potato (*Solanum tuberosum* L.)

Pathogen: *Golovinomyces cichoracearum* (DC.) V .P. Gelyuta

The first field report of the disease was from Washington in 1950, with subsequent reports from Utah and Ohio in USA.

Losses

Powdery mildew causes losses in potato production throughout the world. In the Columbia Basin of Washington, this disease appears most damaging in potatoes grown under furrow irrigation.

Symptoms

Early symptoms comprise small dark areas on the adaxial surface of leaves, along the veins, and at the petioles. Dark lesions consisting of mycelia and conidiophores are also visible on the main stems of affected plants. As the disease progressed, leaves get covered by a gray powdery fungal mass, and older leaves became necrotic (Fig 18.9). Disease symptoms are most severe on cvs. Desiree and Santina. Disease expression was greater along sprinkler lines and in localized areas from which the disease spread to surrounding plants. Severely affected plants began collapsing just prior to water cutoff.

Fig. 18.9: Symptoms of powdery mildew on potato

Pathogen

Golovinomyces cichoracearum earlier reffered as Erysiphe cichoracearum is the causal pathogen. In the Middle East a different pathogen, designated as *Oidiopsi* (anamorphic state of a *Leveillula* species), also was reported on potato to cause powdery mildew.

Conidial chains arising from the hyaline, epiphytic mycelia consist of two to eight conidia. The conidia are cylindric to doliform and measure 16.8 to 22.8 im × 28.8 to 45.6 μm in size. No cleistothecia are reported. Identification of the causal agent as *Golovinomyces cichoracearum* (synonyms *G. orontii* and *Erysiphe cichoracearum*) based on morphology was confirmed by internal transcribed spacer (ITS)-polymerase chain reaction (PCR).

BLAST analysis of the ITS sequence revealed a 99% homology to *E. cichoracearum* from an *Ambrosia* sp. Pathogenicity was confirmed on potato seedlings cv. Red La Soda.

Disease management

Although application of sulfur dust or wettable sulfur for control of potato powdery mildew generally is successful, the applications sometimes fail to prevent development of severe symptoms. The possible explanation for variable effectiveness of fungicide treatments is that the co-occurrence of causal pathogen *L. taurica* and *E. orontii* on potato crops in North America might differ in their sensitivity to fungicides. The factors governing co-infection, and the responses to foliar fungicide applications by the two pathogens needs further studies.

18.10 Powdery mildew of Capsicum and Green house pepper
Pathogen: *Leveillula taurica*

Geographic distribution

Powdery mildew on pepper or capsicum, caused by the fungus *Leveillula taurica* (Lev) *Arnaud* (anamorph = *Oidiopsis sicula* Scalia *syn. Oidiopsis taurica* (Lev) Arnaud], occurs worldwide in Africa, Asia, the Mediterranean and Caribbean regions, and North America. Isolates of *O. taurica* pathogenic to pepper have a wide host range and are usually cross-infective to tomato and egg plant. The disease is a common problem on peppers in greenhouse production. Severe losses are reported when plants are heavily defoliated.

Leveillula taurica first appeared in North America in Florida in 1971. Since the early 1990's it has been a recurring problem in California on chilli and bell peppers. By the late 1990's it had spread to Arizona, Idaho, New York, Oklahoma, Utah, Mexico and Ontario. It was first detected in British Columbia in February 2003 on greenhouse pepper crops and has since spread throughout the greenhouse pepper industry.

Losses

Losses depends on the level of infection.Studies show that higher the level of powdery mildew infection, higher the loss of production. An early, heavy infection with mildew had about 30% loss of production compared to a later, lighter infection. Powdery mildew generally has caused 10-15% yield loss in North American greenhouse pepper crops.

Symptoms

The white, talcum powdery-like growth that is readily apparent for many other powdery mildew diseases is not nearly so obvious on pepper. The signs of the pathogen are best observed on the underside of the oldest lesions. On the top surface of leaves, lesions are yellow with brown necrotic centers. Leaves curl upwards. Premature senescence of the leaves results in defoliation. Both the number of fruit and the size of fruit are reduced in heavily infected plants.

In general, pepper crops become more susceptible to powdery mildew as they mature. Older plants and lower leaves are the first to show evidence of powdery mildew infection. Pepper powdery mildew is different in several ways from the mildews that infect tomato (*Erysiphe, Oidium lycopersicum*), or cucumber, (*Erysiphe cichoracearum, Sphaerotheca fuliginea*). Pepper powdery mildew grows unseen, within the leaf tissue for a latency period of up to 21 days. Unlike tomato and cucumber powdery mildew which is easily seen on the top side of the leaves, pepper powdery mildew grows on the undersurface of leaves (Fig 18.10). In addition, pepper plants can become defoliated and do not recover as quickly as other greenhouse crops when infected with powdery mildew. Pepper powdery mildew does not infect the fruit or stems but can quickly destroy unprotected leaves and eventually the entire pepper crop.The fluffy, white patches of powdery mildew appear on the underside of leave. With time, these patches may turn brown rather than remaining white. The upper surface of the leaf may appear normal or have diffuse, yellow patches which correspond to the mildew colonies on the lower surface. Early powdery mildew infections can be seen more easily by holding the leaf up to the light and looking for developing mildew colonies. Severely infected leaves wither and drop off causing plants to die.

Fig. 18.10: Symptoms of powdery mildew on capsicum leaves

Pathogen

The fungus *Leveillula taurica* (anamorph *Oidiopsis taurica)* is an obligate parasite but unlike other powdery mildews, it appears to have a wide host range. Host specificity of some isolates is observed (Ayesu-Offei 1998, Correll *et al,* 1987). The pathogen has endophytic mycelium. Conidiophores are long and multi-branched. Conidia are dimorphic (pyrifrom and cylindrical) and borne singly or in short chains. The size of the conidia vary according to the isolate.

Mean measurements range from 49-71 µm X 16-24 µm and 44.6-65 µm X 16-23 µm for the pyriform and cylindrical conidia, respectively (Correl *et al,* 1987).

It should be noted that *O. taurica* has been known for many years as the causal agent of powdery mildew of tomato in the western USA but was not the cause of powdery mildew of field tomato in Florida. However, *O. taulica* was identified as the causal organism of powdery mildew on pepper in greenhouse peppers in Florida.

Disease cycle and Weather parameters associated

Pathogen overwinters on the infected leaves as dormant mycelium or cleistothecia. It may also survive on the collateral hosts. Secondary spread is through wind-borne conidia. Conidia are wind blown or dispersed by splashing rain. Temperatures of 25-30°C favor germination of conidia. Some isolates of powdery mildew that infect pepper can apparently develop infection under a wider range of dry to humid conditions compared to other powdery mildews. Relative humidity that is higher at night than during the day and temperatures of less than 30°C are conducive to disease. Free moisture on leaves inhibit spore germination. Throughout the growing season, new infections may develop on new, succulent shoots of plants.

Disease Management

Cultural methods

- Restrict visitor access to the greenhouse
- Follow strict greenhouse hygiene throughout the growing season
- Conduct a through year-end clean up and dispose of all crop debris off-site or by burning or burying in a landfill.
- Control outdoor weeds surrounding the greenhouse
- Keep ornamentals and imported plants out of the greenhouse and immediate area
- Improve greenhouse climate to reduce relative humidity and increase air circulation.

Timely disease detection

Enhance early disease detection by placing suspected leaves in a zip lock bag with some moist paper towelling. After a day or two of incubation in a warm place, use a hand lens (15-30 x) to check the under surface of leaves for white mildew colonies. Be sure to have the disease confirmed by sending a sample to the plant diagnostic lab.

Chemical

In both greenhouse and the field, use registered fungicides to prevent powderymildew. Several systemic and contact fungicides provide good powdery mildew control (Keshwal and Choubay 1983). Preventative spray programs in greenhouses should be established prior to infection. Applications of mono-potassium phosphate as a foliar spray inhibit the fungus on plant leaves and reduce yield losses (Reuveni *et al,* 1998). Bicarbonates ($NaHCO$, or $KHCO_3$) applied to the foliage reduce disease severity, leaf defoliation, and post harvest decay of fruit (Fallik *et al.,* 1997). Auxins (indoleacetic acid, indole butyric acid) applied to infected pepper leaves inhibit defoliation (Reuveni *et al,* 1976).

The disease can be prevented by early application of fungicides (Table.18.2). Apply a protectant fungicide when powdery mildew is first detected, or in a greenhouse with a previous history of powdery mildew. Repeat the treatment by alternating with fungicides in different chemical groups. Continue treatments if your greenhouse has had powdery mildew the previous season or if disease pressure warrants control.

18.11. Powdery mildew of Cluster bean

Pathogen: *Leveillula taurica*

Symptoms

Powdery growth of the fungus is observed on the lower side of the leaves covering the lower leaf surface (Fig 18.11). The upper leaf surface turns yellow. The leaves, petiole and young stems are affected by the fungus. If the infection is severe, the withering of leaves and defoliation takes places.

Fig 18.11: Powdery mildew on cluster bean plant

Pathogen

The fungus *Leveillula taurica* is endophytic and intercellular. Conidiophores emerge through stomata bearing single oidium at the tip. Conidia are pear shaped or rectangular and germinate by germtube. The cleistothecium has definite myceloid appendages.

Table 18.2: A summary of registered fungicides and label information

Product	Chemical/Biological Ingredient	Chemical Group	Mode of Action	REI[1]	PHI[2]	Application
Actinovate SP	*Streptomyces lydicus*	biological	preventative/suppressive	1 hr		use preventatively; apply at 7-14 days interval; use the product within 4 hrs of preparation
Bartlett Microscopic Sulphur	Sulphur	M	preventative/suppressive	24 hrs	NA	use preventatively; apply as required at 14 days interval; do not exceed 10 applications per crop cycle
Agrotek Ascend Vaporized sulphur	Sulphur	M	preventative/suppressive	2 hrs	NA	use preventatively; Run the vaporizer for 1-8 h during night and repeat 2-7 times per week as per label instructions
MilStop	potassium bicarbonate	NC	preventative, non-systemic	4 hrs	0 days	use preventatively; apply at 7 days intervals; treated produce cannot be exported to the USA
Nova 40W	Myclobutanil	3	preventative/some curative action, locally systemic	12 hrs	3 days	use preventatively; do not exceed 3 applications per crop cycle; apply at 10-14 days interval
Pristine	boscalid + pyraclostrobin	7 & 11	preventative/some curative action, locally systemic	12 hrs	1 day	use preventatively; do not exceed 1 application per crop cycle, hence, use in rotation, after 7 days, with other fungicides
Switch	cyprodinil & fludioxonil	9 + 12	preventative/some curative action, locally systemic	1 day	1 day	use preventatively; do not exceed 3 application per crop cycle, hence, use in rotation, after 7 to 10 days, with other fungicides
Timorex Gold	tea tree extract	natural product	preventative	4 hrs	4 days	use preventatively; apply at 7-14 days interval

PHI - pre-harvest interval REI- re-entry interval

The disease cycle and weather parameters

The fungus perpetuates as dormant mycelium or cleistothecia, which serve as primary source of infection. Secondary infection is through air borne conidia.

Disease management

Spray dinocap, tridemorph, triadimefon, penconazole, or myclobutanil @ 0.10 per cent at fortnightly interval.

18.12. Powdery mildew of Eggplant

Pathogen: *Leveillula taurica*

Symptoms

Powdery mildew appears as a dusty white to gray coating over leaf surfaces or other plant parts (Fig 18.12). Powdery mildew, begin as discrete, usually circular, powdery white spots. As these spots expand they coalesce, producing a continuous matt of mildew. Symptoms usually appear late in the growing season on outdoor crops. Injury due to powdery mildews includes stunting and distortion of leaves, buds, growing tips, and fruit. The fungus may cause death of invaded tissue. Yellowing of leaves and death of tissue may result in premature leaf drop and may result in a general decline in the growth and vigor of the plant.

Fig. 18.12: Symptoms of powdery mildew on egg plant leaves

Disease Cycle and Weather parameters

The fungus is favored by periods of high relative humidity or site conditions that promote a more humid environment, such as close spacing of plants, densely growing plants, or shade. Indoors, symptoms may occur at any time of year, but the rate of spread and development will be affected by the relative humidity and temperature.

Disease Management

Before symptoms appear use Kumulos-S 250g/100L. If symptoms appear use Bellis 50g/100L.

18.13. Powdery mildew of turnip and rutabaga (*Brassica* sp.)
Pathogen: *Leveillula taurica*

Symptoms
A powdery, whitish gray color fungal growth and its spores develop on infected leaves (fig.18.13). In advanced stages, leaves are distorted, twisted, and retarded in growth. They ultimately turn yellow and die.

Fig. 18.13: Turnip leaves infected with powdery mildew pathogen.

Pathogen
Leveillula taurica (syn. = *Erysiphe taurica*), overwinters on turnip and rutabaga refuse and other hosts including weeds.

Disease Management

Cultural methods
Plough down cruciferous weeds and volunteers. Follow Crop rotation. If only powdery mildew is a concern, it is enough to rotate out of crucifers just one growing season so that all volunteers and cruciferous weeds are controlled.

Chemical control
Start a spray program when mildew first appears.Cueva at 0.5 to 2.5 L/ 100 L water on 7 to 10day intervals. Fontelis at 500g to 1kg/A on 7 to 14 day intervals. Do not make more than two sequential applications before alternating to a labeled fungicide with a different mode of action. Kumulus DF at 1.360 to 4.5kg/A on 10 to 14day intervals. Microthiol Disperss at 1.360 to 4.5 kg/A. Do not apply if temperature exceed 32°C within 3 days after application. Milstop (85% potassium bicarbonate) at 3 Tbsp/ 10 L water on 7 to 14day intervals. Strobilurin fungicides (Group 11) are labeled for use.

Biological control

Actinovate AG as a foliar spray on 7 to 14 day intervals.Double Nickel LC on 3 to 10day intervals Or Serenade MAX on 7 to 10 day intervals.

19

Powdery Mildew of Fruit Crops

19.1. Powdery mildew of Grapevine

Pathogen: *Uncinula necator*

The known history of powdery mildew of grapevine dates back to 1834 when the fungus causing the disease was first described as ***Erysiphe necator*** in the eastern part of North America. However, it was not considered important until it was reported from England in 1847 and by 1850, it had spread to France and other major grape growing areas of Europe where it caused considerable loss to grapevine growers and the wine industry.

Geographic Distribution

Today, powdery mildew can be found in most grape growing areas of the world including the tropics. It is known to occur in a mild or severe form in North and South America, Europe, parts of Africa, India and in Australia. It appears in almost epidemic form in all the vineyards in India when the conditions are favorable for its development. The disease is much more serious than the downy mildew of grapes and is more dangerous than other powdery mildews of different crops.

Symptoms

The disease attacks the vines at any stage of their growth. All the aerial parts of the plant are attacked. The characteristic symptom of the disease is the appearance of white, powdery patches on affected parts. Cluster and berry infections usually appear first. Cluster infection before or shortly after the bloom results in poor fruit set and considerable crop loss. Young fruits (berries) just after bloom show whitish mycelial growth on the surface (Fig 19.1). When the infection of berries occurs before they attain full size, the epidermal cells are killed and the growth of epidermis is prevented. Because the internal pulp continues to grow, the skin cracks. Such berries either dry up or rot. If the attack occurs when fruits are nearing maturity or beginning to ripen, they fail to color properly, become irregular in form and only few of them ripen, remaining undersized with a blotchy surface. Often the infected berries develop a net-like pattern of scar tissues. *As the fruit approaches maturity it becomes immune to attack.*

Leaf lesions appear late and do not cause much damage. On young leaves, small whitish patches appear on the upper, or sometimes on the lower surface of the leaf. These patches grow in size and finally coalesce to cover large areas on the lamina. Similar floury patches are formed on the stem, tendrils, and flowers. The powdery growth gradually turns gray and finally dark colored. Malformation and discoloration of the affected leaves are common symptoms. The stems turn brown in color. The diseased vines have a wilted appearance and remain stunted in growth. Necrosis of the penetrated epidermal cells and even of adjacent cells is a characteristic reaction of resistance in some varieties of the host. Late in the season, there may be numerous black specks on the surface of the infected spots, but in many regions these are not present. These are the sexual fruiting bodies, the cleistothecia, and are so small that they may be over looked unless one uses a hand lens. Leaves attacked by surface mildew usually retain about their normal green color but may show some discoloration toward the end of the season.

Failure to control the disease results in a chronic reduction in wood maturity (number of nodes, which, are important in subsequent bearing in following year) and losses in the yield of fruits which may be up to 40-60%. The most significant effect of powdery mildew is on berry sugar levels and juice color and acidity. The fungus infection makes the grapes unsuitable for wine making. Berries infected by the fungus tend to be higher in acid than healthy berries. This is not considered desirable by the brewers . The fungus itself produces an off-flavour in wine made from infected grapes. The infected berries tend to crack, thus, providing entry to other pathogens and saprophytes.

Fig. 19.1: *Symptoms of* Powdery mildew on grapevine leaves and fruit berries

Pathogen

The disease is caused by *Uncinula necator* (Schw.) Bur. (anamorph *Oidium tuckeri* Berk.). The fungus was reclassified as *Podosphaera necator* and then as *Erysiphe necator* which is its original name. The mycelium is entirely superficial on the attacked parts to which it adheres by means of bilobate or multilobate appressoria. The hyphae are hyaline, slender, septate, branched, and 4-5 μm in diameter. They turn darker in color when the formation of conidia is over. The conidiophores arising perpendicular to the creeping hyphae on the host surface are simple, multiseptate and erect. They are attached to the mycelial hyphae by a cylindrical foot cell measuring 24-40 μm. Cells of the conidiophores are generally wider than those of the mycelial hyphae, measuring 6.2-7.5 μm. They bear a chain of 3-4 conidia under field conditions. In static humid conditions the chains may contain 8-10 conidia. These hyaline conidia are oval in shape, and measure 25-30 x 15-17 or 27-47 x 14-21 μm. The oldest conidium is at the distal end of the chain. The conidia germinate by a short germ tube terminated by a bilobed or multilobed appressorium.

Cleistothecia of the fungus have been found in North America, Europe, Russia, Peru, and Australia. Under the climatic conditions prevailing in most vine growing areas of India, the Perfect stage of the fungus is not found. The fungus is heterothallic and most populations consist of two mutually exclusive mating types. The sexual compatibility of the fungus is controlled by a single mating type gene (MAT1) with two alleles (MAT1-1 and MAT1-2). These may be equally common in natural populations.

A small percentage of isolates have the capacity to form cleistothecia in protracted association with isolates that initially appear to be of an incompatible mating type. When the mating types are present, the cleistothecia can form on all infected tissues during the later part of the growing season. They are found embedded in the superficial mycelium on leaves or on shoots chiefly at the nodes or in buds among the scales and hairs. They are hyaline and spherical when young but turn yellow due to accumulation of yellow lipid in the ascocarp. When the outer cells of the ascocarp darken, the mature cleistothecium turns black almost round with a flattened top and measures 75-100 or 84-105 μm in diameter.The peridium is covered with 8-25 septate appendages, which appears inserted in the equatorial region of the ascocarp. The appendages are coiled (uncinate) at the distal end and are brown at the base. They are 1-6 times as long as the diameter of the cleistothecium. When the cleistothecium is immature, it remains the functional connection with hypal mass. Each cleistothecium contains 4-6, some time more, ovate to subglobose asci measuring 48-60 x 37-45 μm. In each ascus, there are 4-6 ascospores, which are oval to ellipsoid in shape and hyaline. The ascospores are low in water content.

Disease cycle

The disease cycle of U. necator involves both asexual and sexual overwintering stages in temperate regions. Perennation of the pathogen through vegetative mycelium in dormant buds is more significant than cleistothecia. The conidia can resist desiccation, but it is not known how long they can remain viable. The fungus inside dormant vegetative buds does not affect their survival or vigor.

The dormant buds carrying the primary inoculum develop into flag shoots early in the spring. These flag shoots provide inoculum for secondary spread of the disease early in the season. The shoots commonly have abnormally formed leaves and the borders are often bent upwards. The leaves are covered with whitish growth of the pathogen consisting of hyphae and conidia. Whenever cleistothecia are formed they serve as additional, or may be the only, source of perennation.

In some areas cleistothecia are the only source of primary inoculum, there being no perennation through mycelium in dormant buds. The ascospore infection occurs late in spring and causes development of isolated colonies while the infection by conidia from flag shoots results in uniform and dense colonies. Miazzi et.al, (2003) have hypothesized existence of two genetically separated biotypes in U. necator that are related with its two over wintering modes, a biotype over wintering as conidia and mycelium in dormant buds that causes infection of shoots and leaves early in season and the other biotype overwinter as cleistothecia and infect bunches. The two biotypes are not separated in time and space, and sexual crosses between the two occours in the vineyard. In tropical climate the survival is mainly through mycelium and conidia on green tissues remaining on the vines. Environmental factors are very much responsible for the formation of cleistothecia. Rains disperse the cleistothecia to the bark of the vine where they are retained between leaf fall and bud break in the next season. Density of cleistothecia is higher on fallen leaves than on the bark but their viability is more on the bark than on fallen leaves on soil. After the infection from ascospores or from hibernating mycelium in the host buds, enormous numbers of conidia are produced to carry on the conidia to conidia life cycle.

The conidia on the leaf surface start germination within 90 min. and in about 20 hours about 52% conidia get germinated. Almost all the germ tubes coming in contact with a hard surface form appressoria and in about 5 days after conidial germination on leaves, colonies of U. necator with profuse conidiophores and conidia are seen. In a study of development and adhesion of infection structures of U. necator on grapevine, Rumbollz et al.,(2000) observed that primary

Powdery Mildew of Fruit Crops 273

appressoria appeared 3-5 hours after inoculation followed by hyphal growth on the leaf surface in 14 hours, suggesting successful host colonization. Deposits of extracellular material at the contact zone of fungal structures and plant cuticle ensured firm adhesion of the pathogen. Esterase activity was associated with conidia and infection structures. Temperature appears to be the major determinant of fungal development. It determines the extent of asexual reproduction of U. necator and fluctuating temperatures decide the rate of conidial formation, germination and colony development. In the temperate climate of California (USA), the early warm spring climate usually precedes severe epidemic of powdery mildew.

Weather parameters associated with disease

Rapid germination of conidia, infection, and development of the pathogen occurs at temperatures of 21°- 30°C although the fungus can grow at temperatures from 5° to 30°C. The minimum temperature for germination of conidia is 6°C and for infection and growth 7°C. The optimum for germination is 25°C. Conidial germination ceases at 3°C and above 33°C and at 40°C, conidia are killed. At 25°C, conidia germinate in about 5 hours and time from inoculation to sporulation is 5 days whereas at 23°C and 30°C it is 6 days. Mildew colonies are killed after exposure to 36°C for 10 hours or to 39°C for 6 hours.

Free water causes poor and abnormal germination of conidia. Rainfall is detrimental because it removes conidia and disrupts the mycelium. Atmospheric relative humidity of 40-100% is sufficient for germination of conidia and infection although germination can occur even at less than 20% RH. Humidity has a greater effect on sporulation than on spore germination. Low diffuse light favors development of powdery mildew. In bright sunshine conidial germination is inhibited.

The fungus perpetuates through hyphae inside dormant vegetative buds which are infected in the preceding season. They are most susceptible at the three-to six unfolded leaf stage. Incidence of powdery mildew colonies on the surface of buds is highest at these stages. The colonization of the interior bud tissue from the fungus on the surface is also highest at these stages. Hyphae with haustoria, conidiophores and conidia are present on all internal parts of the buds except the meristems (Rumbolz and Gubler, 2005).

In areas where cleistothecia play a major role in survival of the fungus and in primary infection, it is reported that cleistothecia in the bark of the vines discharge ascospores when rainfall occurs between bud burst and bloom. In a 4-year study in Eastern Washington, Grove (2004) did not find any evidence of perennation of the mildew fungus in dormant buds. Cleistothecia were retrieved from bark fissures and senescent leaves and contained viable ascospores at the time of bud burst. Ascospores could be trapped as late as 70 days after bud burst. Ascospore discharge requires free water. Temperatures within the range of 10° and 25°C have little effect on ascospore release but a temperature of 4°C or lower can suppress ascospore discharge. In laboratory studies, wetting of leaf discs bearing cleistothecia and incubation in a humid chamber for 48 hours at 20°C were found essential for release of ascospores. Storage of cleistothecia on leaves with periodic wetting at 5°C during 110 days were necessary to induce both ascospore release and their germination ability. Ascospores germinate equally well in free water and in saturated atmosphere. Germination declines rapidly as humidity decreases. Appressoria are not formed by ascospore germ tubes at below 10°C and at above 31°C.

It is reported that climatic conditions during October and November in India are ideal for development of Powdery mildew. Warm, dry weather with just enough humidity is very favorable. However, a microclimatic study in Maharashtra, in mid 1990s, showed more rapid spore production and disease development during December and January, when the weather is cool and humid than in November and February. According to this study temperature in the range of 12.2°-30.1°C and relative humidity greater than 57.4% favours sporulation of U. necator. At temperature below 8.8°C and above 34°C and RH below 47.4 % the rate of multiplication is zero. The disease development is retarded in the sunshine. In South India the disease incidence is reported to be significantly influenced by relative humidity and maximum temperature, whereas influence of minimum temperature, rainfall and total rainy days was not significant. Increase of RH by 1 % increased disease incidence by 2.4 %. Increase of temperature by 1°C decreased the disease by 4.4%. Maximum temperature range of 27-31°C along with RH up to 91% favored disease incidence while maximum temperature range of 31-34°C inhibited the development of powdery mildew. Incidence and severity of the disease increases with increasing humidity to an optimum near 85% RH and then plateau or marginal decrease at higher values is seen.

Development of ontogenic resistance to powdery mildew, which varies with host genotype and tissue type, is quite rapid in grape berries (Ficke *et al*, 2003, 2004; Gadoury *et al*, 2003). The fruit becomes nearly immune to infection by *U. necator* within 4 weeks after fruit set. Only fruits inoculated within 2 weeks of bloom develop severe powdery mildew. Rachises of fruit bunches develop severe mildew when inoculated at bloom and disease increases steadily over the next 60 days. The ontogenic resistance does not affect adhesion of conidia, germination and appressorium formation but checks pathogen ingress at the

Powdery Mildew of Fruit Crops 275

cuticle before formation of a penetration pore. As the berries age, hyphal elongation and colony growth lowers down and finally there is no development of secondary hyphae. On aged berries, the fungus produces more appressoria in an attempt to cause infection. This response of older berries is supposed to be due to synthesis of a specific protein. Ficke *et.al.,*(2004) had reported that many factors are responsible for the ontogenic resistance. Cuticle and cell wall thickness, antimicrobial phenolics, pathogenesis related proteins are not the principal causes in halting pathogen ingress on ontogenicaliy resistant berries.The infection is halted by one or more of the following:

(i) A preformed physical or biochemical barrier near the cuticle surface, or

(ii) The rapid synthesis of an antifungal compound in older berries during the first few hours of the infection process . The protein, thaumatin, is reported to be present in grape berries. In a study of effect of grape proteins on *U. necator, B. cinerea* and *Phomopsis viticola,* Monteiro *et.al.,*(2003) have observed that two proteins, osmotin and a thaumatin-like protein, inhibited spore germination and germ tube growth of *U. necator.* Vines infected by *Plasmopara viticola* (downy mildew) show resistance to powdery mildew. While leaves infected with *P. viticola* develop 1.5 powdery mildew colonies per leaf, and the uninfected leaves show 23.9 colonies per leaf. The induced resistance is confined to leaf tissues colo*nized* by *P. viticola* and is not transferred to new developing leaves. Conidia of *U. necator* germinate on downy mildew infected leaves, but failed to produce secondary hyphae and colonies.Protection was reversed by application of 0.1 % sucrose solution, which restored susceptibility of the leaf to powdery mildew.

Disease forecasting

An important part of powdery mildew management depends on the use of disease forecasting models. Several models for forecasting powdery mildew have been developed, but for most of these, information on them is either confidential or difficult to obtain. The exception is the UC model developed at the University of California, Davis, California, which is widely available and thoroughly tested. A major advantage of the UC model software is that many companies that sell weather instruments make it an integral part of the instrumentation. This allows the grower to use it with ease. The model evaluates climatic conditions within the grapevine canopy and assesses the likelihood for the development of powdery mildew (Weber *et al.,* 1997). The ascospore infection forecasts are based on average temperature during the time when the leaves are wet, and utilizes the 0.67 value for conidial infection provided by the Mills table to predict grape infection. The Mills table was originally developed for forecasting infection of apple by *Venturia inaequalis* (Cke.) Wint.and has been revised (MacHardy and Gadoury, 1989). In general 12 to 15 hours of leaf wetness are required between 10 to 15° C for infection of grape by *U. necator*. Once infection has occurred, the model switches to the risk assessment phase which is based entirely on the affect of temperature on the reproductive rate of *U. necator* (Thomas *et. al*,1994). The model generates daily conidial risk indices based solely on temperature in the canopy. In order for the conidial risk index to increase, there must be 6 hours during the day when canopy temperatures are between 21 and 30° C. Once started, the index climbs 20 points on each day that meets the 6 hour temperature requirement. The index falls 10 points on days that are unfavorable for powdery mildew development. By knowing the likelihood that powdery mildew could be developing in their vineyards, growers are able to accurately time fungicide applications. During periods of low risk, fungicide application intervals can be safely stretched.

In order to use the UC model the grower must have a means for recording leaf wetness and temperature. Two types of weather stations have evolved for the use of the UC model software. The most complete system is sold by Adcon Telemetry Inc, Boca Raton, Florida. It utilizes radio telemetry to transmit weather information from field weather stations to a receiving base station. Sensors in the field collect temperature, precipitation, relative humidity and leaf wetness data, usually every 30 seconds and transmit it to a central computer. The data is averaged every 15 minutes. Powdery mildew risk indices are calculated at the base station by a personal computer containing the UC model software. Growers access the weather data by using a computer to download it from the base station computer.

A less expensive alternative is to use miniature weather stations that the grower individually purchases. Sensors measure leaf wetness, temperature, and relative humidity in the grape canopy. Attaching a PC to the weather station downloads the weather data. UC model software that can be purchased with the weather station will automatically predict infection and degree of risk for conidial infection when used in conjunction with the weather station software. This type of configuration can be purchased for less than a thousand dollars.The Model 450 Watch dog dataloggers are reliable and easy to use for this purpose.

Disease Management

Regulatory measures

Due to the wider spread occurrence of powdery mildew in grapevine growing areas of the world, regulatory measures are of no values except where introduction of fungicide resistant strains is feared.

Cultural measures

Clean cultivation of vines is an important part of disease management in grapevine orchards. Pruning after shedding of leaves, thinning out and cutting back of laterals and removal and destruction of all diseased parts constitute clean cultivation. Under no-fungicide treatment conditions, in vertical shoot positioned vines, the disease incidence is higher (30% clusters infected) than in free-positioned, topped vines (5% infected clusters). The difference is better marked under low disease pressure. Excessive nitrogen fertigation tends to promote succulent growth, which is associated with increased incidence of powdery mildew.

Chemical Control

The control of powdery mildew in commercial orchards is generally based on the use of different fungicides (Tab.19.1). Fungicidal control measures should start at an early stages of vine development and repeated at 7-21 days interval depending on the fungicide being used.

Dusting of vines with sulphur (300 mesh) had been an effective control method in the past and is still most extensively used chemical control measure. The first dusting should be done when new shoots are 7-15 cm long, second during .or just before blossoming and a third application can be made 40-50 days later. Sulphur is also applied as a wettable powder. In dry climates sulphur dust is preferred whereas in areas with plenty of rainfall wettable powder or flowable formulations are recommended for their retention qualities. The optimum temperature for sulphur activity is 25-30°C. Above 30°C, there is the risk of phytotoxicity. The activity of sulphur is reduced in humid air compared to dry air. Because sulphur has poor retention qualities, application schedules of 7-10 days are usually required. Pre-bud swell or dormant stage application of lime sulphur delays the development of epidemics.

However, sulphur can leave undesirable residues on table grapes and may taint wine if used within one month before harvest. In addition, sulphur adversely affects beneficial insects. In addition to sulphur dust (15 kg/ha), Sulfex (0.25%), Karathane (0.05%), Calixin (0. 05%), Topsin-M (0.1%), Thiovit (0.25%), all have been found effective against powdery mildew of grapevines.

In most grape-growing areas now, the standard approach is intensive use of sterol demethylation-inhibiting (DMI), quinone outside inhibitor (Qol) and quinoline fungicides. The sterol inhibiting triazole fungicides have proved more effective. Generally, Bayleton (triadimefon) is considered most effective triazole fungicide against grape powdery mildew fungus. Triadimefon at the rate of 40 g/100 lit. or penconazole or cyproconazole at the rate 40 ml/100 liter sprayed in mid-March, last week of April, and first week of May are highly effective against the disease. Fenarimol (Rubigan 12 EC), used in 7 sprays at 0. 05%, at 10 days interval between 15 cm cane length and 60 cm cane length is highly effective against powdery mildew. The fungicide is compatible with wide-spectrum protectants such as mancozeb and its residue persists for 14 days after the last spray but at a very low level. The dissipation rate is high during summer. Under dry temperate conditions of Himachal Pradesh (India) triadimefon (Bayleton), flutriazole (Impact), or hexaconazole (Anival) were more effective than carbendazim (Bavistin), dinocap (Karathane) and sulphur (Sulfex). Single application vapor action treatment with triadimefon, triadimenol (Bayton), flusilazole, mycobutanil and Penconazole.is also recommended at 0.2 to 4.0 g a.i. per vine between bloom and two weeks after shatter. The treatment gave effective control of the disease.

Cyprodinil and related anilopyrimidine fungicides are more recent compounds being used against powdery mildews. One advantage with the sterol-demethylation inhibiting triazole fungicides is that the interval between sprays can be increased and, thus, the number of sprays decreased. Dinocap (karathane) is used at 10-14 days interval while the triazoles can be used at 14-21 days interval.

Among the strobilurins, trifloxystrobin is very effective against conidial germination of U. necator. It shows strong prophylactic and local activity on leaves. Different strobilurin fungicides (azoxystrobin, kresoxim-methyl, Pyraclostrobin and trifloxystrobin) have different intrinsic activity against U. necator.

Polar, a polyoxin B compound, is highly effective against powdery mildew of grapevines.

Table 19.1: Fungicides used in grapes to control powdery mildew

Common name	Type	Disease spectrum	Movement in plant
Benomyl	Organic, benzimidazole compound	Powdery mildew and bunch rot	Systemic foliar
Sulfur	Inorganic, elemental Compound	Powdery mildew	Contact, vapor action
Azoxystrobin	Organic strobilurin compound	Powdery mildew	Systemic foliar with translaminar effect
Kresoxim methyl	Organic strobilurin Compound	Powdery mildew	Systemic foliar with translaminar effect
Trifloxystrobin	Organic strobilurin Compound	Powdery mildew	Systemic foliar with translaminar effect
Myclobutanil	Organic, DMI, triazole compound	Powdery mildew	Systemic protectant and curative
Iprodione	Organic, dicarboximide Compound	Bunch rot	Foliage-contact protectant and curative

Non-conventional chemicals

In order to reduce cost of repeated application of fungicides and to avoid resistance development in the pathogen, attempts to encourage the alternative methods of control of powdery mildew, non-conventional chemicals, plant products and many approaches including biological control have yielded encouraging results. Mineral oil (petroleum derived spray oils, PDSO) and plant oils have been used against grape powdery mildew with positive results. Mineral oils are often as effective as myclobutalin. Petroleum oil applied as emulsion (1% v/v) in water has provided moderate protection, excellent pre-lesion and post-lesion curative action and is antisporulant. Plant oils have also showed significant action in pre-lesion treatment and as antisporulants in treatments applied to established lesions. Generally, the recommendation is to use PDSO in early stages of disease incidence as a prophylactic spray to be followed by reduced number of fungicide sprays. Azam *et al.*, (1998) have reported that a rape oil derivative give good control of grape powdery mildew. Repeated sprays of the rape oil derivative at the rate of 2.0 and 0.5 ml per liter prevented the foliar symptoms as effectively as wettable sulphur.

Potassium bicarbonate, monopotassium phosphate and sodium bicarbonate are hazard free alternatives to fungicides. Spray of 1% sodium bicarbonate 3 times during the growing season, commencing after the appearance of first symptom gives good control of powdery mildew (Sawant and Sawant 2008). The control is as good as that given by a systemic fungicide. Alternating phosphate application with a systemic fungicide like penconazole or myclobutanil improves the control efficacy. The efficacy of these inducers of systemic acquired resistance (SAR) is increased by the addition of 0.1% sulphur.

Organic Measures

Spray of milk and whey is also reported to significantly reduce the mildew on leaves. The milk spray causes collapse of hyphae of the fungus and damage conidia within 24 hours. The effect is attributed to production of free radicals and lactoferrin (antimicrobial compound found in bovine milk). The latter ruptures conidia but has no effect on hyphae (Cisp *et al.*, 2006). Spray of soluble silicom (potassium silicate or metasilicate) inhibits formation of mildew colonies on the leaves. Sprays of aqueous extracts of compost have been used with positive results.

Biological control

In biological control, suppression of grapevine powdery mildew by the mycoparasite Ampelomyces quisqualis Ces. (syn. Cicinnobolus cesatiiDe Bary) is reported. However, under natural conditions, populations of A. quisqualis lag several weeks behind populations of powdery mildew fungi in their development and its colonies are seen late in the season allowing the mildew to develop to damaging levels. Sprays of concentrated conidial suspensions earlier in the season have given partial control of the powdery mildew of grapevines, particularly in wet seasons. The chitinolytic enzymes of Trichoderma harzianum and cells of biocontrol strains of Enterobacter cloacae have antifungal activity against U. necator. They suppress spore germination and germ tube elongation. Mixture of the enzymes and bacterial cells has synergistic effect.

English and Norton (2007) have reported 45% reduction in powdery mildew invasion of leaf area by applying the mite Orthotydeus Iambi. The mite becomes well established in the vines where it is released, however some cultivars of grape support high populations of the mite than others treatments where this mite is released in the vines are as effective as fungicide. Significantly better disease control is found in treatments with both mites and fungicides. The pathogen could develop cleistothecia only on leaves not inhabited by the mites. Application of myclobutanil and strobilurin fungicide azoxystrobin did not greatly affect abundance of the mites but mancozeb and wettable sulphur greatly reduced their population.

Populations of Bacillus, Pseudomonas Serraia, Penicillium, and Trichoderma,all are enhanced on the host surface and they provide biological regulation of the plant pathogens. Sendhilvel *et al*, (2007) reported effective control of powdery mildew on grape leaves by application of a talc- based formulation of Pseudomonas fluorescens strain Pf1 at 2%. Bacillus subtilis is an established biocontrol agent against powdery mildew. Spray of aqueous extract of dry mycelium of Penicillium chrysogenum induces systemic resistance to powdery mildew and downy mildew and gives as good control as copper and sulphur fungicides and the plant activators benzothiadiazole or acibenzolar-S-methyl.

Resistance Varieties

The species of the genus *vitis* and cultivars within the species differ in susceptibility to powdery mildew (Table 19.2.). Under dry temperate conditions of grape growing areas of Himachal *Pradesh, varieteies like* Redsultana, Saint George and No. 1613 are the highly resistant.

Powdery Mildew of Fruit Crops   281

Table 19.2: Susceptibility of some common vine to powdery mildew

Variety	Type	Susceptibility
Chardonnay	White wine	Susceptible
Gewurztraminer	White wine	Susceptible
Chancellor	White wine	Susceptible
Bacchus	White wine	Susceptible
Riesling	White wine	Susceptible
Cabernet Sauvignon	Red wine	Susceptible
Sauvignon blanc	White wine	Susceptible
Concord	Juice	Intermediate

*Source:*Management Guide for Grapes for Commercial Growers 2000-2001 Edition, British

Integrated disease management

Powdery mildew is most effectively managed by using the concepts of integrated Disease management (Gadoury, 1993). Biological control products should be integrated with chemical and cultural controls whenever the opportunity presents itself (Jacobsen and Backman, 1993). The integrated approach to disease control relies heavily on use of cultural practices that reduce excessive vine growth and spread of powdery mildew. Shoot removal and hedging to allow air movement within the vine canopy will slow the spread of powdery mildew. Fertilizer and water should be used judiciously to avoid excessive growth. An open canopy will not only maintains a microclimate less favorable for disease development but also allows better penetration of fungicide (Pearson, 1988). Dormant sprays of fixed copper products or lime sulfur may reduce overwintering inoculum of *U. necator* on grape vines and should be used where powdery mildew was severe in the previous growing season.

19.2. Powdery mildew of Citrus

Pathogen: *Acrosporium tingitaninum* (syn. *Oidium tingitaninum; Oidium citri)*

Geographic distribution

The disease is common in India and is serious on mandarin and sweet oranges in the sub-mountain tracts of Coorg, Nilgris, Pulney, Wynad and Shevaroy Hills in South India and in parts of Assam, Sikkim and West Bengal. Powdery mildew is common during the winter months and is seen in almost all the varieties.

Symptoms

All the aerial plant parts including young and actively growing leaves and twigs are affected. On the upper surface of the leaves, white powdery patches of fungal growth are seen (Fig 19.2). The patches extend and cover the whole leaf. Petioles and stems are also covered by whitish fungal growth. The affected leaves show yellowing and crinkling and they drop prematurely. In severe infections, young fruits show mildew growth and drop off.

Fig. 19.2: Symptoms of Powdery mildew on citrus leaves

Mode of spread and survival

The fungus spreads through wind-borne conidia. The water shoots under the canopy are first infected.

Epidemiology

The disease appears usually from October to March and in higher elevations. Damp mornings with few hours of sunshine are the most favourable conditions for the onset of disease. The disease is common and serious in mandarin and sweet oranges. The disease is noticed on lemon also.

Disease Management

The affected plant parts should be removed and destroyed carefully. Water shoots should be pruned to reduce infection. Dusting finely powdered sulphur gives effective control. Prophylactic spray with wettable sulphur 0.2 per cent or tridemorph 0.1 per cent or carbendazim 0.1 per cent controls the disease. Spraying with carbendazim 0.05 per cent at 20 days interval is also recommended.

19.3 Powdery mildew of Papaya

Pathogen: *Oidium caricae*

Geographic distribution: *Brazil, India*

Papaya powdery mildew is a disease of general occurrence, especially in very shady nurseries and in colder months of the year, as noticed in the regions north of Espiritio Santo and south of Bahia, Brazil,where it occures between the months of June and September. When it occurs with high severity, the disease can cause damage in the leaves, affecting photosynthesis and consequently the commercial quality of the fruits. In nursery seedling plants, a total loss of leaves may occur resulting in death of the plants seedlings.

Symptoms

Symptoms in the leaves cause a light green yellow discoloration of irregular outline and of dark green margins. White powdery like growth of the causal fungus develops on undersides of leaf. With progress of the disease, the discolored patches become covered with a powdery white mass made up of mycelium and conidia of the pathogen and appears on the lower and upper leaf surfaces (Fig 19.3). The leaves may become yellowish (chlorotic), with subsequent drying and leaf fall (Simone, 2002). With the exception of the younger leaves, all of the leaves may be affected by the fungus.The older leaves are more susceptible than younger one. Under severe infection, besides the leaves the pathogen can be observed on the stem, flowers, pedicels and fruits.

The fungus *O. papayae* causes symptoms similar to those incited by *O. caricae*. On the upper surface of the leaf the chlorotic areas that evolve to yellowish spots are observed. These are delimited by principle veins, and are round (or with irregular margins), of approximately 0.5 cm diameter and coalesce covering great foliar area. On the lower surface of the leaf a powdery mass of pale color, corresponding to the yellowish spots on the upper surface, is observed. In contrast to *O. caricae* these signs of the fungus occur normally on the lower surface of the leaf, and are rarely observed on the upper surface. Sometimes the white patches appear on the fruits. In the nurseries seedling plants are especially susceptible to attack and may be seriously affected. Young infected leaves of the seedlings dry up prematurely and drop down.

Fig. 19.3: Symptoms of Powdery mildew on Papaya leaf and fruit

Pathogen

Three species of Oidium have been reported and described as causing papaya powdery mildew. Oidium caricae (conidia elliptical, 24-30 μm x 17-19 μm), Oidium indicum Kamat (conidia barrel shaped, 31-47 μm x 12-33 μm) and Oidium caricae-papayae Yen (conidia 36– 44.4 μm X 15.6-21.6 μm). Another powdery mildew Ovulariopsis papayae van der Bijl. (teleomorph: Phyllactinia sp.), with conidia 14-23 μm x 60-90 μm, has also been described.

O. caricae is known in Brazil since 1898 and reported in other production areas of Papaya in the world.The mycelium is hyaline, septate, with haustoria developing in the interior of the host. The conidia are hyaline and granular, barrel shaped, and formed in chains of three to five or more spores.

Oidium papayae, observed in north Espirito Santo, Brazil, present erect, multiseptate conidiophores originating from cylindncal hyphae. The conidia are large, subclavate and isolated in the apex of the conidiophores.

Disease Cycle and Weather parameters

The disease occurs principally in the Colder and dry months of the year.The disease appears in the northern region of Espirito Santo, Brazil, in the month of May to September, when the average temperature varies from 21-24 °C and the relative humidity of the air is lower than 70 % and the weather is cloudy. For germination of the conidia a brief period of high relative humidity is required, where the presence of water in free state is not required. The great masses of spores produced on infected leaves are readily spread by wind current to healthy plants. Year round production of papaya permits uninterrupted reproduction of the fungus and continuous presence of the disease in an active state. The disease is more serious in orchards with system of drip irrigation, since sprinkle irrigation is unfavorable for the fungus.

Disease Management

The control of powdery mildew is achieved by the application of specific fungicides. The sprays should be applied when conditions are favorable for the occurrence of the disease, principally if these conditions occur for prolonged periods. The product most utilized is wettable sulfur or sulfur dust, applied at bi-weekly intervals after the start of the appearance of the first symptoms. When climatic conditions are highly favorable, the interval may be less, generally weekly. Sprays with wettable sulfur have not been effective when symptoms in the plants are severe, or that is, when the area of the leaf has signs of the fungus of more than 25%. This fact reinforces the importance of carrying out constant monitoring of the disease in the orchard. Sulfur application should be avoided in the hottest periods of the day (temperature greater than 24° C) to

Powdery Mildew of Fruit Crops 285

prevent phytotoxicity.Fungicides of the benzimidazole group, like thiophenate-methyl, benomyl and carbendazim, also have been utilized by producers to control powdery mildew in the northern region of Espirito Santo; however, they have not been effective.

In the nursery, the fungicide triflumizole (15g c.p./100 L), was effective for control of papaya powdery mildew. In the field, triflumizole was less effective, but not significantly different than sulfur (Tatagiba *et al.*, 1998b).

19. 4 Powdery Mildew of Mango

Pathogen: *Oidium mangiferae* Berthet

Berthet first recorded powdery mildew of mango in Brazil in 1914 and named it *Oidium mangiferae* Berthet. Wagle (1928) recorded the disease in India. Earlier; it was considered a disease of minor importance but now it is becoming increasingly important in most of the commercial mango growing countries of the world affecting almost all cultivars, either in severe or mild form (Prakash and Srivastava, 1987a, Prakash and Misra, 1993a,b, Prakash and Raoof, 1994, Prakash etal, 1996, Ploetz and Prakash, 1997). Although infection occurs both on foliage and blossom, the losses mainly occur as a result of blossom infection. In India, incidence of mildew has assumed such devastating proportions that the disease has become the single limiting factor in the expansion of mango cultivation.

Geographical distribution

The disease is known to cause extensive damage mainly up to latitude 40°NS of the Equator. It is reported from India, Myanmar, Bangladesh, Nepal, Pakistan, Sri Lanka, Israel, Lebanon (Asia), New South Wales (Zaire), Queensland and New Caledonia (Australia), Congo, Egypt, Ethiopia, Kenya, Malawi, Mozambique, Mauritius, Reunion, Tanzania, Zambia, Zimbabwe, South Africa (Africa), USA (California and Florida), Mexico, Jamaica, Costa Rica, Guatemala (Central America), Brazil, Venezuela, Colombia, Peru (South America (Prakash *et al.*, 1996). It has also been reported from Canary Island (Hernandez *et al.*, 1955) and Cuba (Padron, 1983). In India, the disease is present in almost all the states of country (Prakash and Raoof, 1994).

Losses

Dropping of unfertilized infected flowers and immature fruits causes serious losses. In Australia, Peterson *et al.*, (1991) reported the incidence of mildew up to 23% on unsprayed trees, whereas it was 11.5% in Mancozeb sprayed trees. Anonymous (1968) reported 20% decreasing yield in Venezuela. The disease can result in yield reduction upto 90% mainly due to its effect on fruit set and development (Schoeman *et al.*, 1995). In India, due to powdery mildew, the yield of mango decreases about 20% in some years (Anonymous, 1930). In one estimate, the loss varied from 22.35 to 90.41% (Prakash and Srivastava, 1987b, Prakash and Misra, 1993a, Prakash and Raoof, 1994).

Symptoms

The symptoms can be noticed on the inflorescence, stalk of the inflorescence, leaves and young fruits. (Fig.19.4). Inflorescences are susceptible during the period beginning when the main axis changes colour and ending at fruit set.

Symptoms on panicles

Infected panicles (flowers, flower stalks, and young fruits) become coated with the whitish powdery growth of the pathogen. Infected flowers and fruits eventually turn brown and dry. The affected flowers fail to open and may fall prematurely. The dead flowers can easily crumble in one's hand. Infection often causes flowers and small fruits to abort and fall off. Fruits that become infected after fruit set have purple-brown blotchy lesions that crack and form corky tissue as the fruitlet enlarges. The full-bloom stage is the most susceptible to infection.

Symptoms on leaves

On some cultivars, new flushes of growth and younger leaves are highly susceptible and may curl up and become distorted. Older leaves are more resistant to infection. Grayish, necrotic lesions or large, irregularly shaped spots may form on leaves .

On very susceptible cultivars, the youngest leaves may become completely covered with fungal spores and mycelium, and eventually die. On some cultivars, the whitish residue of the fungus tends to appear on the lower leaf surface, along the leaf midrib.

Symptoms on Fruits

Young fruits are covered entirely by the mildew. Fruit epidermis of the infected fruit cracks and corky tissues are formed . Purplish brown blotchy areas appear on the skin of older fruits. Recently, the disease has been noticed on young mango fruits in leading varieties *viz.* Dashehari, Kishan Bhog. Such fruits may remain on the tree until they reach up to pea size and then drop prematurely. In normal infection conditions 20-40% flowers and fruits are destroyed but during epidemic, it may result in complete failure of the mango crop.

Fig. 19.4: Powdery mildew symptoms on mango leaves and Panicles.

Pathogen

Powdery mildew of mango was attributed to Erysiphe cichoracearum (Wagle, 1928). On the basis of histological studies (globular haustoria) and the type of conidial germination, the pathogen was kept under E. polygoni group (Uppal, 1937). It was later observed that the pathogen produces saccate *or* lobed haustoria which *is* not charactenstic *of E. cichoracearum* (Uppal *et al.,* 1941). Since the description of the perfect stage *of* the pathogen has not been given (Uppal *et al.,* 1937) the name the stage *Oidium mangiferae* Berthet *is* preferred. Mango is the only known host of the mango powdery mildew pathogen (i.e., only mango can be infected by the fungus).

The morphology of Oidium mangiferae has been described in detail by Uppal *et al.* (1941). The fungus has septate mycelium, which remifies over the surface of the host, forming a white dense coating of branched hyphae measuring 4.1 – 8.2 µm.. From the superficial mycelium, numerous branches arise as conidiophores (pseudoidium type) with two to more basal cells straight ranging from 64-163 µm in length and at tips bear unicellular, hyaline, elliptical conidia having truncated ends. The conidia are produced in a basipetal fashion and are sometimes seen singly or in pairs of two. These are also produced in chains of 20-40 on detached leaves kept in close containers but the conidia easily fall off when mature. Conidia measure 25 to 48.9 µm in length and 16 to 23.9 µm in width, mostly 33 to 42.9 x 18 to 21.9 µm and germinate by a germ tube. The length of germ tube vary depending upon relative humidity, and they terminate in hook like appressoria (Ploetz and Prakash, 1997). Appressoria also develop *on* the underside *of* the hyphae and fix *it* firmly *to* the epidermal wall. Some-times, two appressoria may arise at the same point on the opposite side of a hypha. The hyphae in contact wi*th* the host form saccate haustoria *on* the under surface, which arise *as* slender tubes and pierce the cuticle and cell walls, these tubes then swell up inside the epidermal cells and form sac like structures which are used for the absorption of the nutrients from the host cells. The superficial mycelium form the conidiophores and conidia which gives powdery growth on the infected surfaces.

Perpetuation *of* pathogen

The pathogen does not infect hosts other than mango, *Mangifera indica* (Prakash and Srivastava, 1987, Gupta, 1989a). During off season, the pathogen remains present *in* intact green malformed panicles, mostly hidden under dense foliage. Datar (1985) studied the perpetuation of *O. mangiferae* on leaves and malformed inflorescence and reported that the conidia could not be located after August in malformed bunches, germination was highest in June, decreased *in* July and no germination occurred in August. Studies conducted at Lucknow (India) revealed that mildew pathogen persists on infected leaves of the previous year's flush which are retained on the plant in the succeeding year. During flowering (Jan.-March), the conducive environmental conditions activate the dormant mycelium already persisting *in* necrotic tissue *of* previous year's infected leaves. Abundant conidia are produced in the new flushes of growth or young flowers which provide inoculums for initiating the disease. Out break of disease is initiated either from inoculums harboured on the tree or by air borne conidia from other infection sites. Secondary infection within the tree is mainly due to air borne conidia.

Disease cycle

Wagle (1928) studied the life cycle of the causal pathogen. The fungus is disseminated through wind. The wind borne conidia cause infection after germination which takes 5-7 hours. The germ tube grows and within two days produces branched mycelium. The mycelium spreads profusely on the epidermal cells which are killed by the feeding of the fungus and become brown. On the fourth day, several vertical bodies begin to appear from the mycelium. These conidiophores give rise to conidia in five days. Thus, the life cycle from conidia to conidia is completed in about five days. The attack generally begins from the buds at the tip of the inflorescence, as these being more hairy, very easily catch the spores. Then it gradually extends on the flower-head, only few infected panicles are sufficient to cause a wide spread epidemic under favourable weather conditions. Mildew also appears naturally on the vegetative shoots of the mango but the life cycle is somewhat longer being about nine days. Conidia retain their vitality for 4-5 days only. If kept in the sun without any moisture, conidia shrivel within 4-5 hours. When conidia germinate at lower humidity, they often lose their ellipsoid shape and become rod shaped. Such conidia are always shorter.

Weather parameters

The disease is destructive in the plains of northern India and coastal areas during cold and wet season. The fungus is favoured by cloudy weather and heavy morning mist (Kulkarni, 1924, Prakash & Srivastava, 1987). The disease is most severe during cool, dry weather. Spores are wind disseminated and are released on a diurnal basis (Schoeman *et al*, 1995). Peak spore release, between 11.00-16.00 h, was positively correlated with hourly temperature and negatively correlated with hourly humidity, vapour pressure deficit and leaf wetness. Mildew pathogen persists for a longer period in UP hills. Minimum, optimum and maximum temperatures for germination are 9°, 22° and 30-32°C respectively (Uppal *et al.*, 1941). Conidia germinate best at temperature ranging from 9-32°C (23°C is optimal) and at relative humidities as low as 20%.

Gupta (1979, 1989a,b) found that the atmospheric temperature is important for the appearance and development of the disease. Minimum temperature (10-13°C) and maximum (27-31°C) and RH (82-91%) were most suitable for the development of the disease. He further emphasized that maximum infection occurs at 26-31°C and 100 % RH and rainfall does not play any significant role on disease but dry weather favours development of the disease. Conidia attached to the conidiophores on mildew leaves and inflorescences retain their viability up to 40 days as compared to detached conidia, which lasted 21-35 days on glass slide and host leaf surface. Mishra and Prakash (1988) stated that predominance of susceptible cultivar Dashehari in congenial environmental conditions such as wind velocity for 3-4 days with maximum temperature around 30°C and minimum temperature around 15°C, RH of minimum 23.4-25.5% and maximum 73.3-83.9 % is conducive for the rapid spread of mildew pathogen.

Optimal disease development occurs in the diurnal range of 10-31 °C and 60-90% relative humidity. Conidia germinate in the absence of water within 5-7 hr at 23°C and 20% RH. Since conidial germination occurs at a wide range of relative humidities, the development of mildew is usually independent of these weather parameters. Lonsdale and Kotze (1991) investigated the critical infection period for powdery mildew disease of mango. A sharp increase in infection occurred during the period July-Aug. for cv. Zill and between 3rd Aug-24th Aug. for cv. Keitt. During this period, flowers had begun to open on the panicles. Prior to flowers opening and during fruit set, little infection occurred. Conidia exhibited a diurnal pattern of dispersal and liberated from 12 to 16 hr. Dispersal took 3-4 days to reach pre-rain levels when dry weather followed by rainy period ,however rains reduced the dispersal of pathogen (Gupta, 1988).

Disease Management

Cultural

Choose hot, dry areas for mango cultivation; if possible, avoid areas that consistently have rain during the flowering season. Prune plants to keep flowers within the range of spray equipment and to improve air circulation in the canopy. Pick up fallen mango foliage and destroy it, remove severely infected panicles. Intercrop mango with other fruit trees or forestry species. Keep tall weeds away from mango trees. Prune unwanted or competing adjacent tree species, allowing more sunlight.

Plant nutrition practices

Foliar applications of phosphate fertilizer solutions (i.e., monopotassium phosphate, 0.5% KH_2PO_4), alone or in combination with conventional powdery mildew fungicides, is reported to provide economical and effective results in Israel and South Africa.

Chemical

Periodic spraying /dusting of mango trees with wettable Sulphur/Sulphur dust (Prakash and Raoof, 1994), Dinocap (Ruehle, 1956), Benomyl (Mc Millan, 1973) and Mancozeb (Persley *et al.*, 1989) have been found very effective. In Zimbabwe (Rhodesia), four sprays (when flower clusters have expanded, just before the clusters open) have been recommended (Hopkins, 1941). In India, the disease has been controlled by spraying of wettable Sulphur. Dinocap, Carbendazim, Benomyl, Tridemorph, Tridemephon, Bitertanol, Oxythioquinone. Thiophanate Methyl, Vigil etc. (Datar, 1981, Joshi and Chauhan, 1985). Spray schedule was tested at 3 stages (emergence of panicle, opening of flowers, and fruit set), and Bavistin (0.1 %) at 20 days interval was found quite effective in reducing the disease

The phytotoxic problem with Sulphur during hot conditions is reported in the dry tropics. Probably this may be due to high dose (4.5 g/1) applied over the mango tree. However, wettable sulphur fungicides were evaluated for their repeated sprays on the control of powdery mildew of mango. All the wettable sulphur fungicides tested, significantly controlled the disease without any phytotoxic effect, and reduces the cost of fungicides application (Prakash and Misra, 1986). Modern mildewcides that control powdery mildew, are very effective provided that spraying commences upon the appearance at the first sign of disease and that coverage is thorough. The flowering stage appears to be the most critical for infection, however, little infection occurs before flower opening or during fruit set. Therefore, spraying is utmost important only during flowering and fruit set. Few newer fungicides are Punch C (Flusilazole + Carbendazim) Topas (Periconazole), Afugon (Pyrazophose). In some regions, both mildew and hopper occur together. In such case a combined treatment of fungicide and insecticide have also been recommended (Pal and Prakash, 1984).

Varietals Resistance

Differences in cultivars susceptible to mildew are widely recognized by various workers (Gupta, 1976, Datar, 1983), but none of the cvs. has been found immune or potentially suitable for incorporating in the breeding programme. However, out of 90 mango varieties tested, only Neelum, Zardalu, Banagalora, Totapari Khurd and Janardan Pasand were found resistant (Gupta, 1976).

Datar (1983) reported cv. Totapari with some degree of resistance. Mango cvs. Ghanya, Jahangir, Yellamondela Thiamandi, Bablipunasa, Kharbuja and K.0.7 have also been reported resistant. Zill, Kent, Alphonso and Norn Doc Mai are very susceptible; Haden, Glenn Carrie and Keitt are moderately susceptible, and Sensation, Tommy Atkins and Kensington are slightly susceptible (Johnson, 1994). No differences in resistance were observed among rootstock tested. Grafts were more important for manifestation of resistance. Cultivars *viz.* Extrema Pahiri and Bour-bon were the most susceptible while Oliveira, Imperial and Carlota were more resistant (Simao and Gomes, 1995).

19.5. Powdery mildew of Jujube (Ber)

Pathogen: *Oidium erysiphoides* f.sp. *zizyphi*

This is one of the most predominant disease and causes significant losses due to its infec*tion* on fruits in india (Kapur *et al.,* 1975). It is widely prevalent and reported from Allahabad, Kanpur and Bombay. Disease is now known to appear on both cultivated and wild species. This disease is present in most of the grafted ber plantations causing heavy losses every year both qualitatively and quantitatively.

Symptoms

The disease generally appears by the end of October and prevails from November to April. With the rise in temperature after February, the disease subsides comparatively. The disease first appears on fruits in the form of white floury patches and later cover the whole fruit (Fig 19.5). With the passage of time, the infected area becomes slightly raised and rough.

The infected fruits often become misshapen and may shed from the trees. Besides fruits the symptoms have been observed on leaves and stem by Kumar *et al.,* (1978). The disease appears on developing young shoots in the form of white powdery mass on the leaves which results in shrinkage. Though, primarily the disease appears on flowers and fruits the germinating conidia of the causal fungus are also observed on leaf surfaces.

Fig. 19.5: Powdery mildew on ber leaves and fruits

Pathogen

The disease is caused by the fungus *Oidium erysiphoides* f.sp. *zizyphi* Yen and Wang, which survives in bud wood and on some collateral host during the absence of flowers and fruits and serve as primary source of infection. The secondary spread takes place by air borne spores. The fungal mycelium becomes external on the host. The conidiophores are upright, single, measuring 75.8-139.4x12.6 µm. Conidia are cylindrical, hyaline, catenulate measuring 25.2-37.8 x 16.8-21.0 µm. Powdery mildew of ber caused by *Microsphaera alphitoides* f.sp. *ziziphi* has also been reported by Mehta (1950).

Disease cycle and Weather parameters

The mycelium overwinter in the new shoots and arise annually. It may also survive on some alternate host during the absence of flower and fruits, which become primary source of infection. Air borne spores become secondary source of infection. *M. alphitoides* perpetuation through dormant mycelium in buds or cleistothecia have not been reported so far in India (Parkash *et al.*, 1988). No collateral host of *M. alphitoides* f.sp. *ziziphi* has been reported. Conidia of this fungus germinate and form appressoria after 2-4 hours at 20±2°C in most saturated atmosphere, whereas sporulation starts 96 hours after inoculation on susceptible ber leaves. Temperature range from 10-30°C and relative humidity levels of 32 per cent and above favour disease development. Singh *et al.*(1999) developed prediction model for powdery mildew disease build up based on weather parameters in Haryana. Maheshwari and Singh (1999) assessed the disease severity on fruits in relation to environment during 1991-93 at Kanpur, India.

Disease Management

Chemical

Sulphur dusting in December has been considered enough for satisfactory control by Mehta (1950). Several workers have reported the efficacy of dinocap, carbendazim, fenarimol and triademefon in reducing the disease (Gupta *et al.*, 1977, 1978, Reddy *et al.*, 1990; Das *et al.*, 1994; Singh *et al.*, 1995). Five sprays of dinocap one each from August to December at monthly interval is useful. Sulfex (0.2%), dinocap (0.05%) and carbendazim (0.05%) or wettable sulphur (0.2%) at ten days interval between the second week in October and December are also quite effective recommendations (Yadav *et al*, 1980). Benlate is reported to give 92 per cent control of the disease followed by Afugan (Kapur *et al*, 1975).

Disease can effectively be checked by spraying the fruits with carbendazim (0.1%) first when the fruits are of pea size stage followed by two or three sprays of dinocap (0.1%) at 10 to 15 days interval (Singh and Sidhu, 1985). Triadimefon, dinocap, sulphur and propiconazole have been reported much effective (Thind and Kaur, 1998). Of the six fungicides tested, triadimefon (0.1%) was most effective followed by thiophanate methyl (0.1 %) and dinocap (0.1%) and recorded highest yield (Reddy *et al.,*1997). Both floral and fruit infection can be significantly reduced by giving six sprays of wettable sulphur involving either 1 or 2 sprays of triadimefon 0.1% at the flowering stage (Desai, 1998a). Wettable sulphur is cheaper substitute for powdery mildew control. A schedule of 5 sprays involving first two fortnightly sprays with bayleton (0.1%) commencing from third week of September at flowering and early fruit set stages and remaining three fortnightly sprays with sulphur (0.2%) at pea size and marble size fruit stages was also found very effective in controlling the disease (Anonymous, 2000).

Varietal resistance

The disease can also be averted by growing vars. Safeda Rohtak, Sua, Noki, Chonchal, Sanaur-5, Kathaphal, Sanaur-1, Illachi-Jhajjar, Kakrola Gola, Kala Gora, Pathani and Mirchia, which have been reported resistant by several workers (Jeyarajan and Cheema, 1972). Five ber cultivars *viz.*, Guli, Seedless, Villaiti, Darakhi and Darakhi-2 were found resistant to powdery mildew under artificial epiphytotic conditions at Rahuri, Maharashtra, India (Anonymous, 1985). As such, these varieties may be taken up for cultivation to ward off powdery mildew.

19.6. Powdery Mildew of Apple
Pathogen: *Podosphaera leucotricha*

Powdery mildew occurs in all major apple growing regions of the world, but is especially serious in semiarid regions and in nursery production. Losses from the disease vary depending upon the inherent susceptibility of the cultivar, environmental conditions, and management practices. In New York, the disease can cause extensive foliar infections in dry years, following mild winters, or on highly susceptible cultivars. The disease can be particularly severe during orchard establishment because non-bearing trees continue producing susceptible tissue until late in summer.

History

Bessey in 1877 first reported this disease from Iowa, where it caused stunting of seedling nursery stock. It was reported from the eastern United States by Galloway in 1889 as a serious disease on young nursery trees. The fungus causing the disease was described and named by Ellis and Everhart in 1888 . In 1914, Ballard and Volck gave a full description of the disease as it occurred in the Pajaro Valley in California, where climatic conditions were such as to cause extensive injury to apple orchards in this region. A study on this mildew was made by Fisher as it occurred in the arid regions of the Pacific north west and contributed to the earlier work of Ballard and Volck on its control. While the apple mildew fungus was known to occur in European countries, little attention was given to it, probably because of its minor importance.

Geographic Distribution

Apple powdery mildew appears to be widely distributed over the world wherever apples are grown. It is reported from Japan, Australia, and New Zealand, as well as from most European countries. It is known to occur in South America and Africa also. The fungus has been found practically in all states in the United States and the provinces of southern Canada.

Economic Importance

Damage from mildew attack, results mainly in stunted growth. The foliage is dwarfed and twig growth is reduced. In the nursery, seedling stock as well as young nursery trees are reduced in size and quality so they are sold as "seconds." In the orchard, where the disease is severe, the shoots, blossoms, fruit pedicels, and the fruit may be severely damaged, but this is rare in most apple growing sections.

Symptoms

Powdery mildew infects young green tissues of the plant, as well as young blossoms. Symptoms of the disease are variable and dependent upon the variety, when infection occurred, degree of infection, and weather conditions.

On Leaves

Leaves are most susceptible to infection in the first few days after they open. Initial infections on the underside of the leaf may cause chlorotic (yellow) patches or spots to occur on the upper side of the leaf. This symptom is not unique to powdery mildew, so inspection of the underside of the leaf is necessary to confirm the presence of the fungus. Lesions on the upper leaf surface appear powdery white but eventually turn a darker brown. Infected leaves have a tendency to crinkle, curl, or roll upwards along the edges giving them a narrow appearance (Fig 19.6). Severely infected leaves usually drop prematurely during the summer. When very young leaves are infected, they tend to increase in length but not in width, stunted and become folded longitudinally. Infected foliage is rendered hard and brittle and frequently is killed. Under favorable conditions, the disease will spread over the entire leaf and progress down the petiole onto young, green shoots. Small black fruiting bodies (cleistothecia) form late in the season and are visible to the naked eye.

Fig. 19.6: Symptoms of powdery mildew on apple leaves

On Shoots

Shoot infections are the result of overwintering infections in dormant buds. When the terminal buds begin to grow in early spring, the fungus advances with the new succulent growth. Infected terminal shoots appear stunted and young shoots may be killed outright in the spring, or may survive the season and die in late fall or winter. The internodes of infected twigs are shortened, and the lateral buds, which tend to be elongated and purplish red in color, are bunched together. According to Baudys, severe infections of some varieties result in the production of witches'-broom.

On Blossoms

Blossoms are usually infected from the overwintering mycelium in dormant buds, and as a rule the entire cluster with attendant leaves soon becomes diseased (Fig 19.7). Floral parts are shriveled and blighted, so that no fruit is produced. Fortunately, infected blossoms and leaves do not bear the conidia early enough in the spring to make a general "blossom blight stage of the disease a likelihood.

Fig. 19.7: Symptoms of powdery mildew on apple blossoms

On Fruit

Fruits are often attacked shortly after the blossoming period, but such infection seldom persists after the apple skin hardens, as it usually does by mid-summer. Infection resulting from the germination of conidia are usually localized at the calyx end, whereas stem end infection generally originated from the margin of the pedicels. As a rule, infections of very young apples causes a stunting of their growth, alongwith the typical russeting which results from later fruit infection (Fig 19.8). The russeting is characterized by a network of lines, sometimes so closely woven as to give the appearance of a solid patch. The infected areas harden, and as a result, the growing fruit often cracks.

Pathogen

Powdery mildew of apple is caused by *Podosphaera leucotricha* (Ellis and Ever.) Salm. In eastern United States the

Fig. 19.8: Symptoms of powdery mildew on apple fruits

closely related mildew *P. oxyacanthae* (DC.) De Bary has been reported as the cause of a much less serious but similar disease.

Apple powdery mildew was first described by Bessey in 1877, and he named the organism as *Podosphaera kunzei.* In 1888 Ellis and Everhart described it as *Sphaerotheca leucotricha.* Burrill changed the name to *S. mali* (Duby), identifying the fungus with *Erysiphe mali* (Duby), which was later classified as *Phyllactinia corylea* (Pers.) Karst. The early confusion concerning the identity of the fungus was not cleared up until the work of Salmon was published in 1900. He definitely referred the fungus to *Podosphaera leucotricha.*Each conidiophore bears a chain of oval, hyaline conidia which measure 20-38 x 12-14 µm. They contain distinct fibrosin bodies.

P. leucotricha is heterothallic. The cleistithecia may form in autumn especially on 1-year-old shoots, on petioles and on midrib, but especially on sucker shoots. Up to 1100 cleistothecia per cm of such shoots have been recorded. The cleistothecia are globose, black, partially embedded in the mycelial web and measure 75-96 µm in diameter. They are densely gregarious. Two types of appendages are formed on the surface of these cleistothecia. Some are long, stiff, diverget and apically formed while others are basal, short, and tortuous and serve to anchor the cleistothecium to the substrate. The apical appendages, (usually 3-5, sometimes 11) are 3-7 times as long as the diameter of the cleistothecium. The top of the appendages are usually unbranched and blunt, rarely, they may be dichotomously branched once or twice. The basal appendages are rudimentary, pale brown, more or less tortuous and simple or irregularly branched. Each cleistothecium contains a single ascus measuring 55-70 x 44-50 µm. Eight ascospores, 20-26 x 12-14 µm in size, are produced in each ascus. cleistoithecial stage is rare in India. Physiological races of the pathogen are reported in Europe.

Hosts

Besides apple which is the most important host plant, *P. leucotricha,* attacks the pear and the quince. It has also been listed as a disease pathogen of cherry, plum, hawthorn, and service berry, but in the recent lists of diseases, *P. oxycanthae* is generally given as the cause of powdery mildew on these plants.

Disease Cycle

Powdery mildew survives the winter as fungal strands (mycelium) in vegetative or fruit buds that were infected the previous season. Low winter temperatures (< -11° F or -24°C) can kill the mycelium in the buds or the infected buds themselves, thus reducing this source of infection. Infected buds usually break dormancy later than healthy ones.

As infected buds break dormancy, the fungus resumes growth and colonizes developing shoots and young leaf tissue, causing primary infections. Primary mildew infections may occur on vegetative shoots and blossoms and thereby cause a reduction in yield.

Infected terminal shoots, or flag shoots, may have a silvery gray color, stunted growth, and a misshapen appearance. They are more susceptible to winter kill than healthy shoots. The powdery white appearance on infected shoots consists of many thousands of spores, called conidia, which are responsible for spreading the fungus and causing secondary infections. Secondary infections usually develop on leaves and buds before they harden off and may reduce the vigor of the tree. Secondary infections also result in the infected buds that carry the fungus through winter.

In late summer and fall, the powdery mildew fungus produces masses of small black structures, called cleistothecia on infected leaves and terminals. These spore-producing structures are another form in which the fungus survives the winter; however, they appear to play a limited role in the infection process.

Weather parameters associated with disease

Powdery mildew infections occur when the relative humidity (RH) is greater than 70%. Even on days when RH is low, infections may occur during night or early morning hours when RH usually rises. Infections can occur when the temperature lies between 10 to 25°C. The optimum temperature range for infection is between 19 to 22°C. Unlike other foliar diseases, leaf wetting is not a requirement for powdery mildew infection. Conidia will not germinate if immersed in water, although high RH is required for infection. Under optimum conditions, powdery mildew will be visible 48 hours after infections are initiated; new infections produce spores in about 5 days. Conidia do not germinate below 88.5% relative humidity. High atmospheric humidity is essential for penetration of the leaf by germ tubes of germinating conidia. Following penetration of the cuticle the hypha becomes thin and peg-like, penetrates the epidermal cell wall and forms a haustorium. The penetration of the cuticle is by action of enzymes. No appressorium formation has been reported. The infection process does not occur when the leaf surface is covered with a water film. Incubation period is usually short and several cycles are completed during the season. The ectophytic mycelium grows well at 20°C and is no responsive to humidity. The rate of development of young colonies depends more on temperature than on moisture stress.

Disease Forecasting

Xu (1999) developed a model to simulate powdery mildew epidemics. The model, named Podem™ (short for Podosphaera, East Mailing), has been incorporated into Adem™ (Apple Diseases, East Mailing), a more comprehensive forecaster for assisting growers in managing apple disease. Adem™ also contains forecasters for apple scab, fire blight, and Nectria fruit rot and canker. Podem simulates powdery mildew epidemics on vegetative shoots on a daily time step from vegetative bud break (assumed to occur 1 week before full bloom) through the end of shoot expansion. The model consists of a series of submodels to generate daily forecasts of the severity of new infections and the total amount of infectious disease. To do this, the model calculates the percentage of susceptible host tissue, the percentage of infectious disease, the latent period, the rate of infection, and uses a number of weather variables (temperature, relative humidity, and rainfall). Podem will also generate risks of infection based only on the weather factors i.e. vapor pressure deficit and temperature (current and past) and tree phenology. The model, although complicated in its construction, is simple to use because it has been programmed for use on a PC. Model validation occurred over the course of 4 years in two unsprayed research orchards and performed well under these conditions. As Xu points out, the model does not incorporate the effects of disease management practices on disease development and this limits Podem's use in commercial operations. Clearly, fungicides have a tremendous impact on many aspects of disease development and these must be considered when modeling the development of disease in commercial orchards. However, the Podem submodel that incorporates weather indices to predict the favorableness of the environment for powdery mildew development is still valuable, and this is the part of the model that has been incorporated into Adem.

Disease Management

Cultural practices

Pruning out shoot infections during the dormant season has not proven effective for eradicating over wintering inoculum and is not done in commercial operations.

Resistant Varieties

The use of resistant varieties is the most effective means for avoiding problems with powdery mildew. All commercial varieties of apple are moderately to highly susceptible to powdery mildew.The cultivar Jonathan is highly susceptible (Dar and Kaul, 1981) while cultivars Red Delicious and Golden Delicious are rated susceptible or resistant at some locations. A few apple cultivars with Vf resistance to scab are resistant to powdery mildew also. Cultivars of apple with commercial quality and resistance or immunity to scab, cedar apple rust, and resistance to powdery mildew are now becoming available in USA. Due to existence of races in the pathogen, resistance does not last long. *P12,* a major resistance gene originating from *Maluszumils* mainly used in apple breeding programmes for resistance to powdery mildew but some virulent strains of the fungus are reported to break the resistance (Caffier *et al,* 2005)

The cultivars Baldwin, Braeburn, Cortland, Crispin, Gala, Ginger Gold, Granny Smith, Idared, Jonagold, Jonathan, Monroe, Paulared, and Rome are particularly susceptible to powdery mildew.

Chemical

In commercial orchards, fungicides are almost always necessary to control the disease when powdery mildew susceptible varieties are grown. The major objectives of the spray program are to: 1) reduce the number of spores produced on newly infected tissues in the spring, 2) prevent secondary infections of new shoots, buds, and leaves during the growing season, and 3) prevent fruit infections. Spray programs should include a fungicide for powdery mildew beginning just before bloom and continuing at 7-10 day intervals until terminal buds have set and shoots are no longer producing new leaves that are susceptible to infection. Spray coverage is extremely important because infections can occur in the absence of rain and susceptible tissue not covered with the fungicide may become infected before fungicides can be redistributed by rainfall.

The old chemical control methods included the use of sulphur dust or spray of lime. Lime sulphur was recommended for spray according to following schedules:

1. At green tip stage (when buds are green) 1:5 dilution.

2. At the open cluster stage, 1:35 dilution.

3. At blossoming or full pink stage, 1:60 dilution

4. At petal fall stage when about half the petals have fallen, 1:100

More sprays could be given afterwards if necessary. Cupric hydroxide (125 g Cu/100 lit) reduces the disease but causes fruit russet. Addition of slaked lime (2.2 kg/100 lit) reduces fruit russet.

Fungicides in use as substitute for elemental sulphur are dinocap (Karathane), oxythioquinox, benomyl (Benlate), carbendazim (Bavistin), thiophanate methyl (Topsin-M) and the sterol biosynthesis inhibiting fungicides fenarimol

(Rubigan), triadimefon (Bayleton), bitertanol (Bacor), triforine (Cella, Funginex or Saprol), bupirimate (Nimrod), myclobutanil (Systhane), etaconazole (Vangard), difenoconazole (Score), penconazole (Topas), flusilazole, triflumazole and fluquinconazole. Tetraconazole (1 lit/ha), hexaconazole (0.28 lit/ha), triadimenol (0.18 lit/ha) and bitertanol (0.4 lit/ha) are reported to give good control of mildew on some apple cultivars.

The strobilurin fungicides are also highly effective against the powdery mildew. The compound CGA 279202 of this group at 5-7.5 g/100 L provides high level of protective and curative action. The fungicide inhibits mitochondrial respiration and strongly inhibits conidiophore and conidia formation, spore germination and germ tube elongation. It has high affinity for the waxy layers and stays as a protective reservoir on the leaf surface. It ensures retention against wash-off by rains. Trifloxystrobin, a later formulation of strobilurin fungicides, controls powdery mildews of apple, mango and nectarines trees. In field trials, 0.01-0.015% (v/v) concentration was superior to DMI or sulphur fungicides (Reuveni, 2000).

Combinations of SBI fungicides have also been used to enhance spectrum of disease control. Thus, a mixture of Bacor 25 WP (bitertanol) at 0.05% and Bayleton (triadimefon) at 0.025% gives good control of powdery mildew and scab both. Combination of the SBI fungicide fluquinconazole (Palisade 25 WP) and the anilopyrimidine pyrimethanil (Clarinet 200 SP or Vision) also gives control of both diseases.

In post-symptom activity, Bayleton reduces the number of normal conidia produced 10 days after treatment and etaconazole and sulphur give reductions 20 days after treatment. Four sprays of Bavistin (0.05%) or Morocide (0.1%) starting at bud swell stage and repeated at 15 days interval give control of the disease. Triadimefon (Bayleton) at 0.05% gives the best control by persisting on the host surface for up to 15 days. Baycor, Rubigan and Saprol give a better reduction of conidial production than Karathane. In another study maximum reduction in germination of spores and germ tube growth was obtained by triforine, followed by tridemorph, bitertanol and carbendazim. The best disease control in the field was given by bitertanol followed by carbendazin and triforine. Antisporulant activity of bitertanol and carbendazim was noticed for 21 days. In a comparative study of Bavistin (carbendazim), Bacor (bitertanol) and Karathane (dinocap) four sprays of Karathane or Bacor were found superior to Bavistin. Addition of one dormant or bud swell stage spray of Bavistin to the schedule recommended for apple scab control gives control of powdery mildew also.

The sterol biosynthesis inhibitors are the most advanced mildewcides in use. However, while resistance to benzimidazole fungicides in the fungus is reported, possibility of resistance to SBI fungicides has also been expressed (Singh, 2000). In addition to this problem, the Period for application of fungicides is also important for economy and effectiveness of the treatments. Winter application of any fungicide is considered useless since the fungus in the buds is protected by thick scale leaves. In highly susceptible cultivars the most suitable time for starting fungicide sprays is the tight cluster or pre-pink bud stage and continued till midsummer when terminal shoot growth ceases. Blossoms must be protected as early as pink bud stage to prevent fruit infection. Sometimes, fungicide sprays may be required in the late season to protect unseasonal flush of new growth. Triadimefon can be applied from bloom stage. Phytotoxicity of some fungicides during the growing season affecting pollen germination, or causing fruit russet is reported. The length of interval between two sprays is more important than the fungicide concentration. The control is enhanced when the interval is shortened rather than increasing concentration. Disease forecasting systems should be effective in optimizing timing of sprays. Podem- TM developed at East Malling in UK is one such model which simulates epidemic of secondary mildew on vegetative shoots at daily intervals. This model is incorporated in a commercial PC based software package called Adem TM which also contains models for forecast of scab, blight, Nectria fruit rot and bacterial canker on apple and pear.

Non conventional Chemical and Oils

Plant oils have been more effective against powdery mildew than against scab of apple .Oils of sunflower, olive, maize, soybean, and rapeseed have provided more than 99% control of *P. leucotricha* under controlled conditions when applied to foliage one day before or after inoculation. Mechanically emulsified rape oil is reported comparable to dinocap (Karathane) and gave 99% control when applied 1-7 days after inoculation. Control of powdery mildews *by* sunflower oil results mainly from the inhibition of conidial germination and suppression of mycelial growth of the pathogen. When acetic acid at 10-12 mg/L is applied to apple shoots known to contain infection of Podosphaera powdery mildew fails to appear on the shoots (Sholberg *et. al.,* 2005). Acetic acid treatment of root stock and scion shoot eliminated the microflora from the surface and ensures production of healthy nursery stock.

Biological Agents

The possibility of biological control of apple powdery mildew exists. *Ampelomyces quisqualis*, described in grapevine and cucurbit powdery mildews, is reported to overwinter in mildewed apple buds and its growth is favoured by wet weather. Szentivanyl and Kiss (2003) have studied the survival of *Ampelomyces* in apple and other hosts. On apple trees the mycoparasite overwintered as resting hyphae in the dried powdery mildew mycelia covering the shoots and in the parasitized ascostromata of *Podosphaera leucotricha* on the bark and the scales of the buds. Overwintered structures collected in spring when placed close to fresh powdery mildew colonies started the life cycle of the mycoparasite. The plant pathogen and the mycoparasite start their life cycle during or soon after bud burst but *Ampelomyces* can only slowly follow the spread of its mycohost on infected leaves. Epiphytic yeasts are potential biocontrol agents of *P. leucotricha.* Alaphilippe *et al.,*(2008) have reported an isolate (Y16) that controls not only powdery mildew but also suppresses *scab without inhibiting its conidial germination and leaf penetration.*

Pruning of dormant shoots infected with mildew in the previous season has been recommended as a means of reducing primary inoculums. Removal of these infected shoots on fungicide treated trees reduce secondary infection by nearly half.

19.7 Powdery Mildew of Strawberry

Pathogen: Sphaerotheca macularis

Powdery mildew occurs on strawberry plants in all areas of the world and is one of the earliest recorded diseases of strawberry. The disease was first described in the 1850s. Powdery mildew can affect flowers and fruit as well as leaves. It also can be a serious problem in plant nurseries. Occasionally a powdery or surface mildew causes some damage to strawberry plants, but the disease is of minor importance to the industry. It was first reported in England in 1854 and in the United States in 1886. Outbreaks have been reported at various times since that date in the northeastern and Pacific Coast states.

Symptoms

Symptoms of powdery mildew infection are very distinctive. White patches of mycelium develop on lower leaf surfaces, which may continue to develop to cover the entire surface. The leaf edges characteristically curl upwards exposing the undersurface. Mycelial growth on lower leaf surfaces may be accompanied by purple to reddish blotches. Flowers may get covered with mycelium and are either deformed or killed. Infection may also cause abnormal pollen production. Developing fruit in all stages may become infected; immature fruit become hard and fail to ripen and ripe fruit remain soft and pulpy. Profuse white, powdery sporulation may occur on leaf and fruit surfaces (Fig 19.9). Occasionally, small dark cleistothecia, or fruiting bodies, of the fungus may develop along with mycelium on leaves. The fruit often fails to color at maturity.The powdery appearance is caused by the production of chains of conidia on simple conidiophores.

Fig. 19.9: Symptoms of powery mildew on strawberry leaves and fruits respectively

Hosts

The fungus *Sphaerotheca macularis* (Wallr. ex Fries) W. B. Cooke occurs on a wide range of hosts including many common weeds.However it is probable that a biological form i.e. *Sphaerotheca maculata* f. sp. *fragariae*, is responsible for the strawberry mildew, since in the presence of abundant infections on the weeds the disease is not present on the neighboring strawberry plants. This fungus causes one of the most serious diseases of the hop. *S. macularis* has the usual characteristics of the Erysiphaceae.

Pathogen

The cleistothecia,a sexual fruiting bodies, are usually abundant on the lower leaf surface as small black dots. These have a few long, colorless, unbranched appendages. The cleistothecium contains a single ascus with eight large oval ascospores about 20 μm long which serve as a primary source of infection while conidia helps in spread of the disease and its intensities.

Disease Cycle and weather parameters

Sphaerotheca maculata f. sp.fragariae,is an obligate parasite and survives from season to season only within living tissues of its host (Peries 1961).The cleistothecia survive the winter and the liberated ascospores give rise to new infections in the spring.

Although powdery mildew is inhibited by rainy, wet conditions, it thrives under short day or low light intensities, and relatively cool temperatures (14-27 °C) accompanied by high relative humidity.

Disease Management

The first step in controlling powdery mildew is the use of resistant cultivars, which are adapted to most strawberry-growing areas of the world. Darrow et.al (1954) reported on the relative resistance of strawberry varieties to powdery mildew where mildew was more severe on the plants in "screenhouses" than in the field. Out of 41 varieties tested, 11 were free of mildew and 8 were very resistant. The varieties like Ambato Armore, and Royal Sovereign were killed . The remaining 19 showed slight to considerable leaf rolling and some necrosis.

Nnursery plant and fruit growers should establish plantings with disease free transplants, and fungicides must be applied at the first signs of infection. Fungicidal control is problematical in tunnel, glasshouse, and annual production systems where reentry and harvest limitations are imposed with use of most fungicides. In addition, in field plantings, the powdery mildew fungus has developed populations resistant to many formerly effective fungicides. Since the fungus grows within host tissues, only systemic fungicides will have eradicative effects. Protective fungicides will not eradicate infections, but may slow the progression of powdery mildew in a field and help prevent the occurrence of a potential epidemic. Fungicide coverage for control of flower and fruit infection should be applied as early as possible when flowering begins and should continue at regular intervals throughout the fruiting season in annual plantings and throughout the growing season in perennial plantings. Late season fungicide applications in perennial plantings may help reduce the amount of inoculum carryover from one year to the next.

19.8. Powdery Mildew of Gooseberry and Currant

Pathogen: *Sphaerotheca morsuvae*

This disease is referred to as "American powdery mildew" in order to distinguish it from the much less destructive European powdery mildew caused by a related species. While it is most important as a disease of gooseberry, it occurs in a milder form on red, white, and black currants.

History and Geographic Distribution

As the name implies, this mildew is of American origin, being native on wild species of gooseberry. It is believed to have been carried to southwest Russia in 1890. It was reported from Ireland in 1900, probably having been independently introduced from America. In Europe it soon became established at the beginning of the century and was apparently introduced into England from the Continent where an outbreak in 1906 was recorded.

American mildew is known to occur in most countries in the North Temperate Zone where species of Ribes are grown. It has not been reported from the South Temperate Zone.

Economic Importance

Most damaging to gooseberries in North America, this mildew is of minor importance except on certain susceptible varieties, especially those recently imported from Europe. In the British Isles and in the European Continent severe losses may occur. The thick, webby growth of the fungus occurs over the surface of the leaves and shoots, though superficial, greatly retards photosynthesis and thus stunts the plant. The developing fruit is also often covered with the mildew growth, resulting in dwarfing, roughening, and often cracking of the fruit rendering it worthless.

Symptoms

The lower parts of the bush usually show the first signs of infection. In May or June small superficial white patches first appear on the leaves, shoots, and berries. These spots soon enlarge and coalesce to form large patches . The leaves and tips of the shoots may become distorted. The infected patches take on a white, dusty appearance due to the production of great numbers of conidia. These patches later turn to a rusty-brown color as the superficial layer of mycelium ages. The affected fruit is discolored and has a roughened surface (Fig 19.10). The thick weft of brown mycelium remains on the shoots after the leaves are shed. The black cleistothecia may be seen as very small dots in the mycelial patches.

Fig. 19.10: Symptoms of powery mildew on gooseberry and red current

Pathogen

The imperfect or Oidium stage of the American mildew fungus Sphaerotheca morsuvae is responsible for the white powdery appearance of the patches. The mycelium is entirely superficial, but like other powdery mildews, the off-shoots from the hyphae penetrate the epidermal cells, forming haustoria, from which the fungus derives its nourishment. An unusually thick mat of brown mycelium is formed by this fungus later in the season. Arising from the mycelial mat are numerous, club-shaped conidiophores on which a chains of powdery conidia are borne. These conidia are oval to oblong, hyaline, and easily detached after maturity.

The cleistothecia are closed (cleistocarpic) as in other powdery mildews. They are dark walled, sub-globose, 76 to 100 μm in diameter. The appendages arising from the cleistothecium are few in number (sometimes absent), filamentous, flexuous, and with no distinct terminal branching. Only a single large ascus is produced in each cleistothecium. This contains eight ellipsoid hyaline ascospores measuring 12 to 15 x 20 to 25 μm.

Disease Cycle

Both conidia and ascospores may cause infection. In the spring the cleistothecia are ruptured by the swelling asci, and the ascospores are forcibly discharged. The cleistothecia are abundant on the pruned or broken twigs on the ground, and for this reason infection often appears first on the lower leaves and shoots. Conidia are soon formed on the initially infected spots and are carried by air currents and washing to expanding organs, including the fruit.

Gooseberry mildew is especially destructive in areas having cool, humid, and rainy periods during spring and early summer.

Disease Management

Considerable difficulty has been experienced in England and Ireland in checking this disease. Lime sulfur is effective, but some varieties are sulfur-sensitive. Dusting sulfur is fairly effective if applied often.

Spacing bushes to allow quick drying and cleaning out all debris from under the bushes will reduce chances of infection.

A number of resistant varieties have been listed from time to time, but some of these seem to be quite susceptible in some regions, probably because of biological strains of the fungus. In Austria the Hoflein variety is reported as being very resistant. In England Butler reports Lancer, Crown Bob, and White Smith as fairly resistant. Some resistant American varieties when grown in Ireland were found to be susceptible.

19.9. Powdery Mildew of Peach and Apricot

Pathogen: *Sphaerotheca pannosa* and *Podosphaera oxyacanthae*

Although a disease of major importance, powdery mildew of the peach and apricot may be a serious disease where weather conditions are favorable for infection. Fikry reports it as one of the most serious diseases of peach in Egypt, causing much reduction of fruit and such infected fruit produce has to be sold at a much lower price. It is often a serious problem in nurseries where the seedling stocks are often stunted when attacked early in the growing season. Yarwood reports a serious loss on apricots in California, as a result of rejection of marred fruit by the canners.

Symptoms

S. pannosa attacks leaves, young shoots, and fruit. The young leaves may become entirely coated with a thick layer of the mycelium, curled, and narrowed as they expand. On older leaves, white patches with radiating edges appears as a rule (Fig 19.11). The white, powdery layer may extend over the entire terminal portion of the growing shoots or individual patches as on the older leaves may appear. On the fruit the disease first appears in the form of white round spots which increase in size until a large portion or the whole surface of the fruit is involved. The fruit takes on a pinkish color, which later turns to a dark brown. The epicarp of the fruit becomes leathery and hard, sometimes cracking. If it does not fall off, it serves as endered worthless. This serve as a foci where powdery mildew thrives, as in Egypt, but it is not so common on fruit in most regions. On the apricot fruit, the fungus produces small necrotic areas, which give the mature fruit a "dirty" appearance (Fig 19.12).

Fig. 19.11: Symptoms of powdery mildew on peach leaves and fruit respectively

Fig. 19.12: Symptoms of powdery mildew on apricot leaves and fruit respectively

Pathogen

Two species of the powdery mildew are reported. The destructive one is Sphaerotheca pannosa (Wallr.) Lev., ordinarily known as the "rose mildew." The other one is Podosphaera oxyacanthae, the common species on cherry also recorded as occurring on peach but is much less common and does little damage.

The white, powdery appearance is caused by the great numbers of conidia produced in chains arising from the wefts of hyphae on the leaf surface. These are produced rather early in the season and give rise to secondary infections. Older leaves and fruit are fairly resistant to infection. The perfect stage is rarely found on peach and apricot but is common on such hosts as the rose. The cleistothecia have hairlike, unbranched flexuous appendages and they contain a single ascus with eight asco-spores.

There is some question as to whether the fungus on peach is properly referred to as *S. pannosa*. Some authors regard it as a special strain of this species. However, Yarwood in California submitted evidence that infections on apricots were from mildewed roses in the neighborhood of the apricot trees and that when these were removed the mildew on the apricots did not appear. He was able to infect apricot fruit with conidia from rose. He pointed out, however that there were probably four strains of this mildew giving different reactions on apricot. One strain attacks both peach and apricot, while one from rose does not attack apricot at all. Since it is known that peach mildew can appear in the absence of the disease on roses, there must be other explanations for initial infection. It may be that the fungus is able to live over in the peach buds, as is the case with the powdery mildew of apple, but there is no evidence on this point.

Disease Cycle

This fungus overwinters as mycelia inside the budscales and the primary infection occurs as leaves emerge from these infected buds. Secondary infections occur when conidia produced by primary infections infects the host and subsequent infections occurs when conidia are blown or splashed by rain onto susceptible host tissues. Fruit (before pit hardening) and succulent terminal growth are susceptible to infection.

Powdery mildew is common under similar relative humidity and temperatures as cherry powdery mildew.

Disease Management

Cultural practices

Removal of infected fruit and pruning out terminals with infected buds during normal orchard operations will reduce the amount of infection within the orchard.

Resistant Varieties

Planting resistant cultivars such as Angelis, Walton, Johnson, Halford, and Stuart will further reduce mildew within the orchard.

Chemical

Powdery mildew of peach can be controlled by applications of sulfur fungicides, but where regular sprays are applied for other peach diseases it is rarely necessary to make any additional applications for this disease. Chemical sprays for control are suggested to start at petal fall or shuck split and continue every 7-14 days until terminal growth ceases. The other fungicides used to control this disease are Rally40WP,Orbit, Microthiol Special, JMS Stylet Oil, Amicarb 100, Funginex 1.6EC and Sulfur 92WP.

19. 10. Powdery Mildew of Cherry

Pathogen: Podosphaera oxyacanthae

Powdery mildew is not an important disease, although at times serious outbreaks occur in the nursery planting and seriously stunt the terminal growth. In general, budded sour cherry stocks are most severely attacked, but Mazzard stocks and sweet cherry varieties may suffer some damage.

Symptoms

The fungus attacks young twigs and leaves. The general effect is upward curling of the leaves, with the terminal leaves decreasing in size and a shortening of twig growth. The mycelium grows rather evenly over the surface of the leaf and shoot but sometimes appears in patches. With the production of the conidia on this mycelium the characteristic powdery coating results (Fig 19.13).

Fig. 19.13: Symptoms of powdery mildew on cherry fruits and leaf respectively

Pathogen

The disease is caused by *Podosphaera oxyacanthae* (D. C.) De Bary, one of the common species of the powdery mildew group of fungi. This same species has been reported from cultivated friut plants such as plum, peach, apricot, apple, pear, quince, hawthorn, serviceberry, spirea, and persimmon. The *Podosphaera clandestine* is also reported to cause powdery mildew in cherry.

Disease Cycle

This fungus overwinters as cleistothecia. The cleistothecia drop to the orchard floor and tree crotches or become trapped in bark crevices. They are about the size of the point of a pin and, therefore, not easy to see. Spores released from the cleistothecia in the spring are spread by rain or irrigation to young leaves. The earliest infections are found on leaves of suckers or succulent terminal growth near the crotches. These infections produce conidia in repeated cycles during the summer, resulting in the powdery appearance of infected leaves. Secondary infections from these conidia occur throughout the season when weather conditions are favorable. Later in the season numerous black cleisto *thecia* appear, especially on the under-side of the leaf. The cleistothecia have appendages dichotomously branched and recurved at the tip. Each cleistothecium contains a single *ascus* with eight *ascospores.* These cleistothecia carry the fungus over winter on the fallen leaves, and the ascospores are released to bring about initial infection in the spring.

Like with the most powdery mildews, hot dry weather with sufficient moisture in the form of high humidity, fogs, dews, or intermittent rains, to permit the germination of the spores, seems to be the best environment for the development of the disease. Powdery mildew is most common when the relative humidity exceeds 90 percent and day time temperatures are between 10 -25.5°C although some infections can occur when humidity is quite low. Long periods of rain are not necessary for infections since the spores will not germinate in free water.

Disease Management

Where powdery mildew is a serious disease, as in some nurseries, control can be obtained by rather frequent applications of sulfur dust or liquid sprays of wettable sulfur. Where sprays are applied for the control of cherry leaf spot, this disease is not likely to appear.

Tart cherries must be treated routinely; control is not usually necessary in sweet cherries. Removal of suckers in the center of the tree may eliminate a site for early infections. Fungicides are not necessary early in the spring because the primary infections arise from the cleistothecia, which germinate in late spring.

The first spray is recommended when very subtle mildew lesions develop on leaves. These lesions are difficult to detect at first but ultimately develop into a cottony growth of mycelium. Control will be poor if spraying is delayed until the mildew is obvious. The recommended date to initiate spraying is available from county agents or agricultural experts of the area. Repeat sprays every two weeks with most fungicides except sulfur which requires applications every five to seven days. It is usually necessary to make two to three applications per year. Fungicide sprays are crucial in young orchards, in orchards with vigorous growth, or in orchards with poor air circulation. In older orchards, it may be possible to achieve good control with one to two applications.

19.11. Powdery Mildew of Plum

Pathogen

Sphaerotheca pannosa and Podosphaera tridactyla (Podosphaera oxyacanthae)

Symptoms

Areas of white powdery fungal growth, roughly circular in shape, develop on the fruit in spring (Fig 19.14). These infected areas later become scabby and dry. In late summer and fall, similar fungal growth appears on leaves. Occasionally, symptoms may develop on fruit and leaves in spring.

Fig. 19.14: Symptoms of powdery mildew on plum fruit

Disease Cycle and weather associated

Sphaerotheca pannosa attacks the plum fruit whereas *Podosphaera tridactyla* attacks the foliage. An unidentified species, possibly of *Podosphaera*, attacks fruit and leaves of certain plum varieties like Red Beaut and Black Beaut in spring; other varieties may be affected in some years as well.

Sphaerotheca pannosa is known to survive as mycelium on roses and in infected buds of peach trees, and these plants may serve as a source of inoculum for plum trees. This pathogen is not known to overwinter on plum, but recently cleistothecia were discovered on peach trees, which suggest that this pathogen may also produce cleistothecia and survive on plum trees.

Podosphaera tridactyla overwinters as special spore-forming structures known as cleistothecia on the surface of shoots, on dead leaves on the orchard floor, and on bark. Spores are produced from these structures during spring rains, and they infect the developing foliage on plum trees.

Growth of the pathogen is favored by cool, moist nights and warm days.

Disease Management

Cultural operation

Watching for the disease during routine monitoring helps to determine the need for possible action the following year, but by the time it appears on the fruit it is too late to spray during the current season. If there are roses infected with powdery mildew near the orchard, these bushes are potential sources of inoculum, and it may be beneficial to control the disease on the roses or to remove them.

Chemical Control

Apply a fungicide at full bloom and make additional applications on a 10 to 14day interval as needed. The fruit is thought to be resistant to infection after pit hardening. It is important to alternate fungicides of a different chemistry to prevent the development of resistance.

19. 12. Powdery Mildew on Blueberries
Pathogen: *Microsphaera alni*

Powdery mildew is the most common and most widespread disease of blueberries, occurring from Maine to Florida and westward wherever blueberries are found. Usually it is not a destructive disease, but in hot and dry seasons which is favorable for its development, causes extensive defoliation.

Symptoms

Injury to the diseased plants comes from the devitalizing effect of the fungus on the leaf, often resulting in defoliation by midseason. The fungus shows considerable variation in symptoms on different varieties. On some, the fungus is confined to the upper surface, forming a compact layer or patches of mycelium (Fig 19.15). On others, the hyphae are inconspicuous and confined to the undersurface of the leaf, with reddish areas where the fungus is attached.

Fig. 19.15: Symptoms of powdery mildew on blueberry leaves and fruits

Pathogen

The fungus causing the disease is Microsphaera alni, one of the surface mildews, which appears on a great variety of unrelated plants. The "oidium" (imperfect) stage forms chains of oval, hyaline conidia on separate conidiophores, thus giving the powdery appearance to the mildewed leaf. The small black cleistothecia, barely visible to the naked eye, may be seen scattered over the weft of mycelium later in the season. They are without ostioles (cleistocarpic) and the appendages are dichotomously branched and recurved at the tips. The asci number two to eight, and each ascus contains four to eight ascospores.

Disease Management

Variation in susceptibility of cultivated varieties is marked. Bergman (1939) reported that in Massachusetts, Pioneer, Cabot, and Wareham were most susceptible; Concord, Jersey, and Rubel were intermediate while Stanley, Rancocas, Harding, and Katherine were highly resistant. Sulphur dust is recommended for the disease control.

19.13. Powdery mildew of Stone Fruit (*Almond and others*)

Pathogen: *Podosphaera* spp, *Phyllactinia and Uncinula* spp

The disease is very serious in nurseries where seedling stocks infected early in the growing season remain stunted. Khan *et al*. (1975), Sharma (1985). Kaul (1967) and Pandotra *et al* (1968) reported occurrence of powdery mildew on almond and plum due to Phyllactinia Lev. and Uncinula Lev.

Symptoms

The fungus infects leaves, young shoots and fruits. On the young leaves the disease appears at first as blister like areas that soon become covered with grayish white patches of fungus growth. Infection causes leaf distortion, curling and premature leaf fall. White patches of fungus appear on green shoots, which may become curved at the tips. Buds and flowers may also be directly attacked. In that case buds either fail to open or open improperly while flowers are discoloured, stunted, and eventually dry up. Infected fruits first show circular spots which may spread over the whole fruits (Fig 19.16). The colour of affected fruits is at first pinkish and later becomes dark brown. Epicarp of the fruit becomes leathery and hard.

Fig. 19.16: Smptoms of powdery mildew on almond fruits and leaves respectively

Pathogen

Different species of Podosphaera Kunze ex Lev. Viz. P. clandestina Wallr. Ex Fr, P. leucotricha Ell & Ev., P. oxycanthae and P. tridactyla (Wallr.) de Bary and Sphaerotheca pannosa Wallr. & Lev. causes powdery mildew in almond, and other stone fruits particularly apricot, peach, plum and cherry.

Disease cycle and weather parameters

The fungus over winters as mycelium in the buds. Cleistothecia are only rarely found on peach and apricot but are common on rose. Primary inoculum is therefore, represented by conidia produced by the overwintering mycelium, and sometimes by ascospores.

Conidia and ascospores are carried by wind to green tissues where they germinate, each producing a germtube which produces a fine hypha. This hypha penetrates the host directly through the cuticle and when it reaches the cell lumen of an epidermal cell it forms a globose haustorium by which the fungus obtains its nutrients. The aerial mycelium on the other hand produces short conidiophores each bearing a chain of egg shaped conidia. These conidia are dispersed by wind and infect expanding leaves, young shoots and fruits until they are 2-3 cm in diameter. The optimum temperature for conidia germination lies between 21 and 27°C (Weinhold, 1961).

Disease Management

The disease can be effectively controlled by three sprays of wettable sulphur or carbendazim/thiophanate methyl before opening of blooms, at petal fall and two weeks later (Anonymous, 1995). New fungicides viz myclobutanil, fenarimol, flusilazole, pynfenox, triademefon, captafol and tebuconazole provide good control of powdery mildew in stone fruits

19.14. *Powdery Mildew* of Mulberry (*Morus* species)

Pathogen: *Phyllactinia corylea* (Pers.) Karst

Powdery mildew is common in all the mulberry areas of the India and causes considerable damage to mulberry plants. (Gangwar *et al.,*1994; Biswas *et al.,*1995b). Leaf yield loss to the extent of 12.1-32.5% have been reported from Karnataka during 1995-96 and 24.02% coefficient of disease index in West Bengal. (Srikantaswamy *et al.,*1998). The disease cause severe damage during rainy and winter season in Karnataka, India.

Symptoms

White powdery patches on the abaxial surface of leaves making them unsuitable for feeding to silkworms. Initially, white powdery patches appears on lower surface of leaves which later on cover the entire leaf surface (Fig 19.17) and in due course turn black to brown in colour. Infected leaves turn yellow and fall off.

Fig. 19.17: Symptoms of powdery mildew on leaves of mulberry.

Pathogen

The disease is caused by fungus *Phyllactinia corylea* (Pers.) Karst. The conidia of the pathogen has a typical appearance particularly the surface of the clavate, hyaline conidium is ornamental with evenly distributed spine like protrusions (Kumar *et al*.,1998). The conidia germinate 4 h after inoculation producing a single germ tube mostly from a little behind the distal end of conidium. The hyphae soon produce special branches, stomatopodia, which enter leaf through stomata. The stomatopodia are produced singly or in pairs, and 1 or 2 stomatopodia enter in a stoma. The branched superficial hyphae form a mycelial mass within 72 h after inoculation. The conidiophores are straight at early stages, but appear spirally coiled when the conidia are mature, which may aid for conidial detachment. The disease becomes apparent at the conidial stage.

Weather parameters

High humidity (>70%) and low temperature (24-26°C) favour outbreak of the disease. Disease assessment key for powdery mildew of mulberry that reduce leaf yield by 10-30% during rainy season in Mysore were prepared which will allow the precise quick and consisted assessment of disease in the field. Vidyasagar and Rajasab (1997) indicated that dark brown spots caused by *Phoma mororum* on mulberry leaves occurred only when the leaves were also infected by *P. corylea*. In Darjeeling hills, during 1991-92 maximum disease intensity occurred in August-September followed by May-July with lowest in spring (March-April). Disease intensity increases with increasing temperature and humidity.

Disease Management

Chemical

The disease control is not practically advisable using Bordeaux mixture or lime sulphur since the fungicide affect the health of silkworms. However, where rearing of silkworm is not the consideration, use of fungicides known to be effective against powdery mildew can be resorted too. Of these, Bavistin 0.2% can be successfully used. Bavistin applied once as a spray at 0.15%, 35-40 days after pruning appeared most effective. Two sprays of dinocarp or carbendazim were found most effective and residual toxicity of these fungicides on silkworms and cocoons after 3,6 and 9 days of spraying had no adverse effect.

Mycorrhiza

Disease incidence of powdery mildew, rust, leaf spot, bacterial blight and blight was reduced in plants inoculated with VA-mycorrhizal fungi, *Glomus fasciculatum* and *G. mosseae* in combination with 60 or 90 kg P/ha/year. Both fungi are equally effective for reducing disease incidence.

Botanical pesticides

Extracts of *Adhatodo zeylanica* was most effective in reducing the disease followed by extracts of *Azadirachta indica, Launaea coromandelica* and *Oxalis corniculata* (Biswas*et al*, 1995a). Extracts of plants leaves of *Calotropis jigantea, Ocimum sanctum* (*O.tenuiflorum)* and *Tageles patula,* onion bulbs and ginger rhizomes can be used without causing any adverse effects.

Resistant Varieties

Among the 10 varieties screened, in West Bengal, India, S-799 is most tolerant followed by C-1729 variety. Among the 58 genotypes of mulberry screened 19 genotypes *viz.,* Chinese white, Ichinose, sanish 5, K.N.G.,Tsukasakawa, Limoncina, Serpentina, Italian sarnal, Brantul Kashmir, Zagatul,Chattatul, Botatul, Zanzabud, Lajward, Nadigam, Kokuso 21 Kairyorosa, Kasuga were field resistant to powdery mildew.

19. 15. Powdery Mildew of Passion Fruit

Pathogen: *Oidium passiflorae,*

The occurrence of the disease was recorded on leaves of passion fruit by Sundaram (1961).

Symptoms

The affected leaves show irregular, pale yellow discolouration on the upper surface and whitish growth of the fungus on the lower leaves (19.18). The disease is incited by *Oidium passiflorae, Syn: Leveillula taurica* (Lev.) Arn.

Fig. 19.18: Powdery mildew on passion fruit vine leaves.

Control Measures

Destruction of infected plant parts and spraying of plantation with sulphur fungicide help to reduce the powdery mildew on passion fruit.

19.16. Powdery mildew of Pomegranate

Pathogen: *Erysiphe* sp.

Geographical distribution

Powdery mildew of pomegranate was reported as early as 1987 in Azerbaijan and *Erysiphe. punicae* was reported as its causal agent (Akhundov 1987). It was also reported from Iran (Khodaparast and Abbasi 2009), and Montenegro. A namorphs of powdery mildews on pomegranate were also recorded from Ethiopia, Greece, India, Iraq, Ukraine, and Crimea but it is unclear if they belong to *E. punicae* (Braun and Cook 2012).

In June 2015, a severe outbreak of powdery mildew was observed on 70% of 0.45 million of 2-year-old pomegranate plants cvs. Wonderful One and Ako in a nursery located in Lecce Province, Italy. (Pollastro *et al.,* 2016).

Symptoms

White, dense plaques of mycelium and conidia appears on young leaves covering upto 30 to 40% of the leaf area (Fig.19.19). At later stages of disease development, leaves turned yellow and abscised. No chasmothecia is detected.

Fig. 19.19: Dull powdery growth patches on pomegranate leaves.

Pathogen

The conidiophore foot cell was cylindrical and the appressorium lobed. The conidia were hyaline, ellipsoid to cylindrical, measuring 28.1 to 35.7 × 12.9 to 14.4 µm (average 32.8 × 13.4 µm), fibrosin bodies were absent, and the germ tube was subterminal.

ITS genomic regions were sequenced for molecular identification where BLASTn analysis of the ITS sequence (563 bp) showed a high homology (identity: 99%; e-value: 0.0; coverage: 99%) with species of the *Erysiphe aquilegiae* clade (i.e., *E. aquilegiae* [EU047570.1], *E. sedi* [JX173288.1], *Pseudoidium hortensiae* [JQ669944.1], and *P. neolycopersici* [GU358451.1]) (Takamatsu *et al.* 2015).

Under artificial inoculations, typical powdery mildew symptoms developed within 7 to 10 days only on inoculated leaves. Therefore, on the basis of morphological features, the pathogen was tentatively identified as *Erysiphe* sp. and, according to the BLASTn results, it appears to belong to the still unresolved *E. aquilegiae* clade.

Control Measures

Due to the poor availability of fungicides allowed on the crop, improved disease management methods are needed to prevent heavy yield losses.

19.17. Powdery mildew on Watermelon

Pathogen: Podosphaera xanthii f. sp. *citrullus*

Geographical distribution

Generally watermelon was considered to be an unimportant host for powdery mildew in many areas. However, this disease had been described as a serious problem in Australia, Egypt, India, Japan, USSR, and the Philippines prior to 1970 *(Robinson and* Provvidenti, 1975). Watermelon *more* commonly remain symptomless while nearby planting of other Cucurbits were affected. Sudden occurrence of powdery mildew on all cultivars and breeding lines of watermelon in a new area, as occurred in Sudan and the United States, suggests introduction or evolution of a new strain or race. Powdery mildew on watermelon is now considered as new emerging disease.

Symptoms

Several different types of symptoms are observed. Symptoms on Watermelon can be difficult to recognize as sporulation often is very sparse in sharp contrast with other hosts. Signs of the fungus can be invisible to the unaided eye, thus the disease may go undetected until chlorotic spots develop and leaves begin to senesce prematurely. Additionally, powdery mildew can begin in discrete foci in watermelon fields, but then the pathogen can spread quickly throughout the field like other powdery mildew fungi. Elsewhere, typical powdery fungal growth characteristic of mildews develops on petioles and stems. This typical powdery fungal growth has been observed on both leaf surfaces in other areas (Davis *et al,* 2001). Young watermelon fruit can be infected as well (Fig 19.20) (Cohen *et al.*, 2000).

Fig. 19.20: Symptoms of powdery mildew on watermelon leaves and fruits respectively

Pathogen

The causal pathogen is suggested to be *Podosphaera xanthii* f. sp. *citrullus* based on results from cross-infectivity stud-ies conducted in Israel in which conidia from watermelon were able to infect some other Cucurbits but Conidia from other Cucurbits were unable to infect watermelon (Cohen *et al.,* 2000). All isolates from watermelon in the United States are race 2 of *Podosphaera xanthi* (Davis *et al.,* 2001). However, they appear to be a different strain as they are more aggressive than other race 2 isolates (Davis *et al.*, 2001). Some race 1 isolates from muskmelon are able to infect watermelon under controlled conditions whereas no isolates from pumpkin or squash are virulent on watermelon (Shishkoff and McGratn, 2001).

Disease Management

Elasol (0.5%) is used as a substitute of elemental sulphur. Other effective fungicides are Sulfex (0.2%), Calixin (0.1%), Karathane (0.05-0.2%). Bavistin (0.1%), Mildex, Ovatram, etc. Use of Sulfex is cheaper. One to two sprays of Calixin or 2-3 sprays of Karathane are required. Strobilurin fungicides, azoxystrobin (Quadns), kresoxim methyl (Flint) and trifloxistrobin (Stroby), are also effective against powdery mildew of water melon.

20

Powdery Mildew of Spices Crop

Spice can be a dried seed, fruit, root, bark, or vegetative substance primarily used for flavoring, coloring or preserving food or sometimes to mask other flavors in the food items. These may have other uses, including medicinal, religious ritual, cosmetics or perfume production, or as a vegetable. Spices are distinguished from herbs, which are parts of leafy green plants used for flavoring or as a garnish. Many spices have antimicrobial properties.

The spice trade developed throughout South Asia and Middle East in around 2000 BC with cinnamon and pepper and in East Asia with herbs and pepper. Generally, early Egyptian, Chinese, Indian, and Mesopotamian sources do refer to known spices. In the production, India contributes 70% of global spice production. Different spices are affected by powdery mildew diseases which hamper its quality and production. Some of the spices which are affected by powdery mildew disease are dealt in this chapter.

20.1. Powdery mildew of Black pepper

Pathogen: *Leveillula taurica* (imperfect stage = *Oidiopsis taurica)*

Symptoms

Powdery mildew primarily affects leaves on pepper plants. Although the disease commonly occurs on older leaves just before or at fruit set, it can develop at any stage of crop development. Symptoms include patchy, white, powdery growth that enlarges and coalesces to cover the entire lower leaf surface (Fig.20.1). At times the powdery growth is present on the upper leaf surface as well. Leaves with mildew growing on the undersurface may show a patchy yellowish or brownish discoloration on the upper surface. The edges of infected leaves may roll upwards exposing the white, powdery fungal growth. Diseased leaves drop from the plants and leave the fruit exposed to the sun, which may result in sun burning.

Fig. 20.1: Powdery mildew on black paper leaves.

Disease Cycle

The powdery mildew disease cycle starts with the landing of fungal spores (known as conidia) on a pepper leaf. Spores germinate and infect the host tissues, form haustoria inside the cell to absorb the food nutrients and to support the growth of the fungus which emerge through stomata in the form of conidiophores and conidia. The fungus initially grows unseen within the leaf for a latency period of 18-21 days. Then the fungus grows out of the leaf openings (stomata) on the undersurface of the leaf, producing conidia which are borne singly on numerous, fine strands or stalks called conidiophores. These fungal strands become visible as white patches, i.e. mildew colonies, on the undersurface of the leaf. Air currents within the greenhouse carry these microscopic, infectious spores to other plants. Spores are dispersed further through the greenhouse vents. In addition to dispersal by air currents or wind, powdery mildew can spread on ornamental plants and weeds, and by workers on their clothing. Repeated generations of powdery mildew can lead to severe outbreaks of the disease that economically damage the crop.

Weather Parameter

Infection of plants can occur over a wide temperature range (18° to 33°C) under both high and low humidity. Under favorable conditions, secondary infections occur every 7 to 10 days, and disease can spread rapidly.

Management

Regular monitoring to detect powdery mildew, especially during warm weather, is important to time fungicide applications early enough to prevent damage. Powdery mildew is managed primarily with fungicides.

Cultural Control

The fungi that cause powdery mildew can survive between crop seasons on other crops and on weed species. The degree of survival depends on environmental conditions. Because of the wide host range of the fungus, it is difficult to control the amount of inoculum that overwinters. Thus, simple sanitation methods in and around pepper fields may not provide a sufficient reduction in the primary inoculum to provide disease control.

Chemical fungicides

Strobilurins (e.g., Cygnus, Compass, Heritage), sterol inhibitors (e.g., Systhane, Strike, Terraguard), benzimidazoles (e.g., Cleary's 3336, FungoFlo), coppers (e.g., Phyton 27,), sulfurs (e.g., Microthiol Special), and miscellaneous others (e.g., Daconil, Pipron, Spectro, ZeroTol) are effective in powdery mildew management.

20.2. Powdery mildew of Coriander

Pathogen: *Erysiphe polygoni*

Symptoms

The initial stage of this disease is recognized by the appearance of small, white, powdery patches on young parts of stems, leaves and buds which increases in size, and coalesces to cover entire leaf surface (Fig 20.2). Affected leaves are reduced in size and distorted. If unchecked at this stage the disease gets intensified and the whole plant becomes almost white with the powdery coating (Fig 20.2). Infection, at the early stage, normally does not permit seed formation. However, if the infection occurs at a later stage, the seed formation may occur but the seeds will be small and shriveled, thus affecting the yield and quality of the produce.

Fig. 20.2: Symptoms of powdery mildew on coriander plant

Survival and spread
Fungus can survive in plant debris in the form of cleistothecia and spread long distances by air.

Favorable conditions
Disease emergence is favored by high humidity and moderate temperatures (cloudy weather); infection is most severe in shaded area.

Management
This disease can be effectively controlled by spraying the plants with 0.2% solution of wettable sulphur or 0.2% solution of Karathane or 0.1% Hexaconazole or 0.1% Propiconazole. The spray should be repeated after 10 to 15 days. Dusting the crop with 20 to 25 kg of sulphur dust per hectare would also control the disease. Dusting may be repeated, if necessary.

Resistant Variety
DH 206 is a high yielding variety resistant to powdery mildew.

Cultural practices
Harvesting of the mature crop should not be delayed to avoid powdery mildew attack.

20.3. Powdery mildew of Fenugreek
Pathogen: *Erysiphe polygoni* D.C.

Symptoms
The initial disease symptoms is marked by the appearance of whitish powdery mass on the leaves (Fig 20.3). In this disease, white powdery patches appear on the lower and upper surface of leaves and other parts of plant. If not checked at an early stage, the disease intensifies and the whole plant becomes whitish with the powdery coating. The seed size and the yield are adversely affected.

Fig. 20.3: Symptoms of powdery mildew on fenugreek plant

Disease Cycle

The pathogen survives mostly by developing cleistothecia in diseased plant debris. They survive in soil until the next season. The ascospores first infect the lower and older leaves in the next season. The spores are carried by the wind to new host.

Weather Parameters

It is most commonly found in hot and humid tropical and subtropical weather, as well as in the weather of temperate to sub-temperate region.

Management

The disease can be effectively controlled by Spraying of wettable Sulphur or Dinocap (Karathane or Thiowet) can also be used to control the disease @ 20-25 g per 10 liter of water at the initial stage of this disease. If needed two more sprays should be given at an interval of 15 days after first spray. Dusting the crop with 300 mesh sulphur dust 15-20 kg per hectare also controls the disease. Dusting may be repeated if the disease reappears.

Resistance Varieties

RMt-l, RMt-143 and Rajendra kranti are moderately resistant to powdery mildew diseases while Lam Selection-1 is tolerant.

20.4. Powdery mildew of Fennel

Pathogen: *Erisyphe heraclei*

Symptoms

Symptoms first appeares as white colonies on the leaves as sporadic spots, petioles, flowers stalks and bracts which later spreads on the whole plant, which exhibited abundant mycelial growth on the leaves, often covering the whole surface (Fig.20.4). Infected plants as fennel vegetable are unmarketable mainly due to signs of white fungal growths on leaves and stems. Severe infection often causes leaf withering and premature senescence, which affect the seed set and causes the complete loss in the seed yield.

Fig. 20.4: Powdery patches on fennel leaves

Weather

Cool and high humid weather (20-25Ú C) or cloudy weather favors conidial germination and disease development. High RH > 80% favors disease development.

Conidia germinated throughout a temperature range of 17-31°C, but highest percentage of germination occurred at 25-27°C. Disease development is more severe during periods when the maximum daily temperature remained above 25°C.

Fungus can spread long distances in air and disease emergence is favored by high humidity and moderate temperatures. The infection is most severe in shaded areas.

Survival and spread

Primary: Through soil and seed,

Secondary: Dispersal of conidia through wind, rain splashes

Disease Management

The disease can be controlled by spraying 0.03 to 0.1% solution of Karathane or 0.2% solution of wettable sulphur. Approximately 23 kg sulfur/ha applied as dust in 3 applications beginning just prior to flowering and completed at the time of seed formation effectively controlled the disease.

Plant tolerant varieties and avoid excess fertigation to minimize the disease incidence.

20.5. Powdery mildew of Cumin

Importance of the mildew disease on cumin in India has varied since it was first observed in 1921. The disease was severe in several states of the country in 1953 and caused complete failure in some crops, while losses in yield elsewhere have been estimated at 5-15 percent. Surface Mildew on cumin is known to occur only in India and Bulgaria.

Pathogen: *Erysiphe polygoni*

Symptoms

Symptoms appears first on the lower leaves as small white specks that gradually enlarge and coalesce to cover the entire leaf surface with superficial, external mycelium. Infection may spread upward rapidly to involve young leaves, flowers and stems, and influence on fruit yield. If the infection is not checked at this stage, the disease spreads and the whole plant becomes almost whitish with the powder (Fig 20.5). Sporulation of the fungus on affected plant parts imparts an obvious powdery, white appearance. Infection of the plant in early stage will not permit seed formation but if the seeds are already formed they will be small and shriveled.

Fig. 20.5: Symptoms of powdery mildew on cumin

Pathogen

The pathogen produce septate, superficial hyphae 3.5-6.5 μm in diameters with lobed haustoria. Hyaline, unicellular, erect conidiophores 103-170 μm bear hyaline, unicellular ovoid conidia 10-16 X 28-40 μm, in chains, but conidia tend to fall when mature, so continuous chains of conidia are seldom seen.

Weather Parameters

Disease development is enhanced by warm, moist conditions, and intensity of the disease is severely reduced when maximum temperatures fall below 26^0 C.

330   The Plant Mildews

Cardinal temperatures for conidial germination are 16, 20-22 and 30^0 C, and mature conidia readily germinate in water but will not germinate when dry and maintain viability only for a short period. The fungus survives from one season to another as dormant mycelium on seeds. In cloudy weather during February-March with cool high humid (20-25°C) or cloudy climate with high relative humidity (RH) > 80% favors conidial germination and disease development.

Survival and spread

Disease is both soil and seed borne. Primary spread is through soil and seed, the secondary spread takes place by dispersal of conidia through wind and rain splashes

Disease Management

The disease can be effectively controlled by spraying 400 to 500 litres solution of 0.2% wettable sulphur (2 g/l), 0.3% Karathane (3 ml/L) or 0.05 to 0.1 % calixin (1.5 to 3,0 g/I) per hectare. The spray should be repeated after 10 days.

Crop should be dusted with 300 mesh Sulphur dust @ 25 kg/ha to control this disease as soon as the symptoms are noticed. It may be repeated after 15 to 20 days, if necessary. The control measures should be adopted at the initial stage of the appearance of the symptom to check the disease effectively.

Control is also achieved by dusting sulfur (13.8-23 kg/ha) over the plants during flowering stage, but a second dusting at half of the original rate may be required under environmental conditions favorable for disease development. Perenox as a foliar spray, applied before disease development also provide satisfactory control.

Resistant Variety

GC-2 cultivar is moderately tolerant to powdery mildew.

20.6. Powdery mildew of Dill

Pathogen: *Erysiphe heraclei*

Symptoms

Incidence of powdery mildew is seen on all green parts of the dill plant. The Symptoms first appears as thin white colonies, which subsequently shows abundant growth on the leaves and stems (Fig 20.6). Most diseased plantings are unmarketable and shriveled without being harvested.

Fig. 20.6: Symptoms of powdery mildew on dill

Disease Management

It can be controlled by spraying sulphur 3g/ litre of water twice. First spraying is done as soon as the disease appears in the field and then after 15 days. Occasionally, when the attack of the disease on the crop is at flowering which causes severe damage, the spraying of Bordeaux mixture three to four times at weekly intervals is recommended.

20.7. Powdery mildew of Chilli

Pathogen: *Leveillula taurica* (Lev.) Arn. (anamorph *Oidiopsis taurica*)

Symptoms

The disease appears as small, circular, whitish spot on both the surfaces of the leaves. In later Stages the entire leaf surface is covered by dirty-white growth (Fig 20.7). The infection starts from the older leaves but gradually covers the entire plant. On the top surface of leaves, lesions are yellow with brown necrotic centers. Leaves curl upwards. Premature senescence of the leaves results in defoliation. Both the number of fruit and the size of fruit are reduced in heavily infected plants. In such cases, the lamina develops complete or partial chlorosis

Fig. 20.7: Symptoms of powdery mildew on chilli leaves

and ultimately plant dries up. With age, the mildew mass turn grey and develops numerous minute black bodies distributed all over the surface.

Geographical distribution

Occurs worldwide in Africa, Asia, the Mediterranean and Caribbean regions, and North America.

Yield Loss

Powdery mildew causes heavy yield loss ranging from 14 to 20 per cent. Due to severe defoliation and reduction in photosynthesis, the size and number of fruits per plant are reduced.

Disease cycle and Epidemiology

Conidia are wind blown or dispersed by splashing rain. Temperatures of 25-30^0C favor germination of conidia. Some isolates of powdery mildew that infect chilli can apparently develop infection under a wider range of dry to humid conditions as compared to other powdery mildews. Low relative humidity of less than 50% and temperature of 20-$30°C$ is required for the development of this mildew. Relative humidity that is higher at night than during the day and temperatures of less than $30°C$ are conducive to disease development . Free moisture on leaves inhibit spore germination. Throughout the growing season, new infections may develop on new, succulent shoots of plants. Pathogen overwinters on the infected leaves as dormant mycelium or cleistothecia. It may also survive on the collateral hosts.

Primary source of inoculum: Dormant mycelium in the infected crop debris.

Secondary source of inoculum: Air-borne conidia

Disease Management

1. Spraying of Dinocap (Karthane) 0.1% or Sulphur 0.3% or Carbendazim 0.1% or Tridemorph 0.05%, or Hexaconazole 0.05% at fortnightly interval. Three sprays may be given as per the severity of disease.

2. Spray thrice at 10 -15 days interval with 1 ml Dinocap or 1 g wettable sulphur/litre water.

3. Before flowering, dust sulphur powder at the rate of 8 - 10 kg /acre is also proving very useful.

4. Use resistant varieties like Hisar Shakti, Hisar Vijay, Jawahar Mirch.

21

Powdery Mildews of Flowering and Ornamental Plants

Agriculture is almost entirely dependent on angiosperms, which provide virtually all plant based food, and a significant amount of livestock feed and timbers for many commercial uses in life. Besides these, Agriculture got importance due to different flowering and ornamental plant crops as these crops not only gave pleasure to nature but also play an important role in economy. These ornamentals and flowering crop plants also get affected with Powdery Mildew disease and are not only influenced by reduced market value but better quality also.

Powdery Mildew looks like white fuzzy powder of fungal growth that accumulates on leaves and stems of ornamental and flowering plants predominantly in spring, and again to a lesser degree in fall. It is spread by millions of microscopic spores. It imbeds itself into tender new growth and feeds on the sap of the plant. It damages plants by decreasing photosynthesis and removing nutrients from the host plant cells. Infections weaken plants and increase susceptibility to further pest problems. By the time the naked eye can see the white 'powder,' it has already invaded the plant tissue and is feeding and reproducing at a rapid pace. As it spreads itself on the surface, it eventually kills the cells of the plant leaf, leaving the leaf rippled and curled.

Mildew spores are everywhere in the garden, in the air, in the soil, on debris and on plant surfaces, ready to sprout when the environment is just right. Warm days (10°-26.5°C) and cool nights with elevated humidity and resultant dew provide ideal conditions. Though humidity promotes fungal growth, it grows on dry plant surfaces. Mildews do not require free moisture, but only need warmth and adequate humidity to grow and develop. Dense plantings create microclimates with sufficient humidity to support mildew populations.

Powdery mildews attack many ornamentals and flowering plants. While powdery mildew fungal infection seldom causes plant death, it almost always reduces the plants looks, quality of the produce and its value. Serious economic losses are common in commercial cultivation of ornamental and flowering plants. Powdery mildews disease pathogens are host specific. Important powdery mildews of ornamental and flowering plants are dealts in this chapter.

21.1. Powdery mildew of Rose

Pathogen: *Sphaerotheca pannosa* var. *rosae*

Georgaphical distribution: Worldwide

Symptoms

On garden roses, new shoots in the spring are dwarfed, distorted, and covered with a whitish gray mildew growth. On expanding leaves, mildew first appears on the upper leaf surface as irregular, light green to reddish, slightly raised blister like areas. The typical dense, powdery white growth (mycelium, conidiophores, and spores) of the mildew fungus soon appears (Fig 21.1). Severely infected young leaves become curled or irregularly twisted and are usually covered with enlarged, whitish gray, powdery, mealy, or felt like patches of the fungus.

These leaves often turn reddish purple under the mildew growth, then yellow, dry, and drop prematurely. Older infected leaves are not usually distorted, but develop round-to-irregular areas covered with the flourlike mildew growth. On highly susceptible rose cultivars, the buds, young stems (canes), thorns, peduncles, fruit sepals, and even flower petals may become infected and entirely covered with the typically dense, flourlike growth. Flower petals may be discoloured, dwarfed, and may fail to open properly; the flowers may also die early.

Fig. 21.1: Symptoms of powdery mildew on rose leaves, stem and flower bud

The growing tips and flower buds may be malformed and killed, but the death of an entire plant is rare. Plants can be severely stunted if they are heavily infected early in the growing season. Rose tissue becomes more resistant to infection as it ages. Some resistant rose cultivars may show a hypersensitive reaction where invaded dead cells appear as black-to-rusty specks on the leaf surface, with little evidence of mildew growth.

Under certain conditions, extensive infection of the lower leaf surface may occur with sparse mycelial growth of the fungus and little, if any, production of spores. Such undetected infections sporulate readily when conditions become favourable. An apparent overnight outbreak of the disease may then occur.

Epidemic and losses

Epidemics can be expected any time during the growing season when the rainfall is low or absent, the days are warm and dry, and the nights are cool and damp. Nearly all species and cultivars of roses are susceptible under conditions that are favorable for disease development. Most climbers and small-flowered ramblers, some floribundas, polyanthas, and hybrid tea roses are very susceptible. Wichurainas are reportedly more resistant.

Losses from powdery mildew occur through a reduced aesthetic value that is seen in fewer flowers of poorer quality, a lowered photosynthetic efficiency that results in reduced plant growth, a greater likelihood of winter injury, and a reduced saleability for roses as cut flowers.

Disease Cycle

The powdery mildew fungus overwinters as dormant mycelium in bud scales and rudimentary leaves within the dormant buds. Infected buds break open in the spring and develop into systemically infected shoots. The fungus sporulates on these shoots, producing large numbers of microscopic spores (conidia) in chains that are carried by the wind or other means to healthy rose tissue where they infect the upper and lower leaf surfaces, thus initiating a new disease cycle.

The fungus survives in the winter as cleistothecia, which appear as black specks embedded in the mealy or feltlike mildew growth on rose stems, thorns, and fallen leaves. The minute cleistothecia are formed within the mycelial mat at the end of the growing season. During warm and humid weather in the spring, a cleistothecium absorbs water and cracks open to discharge a single small sac or ascus containing 8 spores (ascospores). The ascospores are carried by the wind or splashing rain to healthy rose tissue and are capable of causing infection. Cleistothecia almost never form on some rose cultivars, especially floribundas and hybrid teas. In greenhouses or mild climates, where roses and powdery mildew both grow continuously throughout the year, cleistothecia are absent and only conidia are formed. New infection cycles are produced more or less continuously.

Conidia and ascospores that land on the surface of a rose germinate and form a holdfast structure (appressorium) on the leaf or stem surface. From the bottom of the appressorium, a fine penetration tube or hypha pierces the cuticle and enters the epidermal cell where a globose feeding structure, or haustorium, is formed. With further growth on the plant surface, the fungus develops a dense, branched network of hyphae. Many additional haustoria form in other epidermal cells. Short, erect branches, or conidiophores, develop at the same time from the surface hyphae, producing a barrel-shaped conidium at the end of each conidiophore. Successive conidia, with one formed each day; commonly remain attached in chains, giving the characteristic powdery white appearance. The conidia eventually break away. The powdery mildew fungus spores are carried by air currents, splashing water, or other means to new infection sites. Handling rose plants, insects, mites, and snails also helps spread conidia. As many as 3 million spores may be formed on one square inch of infected tissue over a period of several weeks.

Weather Parameters

When conidia or ascospores fall on a plant surface, they start to germinate in 2 to 4 hours, reaching a maximum number in about 25 hours. The optimum temperature for germination is about 22°C; the minimum, about 5°C and the maximum is close to 35°C. Spore germination occurs on the surface of a rose over a range of relative humidity from 23 to 99 percent. Free moisture is detrimental to spore germination of the powdery mildew fungus. Once released from the conidiophores, the thin-walled conidia do not live long. At 32°C, and a relative humidity of 70 percent or less, germination reaches 95 to 100 percent in 2 hours and drops to 8 to 20 percent after 5 hours. At 21°C, and a relative humidity of 70 percent or less, germination is only 20 to 40 percent after 5 hours. Although conidia remain viable longer at a relative humidity of 80 to 90 percent, essentially all conidia are dead after 48 hours at 21°C and after 24 hours at 32°C.

The environment most favourable for conidial production, maturation, release and spread, germination, and infection include repeated day-night cycles where the nights are cool (about 16°C) and damp with a relative humidity of 90 to 99 percent, and the days are warm (about 27°C) and dry with a relative humidity of 40 to 70 percent. When spring and summer rainfall is high, epidemics of powdery mildew are most common during the late summer or fall. The disease cycle-production of conidia, release, germination, infection, and production of conidia–can be as short as 72 to 96 hours. If left uncontrolled, powdery mildew can quickly become epidemic when cool, damp nights are followed by warm, dry days.

Disease Management

1. Purchase only top-quality, disease-free plants of resistant cultivars and species from a reputable nursery. Roses with thick, leathery, glossy leaves, for example, wichuraiana hybrids 'Simplicity' and 'Meidiland' roses, *Rosa rugosa* varieties have fair-to-good resistance. Few of the reported resistant cultivars escape infection in seasons that are favourable to the mildew fungus. The presence of different physiologic races in an area greatly complicates the breeding of rose cultivars that are highly resistant or immune to powdery mildew. Most rose climbers, small-flowered ramblers, some floribundas, grandifloras, and hybrid teas are very susceptible.

2. Prune roses in the fall and in early spring, according to type and cultivar. All dead wood should be removed and burned. Drastic pruning to within 1 to 2 inches of the bud union greatly reduces the carryover of powdery mildew and other disease-causing fungi.

3. Maintain rose plants in high vigour.

   a. Plant properly in well-prepared and well-drained soil, high in organic-matter content, where roses will obtain all-day sun (or a minimum of 6 hours of sunlight daily). If possible, avoid planting near large shrubs or trees that will compete with roses for moisture, light, and soil nutrients.

   b. Space plants for good air circulation at the suggested distance for the cultivar, type of rose, and effect desired.

   c. Do not handle or work among plants when the foliage is wet.

   d. Fertilize based on a soil test. Avoid excessive applications of high-nitrogen fertilizers. Newly planted roses should not be fertilized until they are well established and growing steadily. The soil pH should be between 5.5 and 6.5.

   e. Water thoroughly at weekly intervals during periods of drought. The soil should be moist 8 to 12 inches deep. Avoid overhead irrigation and spraying the foliage when waters, especially in late afternoon or evening. Use a soil soaker hose or other methods that will not wet the foliage.

   f. Protect plants for winter by following local recommendations. Winter safeguards provide insulation against extremely low temperatures, alternate periods of freezing and thawing, and protection against damage by wind or heavy loads of snow and ice.

g. Whenever possible, destroy nearby wild or uncared roses.

4. Vaporized sulfur gives excellent control of powdery mildew in greenhouses. A slurry made of 500 ml of water and 500 gm of wettable sulfur is sufficient for each 90,000 cubic feet of greenhouse space. The slurry is painted on two steam pipes in each house, covering sections 3 feet long and leaving equal intervals unpainted. Applications should be made regularly twice a week. If steam is unavailable, the sulfur may be vaporized in small homemade or commercially available vaporizers using light bulbs or small heating elements as a source of heat. One vaporizer should be used for each 1,000 to 1,500 square feet of greenhouse space. Flowers of sulphur (not wettable sulfur) should be used in the vaporizers. The vaporizers should be kept on day and night, with the greenhouse vents open no longer than necessary for proper ventilation. Excessive heat in the vaporizer must be avoided since the fumes of burning sulfur are highly toxic to rose foliage. For homemade vaporizers, a 60-watt bulb should be used. Weaker bulbs will not melt the sulfur. Stronger ones cause the sulfur to ignite. Commercial vaporizers come equipped with bulbs of the proper size. Burned-out bulbs should be replaced by the same size bulb.

21.2. Powdery mildew of Chrysanthemum

Pathogen: *Erysiphe cichoracearum*

Georgaphical distribution

Punjab, Uttar Pradesh, Jammu and Kashmir, Himachal Pradesh, Maharashtra, Haryana and Rajasthan.

Symptoms

White powdery growth appears mainly on older leaves (Fig 21.2). It appears on both sides of leaves as well as shoots, buds and sometimes flowers.

Pathogen

The mycelium is well developed, evanescent but sometimes persistant and effused, superficial with well-developed haustoria in host

Fig. 21.2: Symptoms of Powdery mildew on leaves of Chrysanthemum

cells. Oidiophores arise on mycelial web, and are unbranched, erect, producing oidia ellipsoidal or barrel shaped, 25-45 × 14-26 µm in size and produced in abundance. Cleistothecia gregarious or scattered, globuse becoming depressed or irregular, 90-135 µm in diameter, wall cells usually indistinct 10-20 µm wide. Cleistothecial appendages numerous, myceloid, basally inserted, hyline to dark, interwoven with mycelium, and 1-4 times as long as the diameter of cleistothecium, and rarely branched. Asci 10-25 per cleistothecium, ovate to broadly ovate, rarely subglobose, more or less stalked, 60-90 × 25 µm in size. Ascospores two in each ascus, very rarely three, 20-30 × 12-18 µm in size.

Disease cycle and Weather Parameters

When conidia or ascospores fall on a plant surface, they start to germinate in 2 to 4 hours, reaching a maximum number in about 25 hours. The optimum temperature for germination is about 22°C; the minimum, about 5°C and the maximum is close to 35°C.

Spore germination occurs on the surface of a chrysanthemum leaf over a range of relative humidity from 23 to 99 percent. Free moisture is detrimental to spore germination of the powdery mildew fungus. Once released from the conidiophores, the thin-walled conidia do not live long. At 32°C, and a relative humidity of 70 percent or less, germination reaches 95 to 100 percent in 2 hours and drops to 8 to 20 percent after 5 hours. At 21°C, and a relative humidity of 70 percent or less, germination is only 20 to 40 percent after 5 hours. Although conidia remain viable longer at a relative humidity of 80 to 90 percent, essentially all conidia are dead after 48 hours at 21°C and after 24 hours at 32°C.

The environment most favourable for conidial production, maturation, release and spread, germination, and infection include repeated day-night cycles where the nights are cool (about 16°C) and damp with a relative humidity of 90 to 99 percent, and the days are warm (about 27°C) and dry with a relative humidity of 40 to 70 percent. Powdery mildew thrives when foliage is dry and the weather is warm. Wind spreads the spores to other plants. In fact, the powdery mildew spores can't germinate or grow when the foliage is wet.

When spring and summer rainfall is high, epidemics of powdery mildew are most common during the late summer or fall. The disease cycle–production of conidia, release, germination, infection, and production of conidia–can be as short as 72 to 96 hours. If left uncontrolled, powdery mildew can quickly become epidemic when cool, damp nights are followed by warm, dry days.

340   The Plant Mildews

Disease management

Cultural control

- Keep susceptible cultivars isolated.
- Space plants far enough apart to ensure good air circulation.
- Pick off and burn affected leaves, if practical.
- Keep greenhouse humidity low.

Chemical control

Scout plants regularly and use fungicides just before or when first symptoms appear. Alternate or tank-mix products from different groups that have different modes of action.

- Armada 50 WDG at 85 to 250 grams/375 litres water. Do not use a silicone-based surfactant. Not for nursery or greenhouse use.
- Bayleton 50 T&O at 15 gm/ 100 litre water. Landscape only, not for use on plants for sale.
- Bicarbonate-based products, might be used to supplement a normal program when powdery mildew is first observed. Do not mix with acidifying agents. Thorough coverage is essential. Armicarb "O" (85% potassium bicarbonate) at 1 to 5 kg/ 375 litre water. 4-hr reentry. Monterey Bi-Carb Old Fashioned Fungicide is registered for home use in all states. Cosavet-DF (80% sulfur) at 3 to 5 kg/375 litre water. Group M2 fungicide.
- Cygnus 50 WG at 45 to 90 gm/ 375 litre water plus a non-organosilicone spreader-sticker. Best used before symptoms develop. Group 11 fungicide.
- Eagle 20 EW at 235 ml/375 litre water. Do not apply more than 560 ml/ A/season. Group 3 fungicide.
- Fungaflor TR at 1 can/1000 sq ft of greenhouse. It ignites to form a vapour that condenses on the plants. See label for details. Group 3 fungicide. 24-hr reentry including ventilation after an overnight period.
- Heritage at 30 to 120 ml/ 375 litre water plus a non-silicone-based wetter sticker. Group 11 fungicide.
- Insignia at 120 to 240 ml/ 375 litre water. Do not use with organosilicate-based adjuvants. Use preventively only. Group 11 fungicide.
- JMS Stylet Oil at 30 ml/3 litre water. Do not use with or near a sulfur application. Do not use during freezing temperatures, above 32°C, or

when plants are under heat or moisture stress. Do not use when foliage is wet as good coverage is essential.

- Pageant at 170 to 340 gm/375 litre water. Do not use more than two (2) consecutive applications before switching to a different fungicide group. Group 7 + 11 fungicide.

- Pipron at 120 to 240 ml/375 litre water plus a surfactant. Greenhouse production only. Group 5 fungicide.

- Rubigan AS at 88 to 150 ml/375 litre water.

- Safer Garden Fungicide (Ready to Use 0.4% sulfur) thoroughly sprayed over the entire plant. Do not use when the temperature is over 29.4°C or within a few weeks of an oil spray.

- Spectracide Immunox at 30 ml/3 litre water.

- Strike 50 WDG at 30 to 60 ml/375 litre water for outdoor use. In the greenhouse, use 30 ml/375 litre water in winter and 60 ml/375 litre water in summer.

- Strike Plus 50 WDG at 90 to 255 gm/375 litre water. In the greenhouse, use 35 gm/375 litre in winter and 70 gm/375 litre water in summer.

- Sulfur (90%) at 3 Tbsp/gal water. Avoid using sulfur during hot periods or on open flowers because it may cause bleaching.

- Terraguard SC at 120 to 240 ml/375litre water. Although a 450 gm/375 litre water rate can be used for initial applications to existing infections, use only in conjunction with a regular scouting program that detects initial symptom development.

- Trinity at 110 to 340 gm/375 litre water.

Biological control
- *Bacillus subtilis* strain QST 713 is registered for the home garden. Active ingredient is a small protein.

- Cease or Rhapsody (*Bacillus subtilis* strain QST 713) at 1.8 to 7.5 litre/375 litre water. Active ingredient is a small protein.

21.3. Powdery mildew of Gerbera

Pathogen: Erysiphe *cichoracearum*

Georgaphical distribution: Worldwide

Symptoms

Powdery mildew is easy to identify since noticeable white spots or white patches appear on the upper and lower surfaces of the leaves. These spots gradually enlarge to form a white, powder-like mat that can spread to flowers and stems (Fig 21.3).

Fig. 21.3: Symptoms of Powdery mildew on leaves and flower of Gerbera

The spots are a combination of the growth of the fungus mycelium and the reproductive asexual structure, the conidiophores, which bear the conidia. Severely infected leaves turn pale yellow or brown and eventually die.

Pathogen

The mycelium is well developed, evanescent but sometimes persistant and effused, superficial with well developed haustoria in host cells. Oidiophores arise on mycelial web, and are unbranched, erect, producing oidia Ellipsoidal or barrel shaped, 25-45×14-26 µm in size, produced in abudance and disseminated by wind. Clestothecia gregarious or scattered, globuse becoming depressed or irregular, 90-135 µm in diameter, wall cells usually indistinct 10-20 µm wide. Clestothecial appendages numerous, myceloid, basally inserted, hyline to dark, interwoven with mycelium, 1-4 times as long as the diameter of cleistothecium, rarely branched. Asci 10-25 per cleistothecium, ovate to broadly ovate, rarely subglobose, more or less stalked, 60-90 × 25-50 µm in size. Ascospores two in each ascus, very rarely three, 20-30 × 12-18 µm in size.

Disease Cycle

High humidity (80% - 90%) and moderate temperatures (20 - 27°C) contribute to the development of powdery mildew. Overcrowding and shading of plants will also favour development of this disease. The powdery mildew fungus reproduces primarily through asexual spores (conidia), which are responsible for most of the spread of the disease. It also reproduces through sexual spores (ascospores), which serve primarily for survival of the fungus. The initial infection typically occurs when conidia from infected transplants are carried by air currents to other plants or plant parts. The conidia then germinate and produce an elongated structure (germ tube) that enlarges and penetrates the leaf epidermis and lives on the host plant. Other specific fungal structures develop within the plant tissue to establish a biotrophic (symbiotic) relationship necessary to absorb nutrients. The germ tube continues to grow and spreads across the leaf surface, producing more hyphae and conidia.

Weather Parameters

When conidia or ascospores fall on a plant surface, they start to germinate in 2 to 4 hours, reaching a maximum number in about 25 hours. The optimum temperature for germination is about 22°C; the minimum, about 5°C and the maximum is close to 35°C. Spore germination occurs on the leaf surface over a range of relative humidity from 23 to 99 percent. Free moisture is detrimental to spore germination of the powdery mildew fungus. Once released from the conidiophores, the thin-walled conidia do not live long. At 32°C, and a relative humidity of 70 percent or less, germination reaches 95 to 100 percent in 2 hours and drops to 8 to 20 percent after 5 hours. At 21°C, and a relative humidity of 70 percent or less, germination is only 20 to 40 percent after 5 hours. Although conidia remain viable longer at a relative humidity of 80 to 90 percent, essentially all conidia are dead after 48 hours at 21°C and after 24 hours at 32°C.

The environment most favourable for conidial production, maturation, release and spread, germination, and infection include repeated day-night cycles where the nights are cool (about 16°C) and damp with a relative humidity of 90 to 99 percent, and the days are warm (about 27°C) and dry with a relative humidity of 40 to 70 percent. When spring and summer rainfall is high, epidemics of powdery mildew are most common during the late summer or fall. The disease cycle- production of conidia, release, germination, infection, and production of conidia can be as short as 72 to 96 hours. If left uncontrolled, powdery mildew can quickly become epidemic when cool, damp nights are followed by warm, dry days.

Disease management

Cultural Control

Only plantlets that are free of powdery mildew should be used, or the plantlets should be disinfested before transplanting.

In addition, plants should be grown in sunny areas with good air ventilation and without excessive fertilization. Overhead irrigation may reduce powdery mildew since it washes conidia from the leaf surfaces. However, overhead irrigation may contribute to other disease problems, such as crown rot.

Chemical Control

Labeled fungicides to control powdery mildew in gerberas include the following: Compass (trifloxystrobin), Eagle (myclobutanil), Heritage (azoxystrobin), Milstop (potassium bicarbonate), and Spectro (chlorothalonil + thiophanate-methyl). However, chemical control may not always be completely effective since the fungus may develop resistance to some chemicals. Since powdery mildew goes through many short cycles per season, the potential for development of resistance is great, especially when only systemic fungicides are used to control this disease.

Resistance of powdery mildew has been confirmed with the fungicides benomyl (Benlate) and triadimefon (Bayleton) in the past.

Bio-rational or Bio-fungicidal Control

Bio-fungicides are naturally based microbial or biochemical products derived from animals, plants, bacteria, and certain minerals. Bio-fungicides can affect fungal organisms directly or may stimulate the plant's own defense against the pathogens. These products are generally narrow-spectrum, and they decompose quickly. Because of these characteristics, bio-fungicides are considered to have little potential for negative impact on the environment. Bio-fungicides that have been effective for the control of powdery mildew include biological control agents (BCAs), oils, phosphorous acid, and potassium bicarbonate.

Biological Control Agents

BCAs are organisms that suppress pests or plant pathogens. BCAs that have been used successfully for the control of powdery mildew include the following: *Ampelomyces quisqualis* (AQ10), *Tilletiopsis* spp, Pseudozyma flocculosa (Sporodex, Plant Products Co., Brampton, Ontario, Canada) and Bacillus subtilis QST 713 (Serenade® or Rhapsody®, AgraQuest, Inc. Davis, California). The major impediment to BCA effectiveness is their requirement for high humidity.

Oils

Petroleum or plant oils have not only been effective in controlling powdery mildew, but also have been useful in reducing the development of resistance to fungicides. Reduction of the severity of powdery mildew through use of oils has been reported in other ornamental, fruit, and vegetable crops as well, including roses, apples, cherries, cucurbits, and grapes. In some cases, the efficacy of oils to reduce powdery mildew is comparable to or even superior to results obtained through use of standard fungicides.

Phosphorous Acid

Phosphorous acid (H_3PO_3) is the active ingredient in phosphonate, used in agriculture for disease control. However, H_3PO_3 is not a nutritional source of P for plants. Nonetheless, H_3PO_3 is often confused with phosphoric acid (H_4PO_4) or phosphates, which are effective nutritional sources of P. Phosphite-based products are marketed under several trade names and have all been found effective in reducing the severity of powdery mildew on gerbera daisy and on other crops, such as grapes and muskmelons.

Potassium Bicarbonate

Some potassium-bicarbonate based products are labeled as fungicides and have been approved by the U.S. Environmental Protection Agency (EPA). Kaligreen, Armicarb and Milstop, in particular, are potassium bicarbonate fungicide products that have proven effective to reduce powdery mildew of gerbera daisy.

21.4. Powdery mildew of Marigold

Pathogen: *Oidium sp.*; *Leveillula taurica*

Geographical distribution: Worldwide

Symptoms

The symptoms are in the form of whitish powdery growth on the aerial parts of the plant. Frosty white, powdery spots develop on leaves. Colonies can become so extensive that the entire plant may appear white (Fig 21.4).

Fig. 21.4: Symptoms of Powdery mildew on marigold leaves and flower

Disease cycle

Germination, infection and sporulation follow each other in the life cycle of the pathogen. The life cycle of the fungus is initiated by airborne conidia which are the asexual reproductive spores of the fungus. When the conditions are favourable, the conidia start germination after four hours and produces a single thick germ tube from one corner of the conidia. Germinating conidia produces primary hyphae at about 30 hr after incubation, secondary hyphae after 36 hours and tertiary hyphae after 48 hours of incubation. After about 60 hours the other hyphae elongate. The white patches of infection became visible to the naked eye only after 120 hours. The mature conidia gets released about seven days after inoculation. The conidia get released when the host leaf and the atmosphere are relatively dry during day time.

Detached conidia are carried away by the wind and gets deposited over the fresh host leaves on which they gets germinated again. High humidity (80% - 90%) and moderate temperatures (20 – 27.7°C) contribute to the development of powdery mildew. Overcrowding and shading of plants also favours the development of this disease and causes severe losses.

Weather parameters

When conidia or ascospores fall on a plant surface, under favourable atmospheric conditions, they start to germinate in 2 to 4 hours, reaching a maximum number in about 25 hours. The optimum temperature for spore germination is about 22°C; the minimum, about 5°C and the maximum is close to 35°C. Spore germination occurs on the leaf surface of marigold over a range of relative humidity from 23 to 99 percent. Free moisture is detrimental to spore germination of the powdery mildew fungus. Once released from the conidiophores, the thin-walled conidia do not live long. At 32°C, and a relative humidity of 70 percent or less, germination reaches 95 to 100 percent in 2 hours and drops to 8 to 20 percent after 5 hours. At 21°C, and a relative humidity of 70 percent or less, germination is only 20 to 40 percent after 5 hours. Although conidia remain viable longer at a relative humidity of 80 to 90 percent, essentially all conidia are dead after 48 hours at 21°C and after 24 hours at 32°C.

The environment most favourable for conidial production, maturation, release and spread, germination, and infection include repeated day-night cycles where the nights are cool (about 16°C) and damp with a relative humidity of 90 to 99 percent, and the days are warm (about 27°C) and dry with a relative humidity of 40 to 70 percent.

When spring and summer rainfall is high, epidemics of powdery mildew are most common during the late summer or fall. The disease cycle-production of conidia, release, germination, infection, and production of conidia-can be as short as 72 to 96 hours. If left uncontrolled, powdery mildew can quickly become epidemic when cool, damp nights are followed by warm, dry days.

Disease management

Cultural pratices

- Space plants for good air circulation.
- Remove and destroy infected plants.
- Plant in sunny open locations rather than shady.

Chemical control

- Spraying Sulfex (3g/litre of water) can effectively control the disease.
- Karathane (40 E.C) @ 0.5% or dusting sulphur powder at 15 days interval.

21.5. Powdery mildew of Dahlia

Pathogen: *Erisyphe polygoni*

Geographic distribution: Worldwide

Symptoms

Grayish white, powdery spots generally develop over the entire dahlia leaf surface forming a cobwebby powdery mat of hyphae (Fig 21.5). Conidia production gives it a powdery cottony appearance. Severely affected leaves become distorted and may fall prematurely. The severely affected plants do not produced proper sized flowers, these are distorted. Severe infection destroy the foliage and at young stage, if infected kills the plant.

Fig. 21.5: Symptoms of powdery mildew on leaves of dahlia

Disease cycle

The powdery mildew fungi require living plant tissue to grow. Year-round availability of crop or weed hosts is important for their survival. Special resting spores are produced, allowing overwinter survival of the species.

The powdery mildew fungi grow as thin layers of mycelium (fungal tissue) on the surface of the affected plant part. Spores, which are the primary means of dispersal, make up the bulk of the white, powdery growth visible on the plant's surface and are produced in chains that can be seen with a hand lens. Powdery mildew spores are carried by wind to new hosts where they germinate, and initiate infection in new host tissues, where they grow and form the powdery mycelia colonies. Although humidity requirements for germination vary, all powdery mildew species can germinate and infect in the absence of free water. In fact, spores of some powdery mildew fungi are killed and germination is inhibited by water on plant surfaces for extended periods. Moderate temperatures (15.5 to 26.6°C) and shady conditions generally are the most favourable for powdery mildew development. Spores and fungal growth are sensitive to extreme heat (above 32°C) and direct sunlight.

Weather parameters

Temperature range from 15 to 20 ± 1°C is suitable for the optimum germination of the conidia of *Erysiphe polygoni*. 60-80 percent relative humidity is optimum for the conidial germination. Conidium of Erysiphe polygoni requires presence of light for germination and germ tube elongation. pH from 6.5 to 7.5 is found more suitable for conidial germination.

Disease management

The best method of control is prevention. Planting resistant varieties when available, or avoiding the most susceptible varieties, planting in the full sun,

Powdery Mildews of Flowering and Ornamental Plants   349

and following good cultural practices will adequately control powdery mildew in many cases. However, very susceptible varieties may require fungicide treatment. Several least-toxic fungicides are available but must be applied no later than the first sign of disease.

Cultural Practices

Plant in sunny areas as much as possible, provide good air circulation, and avoid applying excess fertilizer.

Fungicide Application

In some situations, especially in the production of susceptible varieties, fungicides may be needed. Fungicides function as protectants, eradicants, or both. A protectant fungicide prevents new infections from occurring whereas an eradicant can kill an existing infection. Apply protectant fungicides to highly susceptible plants before the disease appears. Use eradicants at the earliest signs of the disease. Once mildew growth is extensive, control with any fungicide becomes more difficult.

Several least-toxic fungicides are available, including horticultural oils, neem oil, jojoba oil, sulfur, and the biological fungicide Serenade. With the exception of the oils, these materials are primarily preventive. Oils work best as eradicants but also have some protectant activity.

Oils

To eradicate mild to moderate powdery mildew infections, use horticultural oil such as Saf-T-Side Spray Oil, Sunspray Ultra-Fine Spray Oil, or one of the plant-based oils such as neem oil or jojoba oil (e.g., E-rase). Be careful, however, to never apply an oil spray within 2 weeks of a sulfur spray or plants may be injured. Also, oils should never be applied when temperatures are above 32°C or to drought-stressed plants. Some plants may be more sensitive than others, however, and the interval required between sulfur and oil sprays may be even longer; always consult the fungicide label for any special precautions.

Sulfur

Sulfur products have been used to manage powdery mildew for centuries but are only effective when applied before disease symptoms appear. The best sulfur products to use for powdery mildew control in gardens are wettable sulfurs that are specially formulated with surfactants similar to those in dishwashing detergent (e.g., Safer Garden Fungicide) To avoid injuring any plant, do not apply sulfur when air temperature is near or over 32°C and do not apply it within 2 weeks of an oil spray.

Other sulfur products, such as sulfur dust, are much more difficult to use, irritating to skin and eyes, and limited in terms of the plants they can safely be used on. Copper is also available to control powdery mildew but is not very effective.

How to Use

Apply protectant fungicides, such as wettable sulfur, to susceptible plants before or in the earliest stages of disease development. The protectant fungicides are only effective on contact, so applications must provide thorough coverage of all susceptible plant parts. As plants grow and produce new tissue, additional applications may be necessary at 7 to 10 day intervals as long as conditions are conducive to disease growth.

If mild to moderate powdery mildew symptoms are present, the horticultural oils and plant-based oils such as neem oil and jojoba oil can be used to reduce or eliminate the infection.

21.6. Powdery mildew of Aster

Pathogen: *Golovinomyces cichoracearum (formerly Erysiphe)*

Geographic distribution: Worldwide

Symptoms

White powdery growth develops on the surface of leaves. Heavily infected leaves wither and die. The adaxial leaf surfaces are heavily covered with white mycelia and conidia, while the abaxial surfaces are less infected. As the disease progressed, infected leaves turned yellow and wilt. Powdery mycelia growth are also observed on stems, petioles, and flower calyxes of inflorescences (Fig 21.6).

Fig. 21.6: Symptoms of Powdery mildew on aster leaves and flower

Pathogen

Mycelium well developed, evanescent but sometimes persistant and effused, superficial with well developed haustoria in host cells. Oidiophores arise on mycelial web, and are unbranched, erect, producing oidia of ellipsoidal or barrel shaped, 25-45 × 14-26μm in size, produced in abudance and disseminated by wind. Cleistothecia gregarious or scattered, globuse becoming depressed or irregular, 90-135 μm in diameter, wall cells usually indistinct 10-20 μm wide. Cleistothecial appendages numerous, myceloid, basally inserted, hyline to dark, interwoven with mycelium, 1-4 times as long as the diameter of cleistothecium, rarely branched. Asci 10-25 per cleistothecium, ovate to broadly ovate, rarely subglobose, more or less stalked, 60-90 × 25-50 μm. Ascospores two in each ascus, very rarely three, 20-30 × 12-18 μm.

Weather Parameters

When conidia or ascospores fall on a plant surface, under favourable atmospheric conditions, they start to germinate in 2 to 4 hours, reaching a maximum number in about 25 hours. The optimum temperature for germination is about 22°C; the minimum, about 5°C and the maximum is close to 35°C.

Spore germination occurs on the leaf surface of aster over a range of relative humidity from 23 to 99 percent. Free moisture is detrimental to spore germination of the powdery mildew fungus. Once released from the conidiophores, the thin-walled conidia do not live long. At 32°C, and a relative humidity of 70 percent or less, germination reaches 95 to 100 percent in 2 hours and drops to 8 to 20 percent after 5 hours. At 21°C, and a relative humidity of 70 percent or less, germination is only 20 to 40 percent after 5 hours. Although conidia remain viable longer at a relative humidity of 80 to 90 percent, essentially all conidia are dead after 48 hours at 21°C and after 24 hours at 32°C.

The environment most favourable for conidial production, maturation, release and spread, germination, and infection include repeated day-night cycles where the nights are cool (about 16°C) and damp with a relative humidity of 90 to 99 percent, and the days are warm (about 27°C) and dry with a relative humidity of 40 to 70 percent.

Disease management

Cultural Control

Only plantlets that are free of powdery mildew should be used, or the plantlets should be disinfested before transplanting. In addition, plants should be grown in sunny areas with good air ventilation and without excessive fertigation.

Chemical Control

Fungicides used to control powdery mildew in aster include Compass (trifloxystrobin), Eagle (myclobutanil), Heritage (azoxystrobin), Milstop (potassium bicarbonate), and Spectro (chlorothalonil + thiophanate-methyl). However, chemical control may not always be completely effective since the fungus may develop resistance to some chemicals. Since powdery mildew goes through many short cycles per season, the potential for development of resistance is great, especially when only systemic fungicides are used to control this disease.

Bio-rational or Bio-fungicidal Control

Bio-fungicides are naturally based microbial or biochemical products derived from animals, plants, bacteria, and certain minerals. Bio-fungicides can affect fungal organisms directly or may stimulate the plant's own defense against the pathogens. These products are generally narrow-spectrum, and they decompose quickly. Because of these characteristics, bio-fungicides are considered to have little potential for negative impact on the environment. Bio-fungicides that have been effective for the control of powdery mildew include biological control agents (BCAs), oils, phosphorous acid, and potassium bicarbonate.

BCAs are organisms that suppress pests or plant pathogens. BCAs that have been used successfully for the control of powdery mildew include Ampelomyces quisqualis (AQ10), Tilletiopsis spp, Pseudozyma flocculosa (Sporodex, Plant Products Co., Brampton, Ontario, Canada) and Bacillus subtilis QST 713 (Serenade® or Rhapsody®, AgraQuest, Inc. Davis,California). The major impediment to BCA effectiveness is their requirement for high humidity.

Oils

Petroleum or plant oils have not only been effective in controlling powdery mildew, but also have been useful in reducing the development of resistance to fungicides. Reduction of the severity of powdery mildew through use of oils has been reported in other ornamental, fruit, and vegetable crops as well, including roses, apples, cherries, cucurbits, and grapes. In some cases, the efficacy of oils to reduce powdery mildew is comparable to or even superior to results obtained through use of standard fungicides.

21.7. Powdery mildew of Anthurium

Pathogen: *Erysiphe communis*

Geographic distribution: Worldwide

Symptoms

The symptoms are in the form of whitish powdery growth on the aerial parts of the plant. Frosty white, powdery spots develop on leaves. Colonies can become so extensive that the entire plant may appear white (Fig 21.7). Economic losses due to powdery mildew infections have so far not been assessed in any countries.

Fig. 21.7: Symptoms of Powdery mildew on Anthurium leaves and flower

Disease cycle

All powdery mildew fungi require living plant tissue to grow. Year-round availability of crop or weed hosts is important for the survival of some powdery mildew fungi. Special resting spores are produced, allowing overwinter survival of the species that causes the powdery mildew disease.

Most powdery mildew fungi grow as thin layers of mycelium (fungal tissue) on the surface of the affected plant part. Spores, which are the primary means of dispersal, make up the bulk of the white, powdery growth visible on the plant's surface and are produced in chains. These powdery mildew spores are carried by wind to new hosts and germinate under favourable environmental conditions to initiate infection. These powdery mildew spores can germinate and infect in the absence of free water. In fact, spores of some powdery mildew fungi are killed and germination is inhibited by water on plant surfaces for extended periods. Moderate temperatures (15.5° to 26.5°C) and shady conditions generally are the most favourable for powdery mildew development. Spores and fungal growth are sensitive to extreme heat (above 32°C) and direct sunlight.

When conidia or ascospores fall on a plant surface, they start to germinate in 2 to 4 hours, reaching a maximum number in about 25 hours. The optimum temperature for germination is about 22°C; the minimum, about 5°C and the maximum is close to 35°C.

Spore germination occurs on the leaf surface over a range of relative humidity from 23 to 99 percent. Once released from the conidiophores, the thin-walled conidia do not live long. At 32°C, and a relative humidity of 70 percent or less, germination reaches 95 to 100 percent in 2 hours and drops to 8 to 20 percent after 5 hours. At 21°C, and a relative humidity of 70 percent or less, germination is only 20 to 40 percent after 5 hours. Although conidia remain viable longer at a relative humidity of 80 to 90 percent, essentially all conidia are dead after 48 hours at 21°C and after 24 hours at 32°C.

The environment most favourable for conidial production, maturation, release and spread, germination, and infection include repeated day-night cycles where the nights are cool (about 16°C) and damp with a relative humidity of 90 to 99 percent, and the days are warm (about 27°C) and dry with a relative humidity of 40 to 70 percent. The disease cycle–production of conidia, release, germination, infection, and production of conidia–can be as short as 72 to 96 hours. If left uncontrolled, powdery mildew can quickly become epidemic when cool, damp nights are followed by warm, dry days.

Disease management

- Treatment with 0.1 % benomyl easily controls the disease.

21.8. Powdery mildew of Begonia

Pathogen: *Oidium spp.*

Geographic distribution: Worldwide

Symptoms

Symptoms are in the form of whitish powdery growth on the aerial parts of the plant. Frosty white, powdery spots develop on leaves (Fig 21.8). Colonies can become so extensive that the entire plant may appear white powdery.

High humidity (80% - 90%) and moderate temperatures (68 - 82°F) contribute to the development of powdery

Fig. 21.8: Symptoms of powdery mildew on Begonia leaves

mildew. Overcrowding and shading of plants will also favour development of this disease and causes severe losses.

Disease cycle

Germination, infection and sporulation follow each other in the life cycle of powdery mildew disease. The life cycle of the fungus is initiated by airborne conidia which are the asexual reproductive spores of the fungus. When the conditions are favourable, germination of the fungus is observed as direct germination of conidia. The conidia start to germinate after four hours and produces a single thick germtube from one corner of the conidia. Germinating conidia produces primary hyphae at about 30 h after incubation, secondary hyphae after 36 hours and tertiary hyphae after 48 hours of incubation. After about 60 hours the other hyphae elongate. The white patches of infection became visible to the naked eye only after 120 hours. The mature conidia get released about seven days after inoculation. The conidia get released when the host leaf and the atmosphere are relatively dry during day time. Detached conidia gets carried away by the wind and gets deposited over the fresh host leaves on which they gets germinated again to cause new infection and symptoms.

Weather parameters

When conidia or ascospores fall on a plant surface, they start to germinate in 2 to 4 hours, reaching a maximum number in about 25 hours. The optimum temperature for germination is about 22°C; the minimum, about 5°C and the maximum is close to 35°C. Spore germination occurs on the surface of a begonia over a range of relative humidity from 23 to 99 percent. Free moisture is detrimental to spore germination of the powdery mildew fungus.

Once released from the conidiophores, the thin-walled conidia do not live long. At 32°C, and a relative humidity of 70 percent or less, germination reaches 95 to 100 percent in 2 hours and drops to 8 to 20 percent after 5 hours. At 21°C, and a relative humidity of 70 percent or less, germination is only 20 to 40 percent after 5 hours. Although conidia remain viable longer at a relative humidity of 80 to 90 percent, essentially all conidia are dead after 48 hours at 21°C and after 24 hours at 32°C.

The environment most favourable for conidial production, maturation, release and spread, germination, and infection include repeated day-night cycles where the nights are cool (about 16°C) and damp with a relative humidity of 90 to 99 percent, and the days are warm (about 27°C) and dry with a relative humidity of 40 to 70 percent. The disease cycle-production of conidia, release, germination, infection, and production of conidia-can be as short as 72 to 96 hours. If left uncontrolled, powdery mildew can quickly become epidemic when cool, damp nights are followed by warm, dry days.

Disease management

Cultural control
- Space plants for good air circulation.
- Remove and destroy infected plants.
- Plant in sunny open locations rather than shady.

Chemical control
- Spraying Sulfex (3g/litre of water) can effectively control the disease.
- Karathane (40 E.C) @ 0.5% or dusting sulphar powder at 15 days interval

21.9. Powdery mildew of Delphinium

Pathogen: *Erysiphe polygoni*

Geographic distribution: Worldwide

Symptoms

Dry, white, mealy, fungal growth occurs on the upper surface of leaves (Fig 21.9). Leaves turn yellow; die from the base of the stem upward. White or grayish patches on leaves, stems and sometimes on closed delphinium buds appears late in summer. Plant vigour and stem quality can be reduced.

Fig. 21.9: Symptoms of Powdery mildew on leaves of Delphinium

Disease cycle

All powdery mildew fungi require living host plant tissue to grow. Year-round availability of crop or weed hosts is important for the survival of powdery mildew fungi, however; some special resting spores are produced, allowing overwinter survival of the pathogen that causes the powdery mildew disease.

Powdery mildew spores are carried by wind to new hosts. Although humidity requirements for germination vary, all powdery mildew species can germinate and infect in the absence of free water. In fact, spores of some powdery mildew fungi are killed and germination is inhibited by water on plant surfaces for extended periods. Moderate temperatures (15.5° to 26.5°C) and shady conditions generally are the most favourable for powdery mildew development. Spores and fungal growth are sensitive to extreme heat (above 32°C) and direct sunlight.

Weather parameters

Temperature range from 15 to 20 ± 1°C is suitable for the optimum germination of the conidia of *Erysiphe polygoni*. 60-80 percent relative humidity was optimum for the conidial germination. Conidia of *Erysiphe polygoni* required presence of light for germination and germ tube elongation. pH from 6.5 to 7.5 was found more suitable for conidial germination.

Disease management

The best method of control is prevention. Planting resistant varieties when available, or avoiding the most susceptible varieties, planting in the full sun, and following good cultural practices will adequately control powdery mildew in many cases. However, very susceptible varieties may require fungicide treatment. Several least-toxic fungicides are available but must be applied no later than the first sign of disease.

Cultural Practices

Plant in sunny areas as much as possible, provide good air circulation, and avoid applying excess fertilizer.

Fungicide Application

In some situations, especially in the production of susceptible varieties, fungicides may be needed. Fungicides function as protectants, eradicants, or both. A protectant fungicide prevents new infections from occurring whereas an eradicant can kill an existing infection. Apply protectant fungicides to highly susceptible plants before the disease appears. Use eradicants at the earliest signs of the disease. Once mildew growth is extensive, control with any fungicide becomes more difficult.

Several least-toxic fungicides are available, including horticultural oils, neem oil, jojoba oil, sulfur, and the biological fungicide Serenade. With the exception

358 The Plant Mildews

of the oils, these materials are primarily preventive. Oils work best as eradicants but also have some protectant activity.

Oils

To eradicate mild to moderate powdery mildew infections, use a horticultural oil such as Saf-T-Side Spray Oil, Sunspray Ultra-Fine Spray Oil, or one of the plant-based oils such as neem oil or jojoba oil (e.g., E-rase). Be careful, however, to never apply an oil spray within 2 weeks of a sulfur spray or plants may be injured. Also, oils should never be applied when temperatures are above 32°C or to drought-stressed plants.

Sulfur

Sulfur products have been used to manage powdery mildew for centuries but are only effective when applied before disease symptoms appear. The best sulfur products to use for powdery mildew control in gardens are wettable sulfurs that are specially formulated with surfactants similar to those in dishwashing detergent (e.g., Safer Garden Fungicide) To avoid injuring any plant, do not apply sulfur when air temperature is near or over 32°C and do not apply it within 2 weeks of an oil spray. Other sulfur products, such as sulfur dust, are much more difficult to use, irritating to skin and eyes, and limited in terms of the plants they can safely be used on.

21.10. Powdery mildew of ornamental plants

21.10.1. due to *Erysiphe cichoracearum*

Host infected

Acalypha, Achillea, African violet, Ageratum, Anchusa, Arnica, Artemisia, Artillery plant, Aspen, Aster, Bachleor's button, Balm, Balsam apple, Balsam pear, Basket flower, Bedstraw, Begonia, Black-eyed Susan, Boltonia, Boneset, Bugleweed, Buttercup, Butterfly weed, Calendula, Campanula, China aster, Chrysanthemum, Cineraria (Florist's), Clematis, Coralbells, Coreopsis, Cosmos, Dahlia, Delphinium, Dusty Miller, Erigeron, Eucalyptus, Eupatorium, Feverfew, Forget-me-not, Gaillardia, Gayfeather, Germander, Globeflower, Gloxinia, Goldenglow, Golden aster, Goldenrod, Hebe, Helenium, Hollyhock, Houndstongue, Inula, Ivy (English), Joepye weed, Larkspur, Leopardsbane, Liatris, Linaria, Lithospermum, Mallow, Marguerite, Mertensia, Mockcucumber, Monkey flower, Nemophila, Oswego tea, Penstemon, Phlox, Prairie coneflower, Romanzoffia, Rudbeckia, St.-Johns-wort, Salpiglossis, Salvia, Smoketree, Snapdragon, Spirea, Stachys, Sunflower, Sweet William, Tansy, Tidytips, Transvaal daisy, Trumpetvine, Verbena, Veronica and Zinnia.

Powdery Mildews of Flowering and Ornamental Plants 359

Symptoms

Powdery mildews first appear as superficial, white, powdery patches on the leaves, young stems, buds, and flowers. These patches may enlarge until they cover the whole leaf on one or both sides. The white powdery growth is composed of mycelium of the fungus and chains of colourless spores (conidia) borne on upright stalks (conidiophores) arising from the white mycelium on the surface of the host plant. Later in the season, these patches may become mealy or felt-like, turn gray to tan in color, and become dotted with minute, dark brown-to-black cleistothecia. A stunting or dwarfing, curling of leaves, chlorosis, premature leaf drop, and deformation of flower buds frequently follow mildew infection. Many powdery mildews, especially those that attack trees and shrubs, are much more unsightly than harmful.

Disease cycle

Mycelium well developed, evanescent but sometimes persistent and effused, superficial with well developed haustoria in host cells. Oidiophores arise on mycelial web, and are unbranched, erect, producing oidia in chains at their apex. Oidia ellipsoidal or barrel-shaped, 25-45 × 14-265µm, produced in abundance and disseminated by wind. Cleistothecia gregarious or scattered, globose becomimg depressed or irregular, 90-135 5µm in diameter, wall cells usually indistinct 10-20 5µm wide. Cleistothecial appendages numerous, myceloid, basally inserted, hyaline to dark, interwoven with mycelium, 1-4 times as long as the diameter of cleistothecium, rarely branched. Asci 10-25 per cleistothecium, ovate to broadly ovate, rarely subglobose, more or less stalked, 60-90 × 25-50 5µm. Ascospores two in each ascus, very rarely three, 20-30 × 12-18 5µm.

21.10.2. due to *Erysiphe lagerstroemiae*

Host infected: Crapemyrtle

Disease cycle

The powdery mildew fungi require living plant tissue to grow. The powdery mildew survives from one season to the next as vegetative strands in buds or as spherical fruiting bodies, called cleistothecia, on the bark of branches and stems.

The powdery mildew fungi grow as thin layers of mycelium on the surface of the affected plant parts. Spores are part of the white powdery growth of the fungi and are produced in chains on upper or lower leaf surfaces or on flowers, or herbaceous stems. Environmental conditions that favour the growth of powdery mildew include low temperatures of 10° to 21°C, a relative humidity of 90% or higher and free moisture. Wind carries powdery mildew spores to

360   The Plant Mildews

new hosts. The powdery mildew species can germinate and infect in the absence of free water. In fact, water on plant surfaces for extended periods inhibits germination and kills the spores of most powdery mildew fungi. Moderate temperatures of 15.5° to 26.5°C and shady conditions generally are the most favourable for powdery mildew development. Powdery mildew spores and mycelium are sensitive to extreme heat and sunlight, and leaf temperatures above 35°C may kill the fungus.

21.10.3. due to *Erysiphe polygoni*

Host infected

Acacia, Aconitum, Alyssum, Amelanchier, Amorpha, Anemone, Arenaria,, Astilbe, Bean (Sclarlet runner), Begonia, Bladder senna, Brassica, Bundle flower, Calendula, California poppy, Candytuft, China aster, Cinquefoil, Clematis, Columbine, Cuphea, Dahlia, Delphinium, Erigeron, Evening primrose, Gardenia, Genista, Geranium, Heath, Hollyhock, Honeysuckle, Hydrangea, Kalanchoe, Larkspur, Locust (black), Lupine, Matrimony vine, Meadow rue, Pansy, Peony Poppy, Primrose, Sandwort, Scabiosa, Sedum, Serviceberry, Sophora, Sweet alyssum, Sweet pea, Teasel, Tulip and Wallflower (Western).

Disease cycle

The powdery mildew fungi require living plant tissue to grow. On perennial hosts powdery mildew survives from one season to the next as vegetative strands in buds or as spherical fruiting bodies, called cleistothecia, on the bark of branches and stems.

The powdery mildew fungi grow as thin layers of mycelium on the surface of the affected plant parts. Spores are part of the white powdery growth of this fungi and are produced in chains on upper or lower leaf surfaces or on flowers, fruits, or herbaceous stems..

Wind carries powdery mildew spores to new hosts on which these germinate and initiate infection under favourable weather condition. The powdery mildew species can germinate and infect in the absence of free water. In fact, water on plant surfaces for extended periods inhibits germination and kills the spores of most powdery mildew fungi. Moderate temperatures of 15.5° to 26.5°C and shady conditions generally are the most favourable for powdery mildew development. Powdery mildew spores and mycelium are sensitive to extreme heat and sunlight, and leaf temperatures above 32°C may kill the fungus.

Powdery Mildews of Flowering and Ornamental Plants   361

21.10.4. due to *Microsphaera alni*

Host infected

Alder, Azalea, Beech, Birch, Bittersweet, Buckthorn, Burning bush, Buttonbush, Catalpa, Checkerberry, Chestnut, Chinquapin, Cranberry bush, Dahoon, Dogwood, Elder, Elm, Euonymus, Forestiera, Hackberry, Hazelnut, Hickory, Holly, Honey-locust, Honesuckle, Hophornbeam, Hornbeam, Laborador-tea, Lilac, Linden, Lyonia, Magnolia, Moonseed, Mountain holly, Mountan laurel, Oak, Pecan Planetree, Privet, Rhododendron, Shallon, Snowball, Snowberry, Spindle tree, Spirea, Sweet pea, Sycamore, Trailing arbutus, Trumpet vine, Viburnum and Walnut.

Disease cycle

The powdery mildew fungi commonly overwinter as mycelial mats in rudimentary leaves within dormant buds, especially on woody plants. Infected buds break open in the spring and may develop into systemically infected shoots. The fungi sporulate on these shoots, producing large numbers of barrel-shaped conidia that are carried by the wind, splashing water, or other means to healthy plant tissue, where they infect the upper and lower leaf surfaces, thus initiating a new disease cycle. Another means of winter survival for powdery mildew fungi in the Midwest is as cleistothecia embedded in the mealy or felt-like mildew growth on plant stems and fallen leaves. The minute cleistothecia are formed within the mycelial mat as the host tissues mature. During warm and humid weather in the spring, a cleistothecium absorbs water and cracks open to discharge one or more asci. The microscopic ascospores are carried by the wind or splashing raindrops to healthy plant tissue where they germinate and may cause infection. A mycelial mat is formed and chains of conidia are evident within a few days, completing the disease cycle.

21.10.5. due to *Podosphaera leucotricha*

Host infected: Crabapple, Photinia and Quince.

Disease cycle

P. leucotricha overwinters as mycelium in dormant flower and shoot buds infected the previous year. In spring, the infected buds break dormancy and the fungus resumes growth, colonizing the developing shoots and young leaf tissue. From these primary infections, asexual conidia are produced on conidiophores and dispersed by wind. Conidia germinate at high relative humidity (greater than 70%, which is commonly available in the microclimate of the lower leaf surface) at temperatures between 10 and 25°C; in contrast to most foliar fungal pathogens, leaf wetting is a deterrent to infection. The youngest leaves are the most susceptible, but become increasingly resistant as they mature.

Mildew colonies generally appear first on the lower leaf surface as white felt-like patches. Conidia germinate to form hyphal outgrowths, which traverse the leaf surface, swell and then flatten to form appressoria. These structures release enzymes, which allow fungal infection pegs to penetrate the plant's epidermal cells and then enlarge to form haustoria. Haustoria are specialized organs formed inside living plant cells, which absorb nutrients and anchor the fungus. As the mildew colony expands or as secondary infections lead to new colony formation, the infection process (hyphal outgrowth > appressorium > infection peg > haustorium) is repeated until susceptible tissue is no longer available. Late season growth may result in a sudden increase in mildew activity. In addition to contributing toward a rapid inoculum buildup, secondary disease cycles are also responsible for infecting lateral and terminal buds that will carry the fungus through the winter.

P. leucotricha also produces sexual spores (ascospores) in sac-like asci enclosed in fruiting bodies (ascocarps). Each ascocarp contains a single ascus with eight ascospores, each of which is elliptical and measures 22-36 x 12-15 μm. Ascocarps are recognized as distinct black dots on the surface of a mycelial mat. Ascocarps are densely grouped together, measure 75-96 μm in diameter and have apical and basal appendages. Ascocarps form late in the growing season and serve as overwintering structures, but don't play any known role in initiating new epidemics, as the ascospores fail to germinate readily. In the past, the ascocarps of *P. leucotricha* were called cleistothecia (reflecting the closed nature and lack of a preformed opening), perithecia (reflecting the arrangement of the asci of many powdery mildew fungi in a layer [hymenium]), and most recently, chasmothecia. All three terms can be found in the literature. Since cleistothecia in other groups of ascomycetes lack a hymenium (*i.e.*, the asci are randomly scattered throughout the enclosed structure), the term *chasmothecia* has been recently introduced to distinguish powdery mildew ascocarps from other cleistothecia. The word is derived from the vertical chasm that is formed during ascospore discharge.

21.10.6. due to *Phyllactinia corylea* (*P. guttata*)

Host infected

Alder, Amelanchier, Aralia, Ash, Barberry, Beach, Birch, Boxelder, Buckeye, Buttonbush, Calycanthus, Catalpa, Chestnut, Chinaberry, Crabapple, Grape myrtle, Dogwood, Elder, Fringetree, Hazelnut, Hickory, Holly, Holodiscus, Hophornbeam, Hornbeam, Horse chestnut, Linden, Locust(black), Magnolia, Maple, Mock orange, Mulberry, Ninebark, Oak, Oso berry, Painted tongue, Plane tree, Quince, Rose, Sassafras, Silver berry, Sycamore, Tulip tree, Walnut Willow, Witch hazel and Yellow wood.

Disease cycle

The fungus occurs most freely by forming a powdery coating of conidia on the leaves, and at a later stage forms perithecia abundantly on detached fallen leaves. The conidia are oval and pointed, and germinate in about 2 hours on the leaf surface. The germ tube elongates and enters the leaf through the stomata. Hyphae grow in the intercellular spaces of the leaf and derive nutrition from the host cells through globular haustoria. The conidiophores emerge through the stomata and form conidia. A single conidium is borne by the conidiophores at one time, which is detached before another formed, a feature characteristically shared only by the other internally-growing genus *Leveillula*. The conidia spread the fungus until defoliation starts with the onset of autumn.

The fungus is heterothallic and the two uninucleate, gamentangia are formed on two closely lying hyphae. The female gamentangium is swollen and a little curved. Nuclear migration and the role of antheridium are controversial. Detail of ascus and stroma formation is same as described earlier. The mature ascoma is brown or black and bears appendages characteristically swollen at the base. In addition, there are "secretary appendages", which have a wide base and form small apical branches with swollen tips. These secrete mucilage, which help in adhesion of the ascoma to the substratum. The ascoma, after completing the dormancy, bursts by elongation of the asci. The asci usually contain only two hyaline ascospores, which on germination initiate an internal mycelium in the host.

21.10.7. due to *Sphaerotheca humuli (S. macularis)*

Host infected

Agastache, Agrimony, Betony, Buffalo berry, Cinquefoil, Delphinium, Elder, Epilobium, Erigeron, Filipendula, Foamflower, Gaillardia, Geranium, Geum, Gilia, Hawksbeard, Hawkweed, Heuchera, Hydrophyllum, Kalanchoe, Larkspur, Matricaria, Meadow sweet, Mitella, Ninebark, Pansy, Phlox, Polemonium, Rose, Saxifrage, Spirea, Sumac, Tamarisk, Veronica and Violet.

Disease cycle

Sphaerotheca humuli uses haustoria to gain access to the leaf epidermal cells. The fungus is usually spread during the spring through mycelium from infected plant, or through ascocarps. Signs appear after 3–7 days of infection if conditions are favourable. The mycelium grows rapidly during the warm summer months with an optimum temperature of about 10-32^0C. The leaves are most susceptible 16–23 days after unfolding. High humidity favours the development of disease, but infection can occur at relative humidity as low as 50%. The conidia of the fungus are spread through the air and thus can travel over great distances. The mycelium can also overwinter in the buds of infected plants.

21.10.8. due to Uncinula *necator*

Host infected

Actindia, Ampelopsis, Ivy (Boston) and Virginia creeper.

Disease cycle

Powdery mildew is a polycyclic disease (one which produces a secondary inoculum) that initially infects the leaf surface with primary inoculum, which is conidia from mycelium, or secondary inoculum, which is an overwintering structure called a cleistothecium. A young clesitothecium, is used to infect the host immediately or overwinter on the host to infect when the timing is right (typically in spring). To infect, it produces a conidiophore that then bears conidia. These conidia moves along to a susceptible surface to germinate. Once these conidia germinate, they produce a structure called haustoria, capable of "sucking" nutrients from the plant cells directly under the epidermis of the leaf. At this point, the fungi can infect leaves, buds and twigs that then re-infect other plants or further infect the current host to produce more white powdery growth as a signs of powdery mildew, on which secondary inoculums is produce to re-infect the host.

For germination to occur using a cleistothecium, the cleistothecium must be exposed to the right environmental conditions to rupture the structure to thereby release spores in hope that they will germinate. Germination of conidia occurs at temperatures between 7 and 31°C and is inhibited above 33°C. Germination is greatest at 30-100% relative humidity.

Symptoms of powdery mildew on some of the important flowering and ornamental plants are given in Fig. 21.10.

Powdery Mildews of Flowering and Ornamental Plants 365

Aralia

Fig. 21.10: Symptoms of powdery mildew on some of the ornamental plants.

Disease management

An effective control program must take into account the interaction of the host, environment, and fungus. As an alternative to exclusive pesticide use, we recommend the coordinated use of multiple tactics—an approach known as integrated pest management (IPM). This approach requires careful monitoring of plants to determine when controls are necessary, and then designing a control program based on appropriate cultural and/or chemical action.

Cultural

Choosing the right plants for the location is a good way to minimize powdery mildew problems for the home gardener. Avoid growing mildew-prone crops, such as phlox or roses, where it is somewhat shady and air circulation is poor. Increase aeration where possible by pruning and thinning plant materials. Some varieties differ in their susceptibility to powdery mildew.

Even varieties of rose, zinnia, and other crops often advertised as "powdery mildew-resistant," sometimes show infection. This is because there are many fungal strains, and the development of powdery mildew-resistant ornamental plants has not been very successful.

Mildew is most severe on young, succulent growth that excessive nitrogen fertilizer promotes. You can sometimes reduce the potential for powdery mildew on susceptible plants by lowering nitrogen and increasing potassium fertilization. Powdery mildew may be the most common foliage disease affecting several greenhouse grown plants, because it is favoured by the indoor environment. These plants include begonia, chrysanthemum, ivy, hydrangea, Kalanchoe, and roses. A high relative humidity (80-100%) occurs frequently in the greenhouse, especially at night, and it benefits spores and fungal infections. High night time humidity is often followed by lower relative humidity (50-80%) during the day. This benefits spore maturation and release for rose powdery mildew and probably other powdery mildew fungi as well. Air currents during periods of low humidity increase the number of spores in the air and aid in their distribution. Greenhouse operators can sometimes manipulate the environment by ventilating and turning up the heat in late afternoon to drive out moisture before nightfall. However, because of fuel costs, this is not a common practice.

Lilac, zinnia, and several other ornamentals often become heavily infected in the fall when there are large temperature fluctuations. Perennials are seldom damaged by fall infection and require treatment only if their appearance is objectionable.

Chemical

If cultural controls fail to prevent disease build-up or if the disease pressure is too great, fungicide spraying may be necessary. The best course of action is to combine both approaches, using cultural methods as well as following a good spray schedule.

Due to the fact that a large part of the fungal organism is within easy reach, powdery mildews are rather susceptible to fungicides.

On the other hand, powdery mildews may develop decreased susceptibility, if substances with only one single mode of action are used repeatedly. Therefore, to avoid the development of resistant powdery mildew strains, it is absolutely necessary to vary the fungicide (active compound) during the course of chemical control. Spraying with fungicides may be necessary as soon as infections are to be expected. Therefore, it is advisable to inspect the plants regularly for the typical symptoms. Especially plants close to windows and doors, where draft is likely to occur, should be checked. Experienced Euphorbia enthusiasts will know the most susceptible plants and check those with particular attention. At the latest when the first white powdery spots are visible, spraying with certified fungicides is indicated. Once the white cover becomes more extensive, the effect of fungicides decreases.

If contact or preventive fungicides are used, it is crucial to cover the whole plant with the spray in order to get an even fungicide layer. Since these pesticides protect the plants against infections (they prevent or stop the germination and germ tube elongation), they cannot work when the fungus has already entered the plant tissue. In some cases it is not clear, whether there is a fungicidal or a plant strengthening effect. The active compounds like hydrogen carbonates and lecithin are often regarded as plant strengtheners.

Their effectiveness against powdery mildews is not without controversy. In some trials - especially with low disease pressure - these components performed quite well, in others they failed. Moreover, they have to be applied frequently and in an early stage of the disease. However, as they at least have some effect directly against the fungus, they are listed here.

Hydrogen carbonates

Baking soda (sodium bicarbonate, $NaHCO_3$) and potassium hydrogen carbonate ($KHCO_3$) are recommended as preventive fungicides. The suggested concentration is between 0.5 and 2 %.

Often the carbonates act better in mixtures with oils plus emulsifier (surfactant). As plants may be damaged, hydrogen carbonates should be used with care. Also important soil parameters are affected by immoderately applied carbonates (pH, sodium content).

Lecithin

As a ubiquitous component in organisms and a common food additive, lecithin is regarded as a low risk biological fungicide. Frequently applied it affects the elongation of hyphae and spore formation. Therefore, the spread of powdery mildew on the plants is retarded.

Sulphur

Sulphur is one of the oldest active ingredients used for plant protection purposes. For more than 150 years it has been applied worldwide in agriculture especially to protect plants against attacks of powdery mildews. As an inorganic compound it is a so-called "multisite inhibitor", which means that sulphur does not bind selectively to a special enzyme, but inhibits or inactivates many different steps of the metabolism at the same time. Therefore, the risk of causing resistance even by frequent application of sulphur is very low.

Wettable Sulphur is the most suitable form of sulphur for plant protection purposes. The active ingredient is formulated with surfactants, which makes the sulphur more easily mixed with water. A traditional alternative to using suspended sulphur is the application of evaporated sulphur. In special devices, pure sulphur is evaporated and distributed in the greenhouse. The sulphur condenses on plants (and elsewhere) and is active against powdery mildew. Although sulphur is a "natural" (which often is wrongly understood as "safe") fungicide, it may cause plant damage when used at temperatures higher than approx. 28 °C.

Strobilurines (Azoxystrobin amongst others)

Fungicidal compounds of the group strobilurines are comparably new in plant protection. They are derivatives of a substance produced by a certain fungus in order to suppress other fungi. The synthetic strobilurines, available since the 1990's are more stable and affect a wide range of fungi.

Although strobilurines are absorbed by the cuticle and diffuse into the cells, they do not provide full systemic activity. Therefore they offer the best effect when used as preventive fungicides. The spraying should be executed as early in disease development as possible.

In contrast to preventive fungicides, some active ingredients – often referred to as curative fungicides – might stop the fungus in a later phase of the disease. Though, as stated above, the earlier control measures are taken, the better the chances of success.

Curative and eradicative

Oils

Horticultural oils (formulated mineral or plant oils) are typically used as contact insecticides against scale insects. Although the effect against powdery mildew has been known for many years, it is not widely used in professional horticulture. May be the effect is too low or uncertain, but there are trial reports promising sufficient efficacy to eradicate at least mild infections. Gardeners should be aware of the risk of harm to susceptible plants especially at high temperatures and in some mixtures with other compounds (especially sulphur) with oil applications.

Azoles (Myclobutanil, Tebuconazole amongst others)

In contrast to all other fungicides mentioned above, azole fungicides are systemically active, meaning that they get into the plant tissue and are transported within the plant. However, most of the systemic compounds are mobile in the xylem (water conducting system) only. Therefore the distribution is always from base to top. From this it follows that branches or leaves located lower on the plant are not protected, when application does not cover them. Azoles are a big group of fungicides with different ranges of efficacy. Myclobutanil and Tebuconazole are two azole compounds active against powdery mildew.

Common names or trade names of fungicides that can be used against powdery mildew on ornamentals.

Dinocap: Karathane

Dithiocarbamates : Duosan, Zyban

Myclobutanil Systhane : "sterol inhibitors"

Triadimefon : Bayleton

Triforine : Funginex, Triforine

Sulfur : Sulfur, wettable sulfur

Thiophanate-methyl: Cleary 3336, Domain, Fungo

Biological Fungicides

Biological fungicides (e.g., Serenade) are commercially available beneficial microorganisms formulated into a product that, when sprayed on the plant, destroys fungal pathogens. The active ingredient in Serenade is a bacterium, *Bacillus subtilis*, that helps prevent the powdery mildew from infecting the plant. These products have some effect in killing the powdery mildew organism but are not as effective as the oils or sulfur in controlling it.

22

Powdery Mildew of Medicinal and Aromatic Plants

Medicinal plants have been used as a major source of therapeutic agents by human beings for thousands of years. Ancient man obtained more than 90% of his medicaments from higher plants. However, the importance of plants as a source of drugs decreased to certain extent with the advancement of knowledge in modern medicine and especially because of the phenomenal development in synthetic drugs and antibiotics. A large number of active constituents of plants were synthesized and it was more or less assumed that most of the plant drugs would be sooner or later obtained from synthetic sources. In spite of the fact that synthetic drugs and antibiotics brought out a revolution in controlling different diseases of man and animals, plants continued to occupy an important place as raw material for some most important drugs used in modern medicine throughout the world (Husain, 1993).

Plant growth and yield depend on protecting the plants from diseases. Anything that affects the health of plants is likely to affect their growth and yield and also seriously reduce their usefulness to mankind. Plant diseases are thus important to man because they not only damage plants but also the plant products. The powdery mildew affects various medicinal plants to lower down their medicinal value.

22.1. Powdery mildew of Plantago

The disease is of minor importance. The disease appeares at the time of flowering (Shankhela *et al.,* 1985).

Pathogen: *Erysiphe cichoracearum* D. C. (Kumawat, 1979).

Symptoms

Small, white or greyish spots appeares on the leaves which gradually enlarge and eventually covere the entire plant surface with white powdery mass (Fig 22.1).

Fig. 22.1: Symptoms of powdery mildew on plantago

Disease Management

The disease can be controlled by spraying Karathane W.D. (0.2%) or wettable sulphur compounds immediately after the appearance of the disease (Shankhela, et.al, 1985).

Powdery mildew can be brought under control by spraying the crop with wettable sulphur, e.g. 'Karathen' and 'Sulfex' two or three times at 15-day intervals after the appearance of infection.

22.2. Powdery mildew of Rauvolfia

The disease was found prevalent on *Rauvolfia serpentina* in the plantations in India (Ganguly and Pandotra, 1962).

Pathogen: *Leveillula taurica* (Lev) Atz.

Symptoms

The characteristic symptom of the disease was rolling of infected leaves enclosing the lower surface of the leaves which is covered with whitish growth of the fungus in patches, consisting of conidia and conidiophores.

Cleistothecia are not found in natural conditions.

Fig. 22.2: Powdery mildew of Rauvolfia serpentine plant.

Disease Management

The disease can be controlled by one application of any wettable sulphur preparation (0.3%) or Karathane WD 0.05% at the time of first appearance of the disease (Pandey 1995).

22.3. Powdery mildew of Opium poppy

The powdery mildew disease of opium poppy was first observed in Pakistan (Zaman and Khan, 1969). The disease has been found in the severe from in many parts of India viz. Rajasthan, Madhya Pradesh and Uttar Pradesh. Poppy cultivation suffers heavy losses on account of this disease. Kothari and Prasad (1972) reported that powdery mildew caused severe damage to the crop in Rajasthan.

Pathogen: *Erysiphe polygoni* DC.

Symptoms

Powdery mildew disease symptoms are observed on 14-16 weeks old poppy plants. The first appearance of the disease in India is in February or early March. The base of the stem attacked first, where it appeared as small circular patches. Several new colonies of the pathogens continue to appear covering large part of stem. Blackening develops in the affected parts; often the stem appeared black interspersed with green patches. The mycelia spread all over the leaves. The luxurious growth of the fungus is found on the leaf surface in case of severe infection. The entire leaf surface is covered by white powdery masses of mycelium (Fig 22.3).

Fig. 22.3: Symptoms of powdery mildew on Opium poppy

The age of host is a strong factor influencing the disease development. The disease is never found on plants less than 80 days old. The environmental factors also affect the disease development. Temperature is one of the main factors; a temperature below 27°C is conducive to the disease development in India (Kothari and Prasad 1972).

Disease Cycle

The fungus is chiefly extrametrical on the host, penetration and infection being confined to epidermal cells where haustoria are found. Mycelium ,white, dense, well developed on leaves and stems. Conidia born singly or in short chains, cylindrical or ovoid, 30-45 x 10-20 µm. Cleistothecia scattered, globose, first yellow becoming brownish-black with maturity, 90-150 µm dia, cells of cleistothecium irregular, brown. Appendages, numerous, basally inserted, narrow, more or less brown, simple or sometimes branched, 1-2 times as long as the diameter of the cleistothecium and forming a dense web around it. Asci 3-12, obviate, 2-4 spored, 50-75 x 26-40 µm. Ascospores elliptic to ovate, hyaline, 20-30 x 10-12 um.

Disease Management

The disease can be effectively controlled by spraying 0.5% wettable sulphur at the time of the appearance of the disease or when the crop attains the age of 80 days. Other fungicides commonly used to control powdery mildew of various crops are also effective for controlling poppy powdery mildew. Carbendazim (@ 0.05 and 0.1%) spray once before (protective) or after (curative) the appearance of disease check the disease.

The diseased plants in the early stage should be uprooted and burnt.

22.4. Powdery mildew of Ginseng

A powdery mildew disease was observed on American ginseng *(Panax quinquefolium)* plants in British Columbia, Canada. The disease was found where three year old plants were growing on raised beds under artificial shades.

A powdery mildew disease was reported in China on both Asian and American ginseng. The pathogen was referred to *Erysiphe panax* Bai & Wang. However, it is not know that the species reported from the American is identical to *E. panax* (Sholberg *et al.,* 1996).

Pathogen: *Erysiphe panax* Bai & Wang.

Symptoms

The upper surface of 5 to 10 leaflets of the plants is covered with colonies of extensive white, superficial mycelium (Fig.22.4).. The leaves became yellow and fell prematurely. No infection of berries or stems is observed.

The disease is believed to cause severe damage to seed production, root weight and winter hardiness of ginseng plants (Sholberg *et al.*, 1996).

Fig. 22.4: Powdery mildew of Ginseng plant.

Disease cycle

The fungus produce abundant cylindrical to oblong arthroconidia, 60-100 μm long and 8.0-8.5 μm in diameter. Yellow to brown spherical cleistothecia like ascocarps are found in groups of 5 to 20 on heavily infected leaves. There are two or more asci in each ascocarp and appendages are hyphoid. The number of ascospores in each ascus is uncertain because few mature asci are seen with seven elliptical immature ascospores (approximately 22 x 13 μm).

Disease Management

The disease can be controlled by application of wettable sulphur (0.3%) or Karathane WD 0.05% at the time of first appearance of the disease.

23

Powdery Mildews of Forest Trees

Forests provide multipurpose benefits such as timber, fodder, fuel and minor forest produces. They also help in conserving soil and water, offering food and shelter for wild life, adding to the aesthetic value and recreational needs of man. Forestry has the prime objective of developing and protecting forests for their maximum productivity. Diseases and insect pests constitute major biological determinants of forest productivity, particularly in nurseries and plantations. They cause heavy damage to seedlings and hence reduce both quantity and quality of planting stock. Large-scale mortality in the nursery due to disease and pest problems could seriously affect the plantation program by reducing the stock of seedlings. In plantations, they cause major problems resulting in the reduction of biomass production or loss of valuable germplasm collections. Further, the infected seedlings are weakened and unable to withstand the adverse field/plantation conditions. Thus, the economic loss resulting from nursery diseases and insect-pests are considerable. Therefore, raising disease free, healthy tree seedlings is not only important for maintaining a good nursery stock but also essential in establishing a healthy stand in the field for better productivity. In forest major diseases problems are like damping-off, leaf spot, leaf blight, leaf rust, powdery mildew, stem rot, seedling wilt, root-rot and collar-rot caused by various pathogens on economically important fast growing tree.

Powdery mildew is one of the important disease of forest trees which reduce the plant growth and economic value of the forest product. The important powdery mildews on forest trees are:

23.1. Powdery mildew of Eucalyptus
Pathogen: *Erysiphe cichoracearum*

Geographical distribution
The disease has been reported on Eucalyptus in Argentina, Australia, Brazil, Denmark, Germany, India, Italy, New Zealand, Poland, Portugal, South Africa, UK and USA (Sankaran *et al.* 1995). Formal reports of powdery mildew on Eucalyptus spp. in nurseries in South-East Asia are few although Kobayashi

(2001) reported that Oidium is often prevalent on Eucalyptus seedlings in nurseries in the region.

Symptoms

It attacks leaves and young shoots of Eucalyptus, producing a thick layer of densely inter-woven white mycelium on the surface of leaves (Fig 23.1) and shoots (Crous *et al.* 1989), sometimes causing spotting and malformation of older growth (Gibson 1975, Boesewinkel 1981). Both sides of leaves of affected plants are covered with mycelia and conidiophores. The ellipsoid conidia are in chains, does not contain fibrosin bodies, and are confirm to those of *Erysiphe cichoracearum* DC.

Fig. 23.1: Symptoms of powdery mildew on eucalyptus leaves

Disease Cycle

Spores of Oidium spp. germinate on the surfaces of the leaves producing germ tubes that penetrate the walls of the leaf epidermal cells. The fungus forms absorbing structures known as haustoria through which it obtains nourishment from the host cells. The fungus proliferates over the leaf surfaces, producing abundant conidia, which result in the powdery white appearance from which the name of the disease is derived. The spores, which are produced successively on specialized hyphae arising from the superficial mycelium, are dispersed by wind to other susceptible hosts, initiating new infections. When the perfect stages are present, small black dots (cleistothecia) are observed immersed in white mycelium layers. The cleistothecia of Erysiphe and Sphaerotheca have hairlike, unbranched flexuous appendages and contain one to several asci with eight ascospores.

Disease Management

Early recognition and prompt removal of infected plants are important in preventing disease spread and fallen leaves should be destroyed to reduce inoculum potential. Chemical treatments are seldom necessary, but sometimes the control of powdery mildew relies primarily on the use of fungicides (Wardlaw and Phillips 1990) and can be achieved by spraying with a fungicide, such as benomyl, chlorothalonil, triademefon, maneb or zineb (Sehgal *et al.* 1975).

Biological control is an alternative means of management of foliar diseases including powdery mildews, especially in greenhouses. Commercial biocontrol products containing *Trichoderma harzianum* T39, *Ampelomyces quisqualis*, *Bacillus* and *Ulocladium* have been developed for greenhouse crops (Paulitz and Belanger 2001). Foliar sprays with other substances such as JMS Stylet-oil for cucurbit powdery mildew (McGrath and Shishkoff 2000) and non swelling chlorite mica clay for cucumber powdery mildew (Ehret *et al.* 2001) have been proved to significantly reduce the severity of the disease in greenhouses.

23.2. Powdery mildew of teak

Powdery mildew of teak caused by the fungus *Uncinula tectonae* Salmon occurs in teak plantations of all age groups but mostly in plantations of age above 15 years.

Pathogen: *Uncinula tectonae* E. S. Salmon

Symptoms

The symptoms include irregular white, powdery coating on teak leaves forming minute, globular dark coloured cleistothecia over the white fungus weft (Fig 23.2). It attacks the lower surface of leaves of teak and many broad-leaved trees listed by Spaulding (1961) resulting in defoliation. The white patches, consisting of mycelium and asexual conidia, developed on the upper leaf surface towards November /December just before the senescence. These patches coalesce and cover the entire surface of the leaf giving greyish-white powdery appearance. However, the mildew may not appear to be seriously damaging as the infected leaves are shed along with healthy leaves during normal leaf-fall. Due to heavy coating of leaves by fungus growth, photosynthetic activity is retarded which may affect growth. Severely infected leaves are defoliated prematurely. The powdery growth is a mass of conidia borne on conidiophores.

Fig. 23.2: Symptoms of powdery mildew on teak plant leaves

Geographical Distribution

The fungus is recorded from North America, Europe and Asia. The primary source of infection is ascospores (sexual); the secondary source is conidia (asexual).

Disease cycle

Conidia which develop abundantly on the leaf surface are wind-dispersed to cause fresh infection. Hot days and cool nights favour conidial germination and growth of the fungus (Damle, 1960). The conidia germinate and enters the epidermal cells to form haustoria which are restricted in the upper epidermal cells only, where it grows and emerge as conidiophores through stomata bearing conidia. The mildew begins to appear soon after the rains. The mycelium bears conidiophores bearing conidia in chains. In November, conidial production declines and perfect state in the form of cleistothecia develops. Appendages develop profusely all over the cleistothecium except on its basal side and are unbranched with the tips uncinate or helicoid. Sclerotial bodies with few or no appendages and resembling cleistothecia in appearance develop on the host. However, no asci develop in them. These sclerotia are meant for over-summering and germinate under favourable conditions (Damle 1960).

Disease Management

Sulphur dust is found to be the most effective fungicide in controlling *Uncinula tectonae* on 2-year-old seedlings followed by Baycor (triadimenor), Morestan (quinomethionate) and Calixin (tridemorph) (Kulkarni and Siddaramaiah, 1979).

23.3. Powdery mildew of oak

Pathogen: *Microsphcera quercina* (Schw.) Burr.

Symptoms

This fungus produces white, cobweb-like patches of varying sizes on the upper surface of the leaves of the oak (Fig 23.3). The patches are isolated at first, but they soon spread and finally cover the whole of the upper surface with a white, powdery coating which adheres firmly to the leaf and cannot be wiped off.

In many cases the outbreaks of the disease occur at two different periods ; the first, which is slight and less noticeable, is in spring (May) shortly after the leaves appear on the one-year-old shoots, and the second in summer or autumn (August and September) on the young, tender leaves of the summer shoots. This latter outbreak appears mainly on young trees and on stool-shoots, especially in places where the ground has been cleared of trees and replanted. A severe outbreak can be recognized from a distance, as the leaves have lost their green colour almost entirely and the trees look as though they had been sprayed with lime-wash. The leaves finally become brown and shriveled and fall pre-maturely. The tips of the shoots do not ripen but wither and die back during the winter. As the fungus has a preference for one to six-year-old plants it is especially dangerous in nurseries.

The fungus was first observed in Europe (France) in 1907; and in the following year patches of mildew appeared on oak leaves in May. The disease spread with great rapidity, and by the autumn of the same year it was known in almost all parts of France.

The white coating consists of conidiophores and conidia; the conidia are able to germinate at once, and so the disease is quickly spread. Efforts to discover the cleistothecial stage were fruitless until 1911, when it was observed in France on small, sickly-looking, but still living leaves of *Quercus sessiliflora*. The cleistothecia appeares as small, black dots in the white mycelium. The appendages are long and

Fig 23.3: Symptoms of powdery mildew on oak leaves

382   The Plant Mildews

much branched at the tips, producing a coral-like appearance. On account of the structure of the cleistothecia the fungus is placed in the genus *Microsphaera*. The cleistothecial stage was later discovered in other countries viz. 1919 in Italy (Bologna, Sicily), 1921 in Germany (Hildesheim), and 1920-22 in Central and Northern Russia. The cleistothecia as a rule only appear on older leaves, and they are found there in large numbers.

Geographical Distribution

The fungus is found in 1908 in Spain, Portugal, Algiers, Italy, Switzerland, Austria, Hungary, Germany, Holland, Belgium, Luxemburg, England, Sweden, and Denmark, but only in the conidia stage. How the mildew came to appear to suddenly throughout Europe is not known; possibly it migrated from North America or it may have previously existed in Europe in a harmless and unnoticed form which then, through unknown circumstances, became epidemic.

In view of the rare appearance of the cleistothecial stage in Europe the question of the over-wintering of the fungus was not decided until it was observed in Germany (Laubert, 1910; Neger, 1911) and in Italy (Peglion, 1911), that the mycelium can hibernate in the winter buds of the oak. The mycelium passes the winter between the scales of the buds and in spring puts out conidiophores from which conidia arise and infect the young tender leaves.

The fungus shows a preference for European species of the oak such as *Quercus robur, Qu. sessiliflora, Qu. pubescens, Qu. Cerris, Qu. tozza, Qu. crispula,* etc.

Disease Management

Spraying with lime sulphur (1 : 20) or with brine 4.5 kg salt to 19 litre of water for young leaves and 9 kg. to 11 litres for older leaves is recommended, or the trees can be dusted with sulphur, once when the leaves have appeared and a second time after the development of the summer shoots.

American varieties, *Qu. rubra, Qu. coccinea, Qu. palustris,* etc., are less susceptible.

23.4. Powdery mildew of Arjuna

Pathogen: *Phyllactinia terminalae*

Symptoms

The infection is observed on leaves of all ages. White powdery patches appeares uniformly on the ventral surface of the leaf (Fig.23.4). Black-dotted cleistothecia later appear, followed by chlorosis and darkening of dorsal surface, slight curving of the lamina, and ultimately resulting in defoliation.

Fig. 23.4: Powdery mildew on arjuna tree.

The incidence of disease is generally recorded during Oct –Nov. It was recorded in 1976 in an epidemic form. A survey of *T. arjuna* plantations revealed heavy infections in 32.8% and light infections in 50% of plants.

23.5. Powdery mildew of khair (Acacia catechu)

Pathogen: *Erysiphe acaciae* Blumer

The disease is recorded on foliage of khair from Poona. The disease is uncommon and unimportant (Browne 1968). *Phyllactinia acaciae* Syd. is recorded on foliage of khair from Sagar, Madhya Pradesh (Tandon and Chandra 1963-64).

Symptoms

It produces grayish or yellowish, persistent dirty dense mycelium on the lower surface of leaves (Fig.23.5). Later the fruiting bodies appear on these spots as black dots.

23.6. Powdery mildew of Sisso

Dalbergia sisso Roxb. is one of the important Indian timber trees. It is locally know as "Sheesham."

Fig. 23.5: Powdery mildew on Acacia catechu plant.

Pathogen: *Phyllactinia dalbergiae (Pirozynski).*

Symptoms

At the advent of the cold season, some of the leaves of sheesham become coated on the lower surface with the mycelium of a powdery mildew. Initially the mycelium is white, and with maturity, after it has produced abundant conidiophores and conidia, the color changes from grayish white to pale yellow (Fig.23.6). By the end of December the under surface of most of the leaves becomes covered with the ectophytic mycelium, and cleistothecial formation is initiated. The young cleistothecia are orange-brown, and then change to brown and ultimately turn black at maturity. The leaves of the infected trees fall before spring, when new leaves appear. fruiting bodies appear on these spots as black dots.

Fig. 23.6: Powdery mildew on sisso leaves.

Geographical Distribution

The fungus is recorded from Dehra Dun, Allahabad (U.P.), Pusa (Bihar), Poona, Bombay and Nagpur (Maharashtra) (Pirozynski 1965).

Disease Management

Karathane and Topsin-M are most effective for control of the disease. Similarly Sulfex have a good protection against this disease (Banerjee *et al.*, 1995).

23.7. Powdery mildew of Tamarind (*Tamarindus indica*)

Pathogen: *Erysiphe polygoni* (DC)

Symptoms

It occurs as white, powdery patches on upper surfaces of leaves (Fig 23.6). The patches coalesce and cover the entire surface, resulting in drying of the leaf. The fungus then spreads to surrounding plant parts. The fungus produces characteristic chasmothecia with mycelial appendages in dried leaves. The primary source of infection is ascospores (sexual spores), and the secondary source of infection is airborne conidia (asexual spores).

Fig. 23.7: Symptoms of powdery mildew on tamarind leaves

Disease Management

Sulphur dust is most effective followed by Karathane (0.1%) and Calixin(0.1%) to control the disease.

23.8. Powdery mildew of Sandal wood

Pathogen: *Pseudoidium santalacearum*

Geographical distribution: India, China

Symptoms

Floury small, circular white patches appears on both sides of leaf surface in initial stage (Fig.23.8); spots increases in size and spread all over the leaf surface.

Pathogen

The mycelium of *Pseudoidium santalacearum* are epiphytic, amphiphyllous, thick coating, hyaline, conidia formed singly (not in chain).

Fig. 23.8: Powdery mildew on sandal wood leaves.

Control measures

Control measures adopted for other powdery mildew of forest trees may control the disease.

24
Powdery Mildew of Forage Crop

24.1. Powdery mildew of Clover
Pathogen: Erysiphe *polygoni* DC. (syn. *Erysiphe trifolii* Grev. Or *E. trifoliorum*)

Powdery mildew of clover is caused by *Erysiphe polygoni* DC. (Syn. *Erysiphe trifolii* Grev.), a member of the *Erysiphales* in family *Pyrenomycetes*. *E. polygoni* can be regarded as a collective species (O'Rourke, 1976), as it has a very wide host range, being recorded on at least 582 species including 212 legumes (Stavely and Hanson, 1966a). Kapoor (1967) suggested a division of the species into two -*E. trifolii* and *E. pisi* on the basis of host range and cleistothecial morphology. Sivanesan (1976) has suggested that the individual 'species' making up the collective *E. polygoni* are in fact *formae speciales*.

Geographical distribution
Powdery mildew caused by *E. trifoliorum* on white clover has been recorded in Scotland, New Zealand, and Australia (Farr and Rossman 2014), in Korea (Lee and Thuong, 2015) and in China, USA, Japan and India.

Symptoms
Symptoms are characterized by off-white, powdery areas of conidia and mycelium on the upper surfaces of leaves (Fig 24.1) often combined with a yellow mottling. Severe infections result in death and browning of the leaves. Mika and Bumeri (1984) noted that attack of red clover by *E. polygoni* resulted in a marked increase in wax secretion by the leaves. An abundance of white powdery mycelial growth appears on leaves of white clover (*Trifolium repens*). The disease frequently occurred even at low temperature (below 7 to 10°C).

Fig. 24.1: Symptoms of powdery mildew on clover leaves

The disease generally appears from the middle of the growing season onwards (O'Rourke, 1976).

Disease cycle

Conidia land on leaves and germinate by producing germ tubes, appressoria formation takes place, infection pegs forms which enter into the leaf cells and bulbous haustoria is formed; with a weft of prostrate mycelium which gives rise to secondary appressorial swellings, further penetration and haustoria formation (Sampson and Western, 1941).This was contradictory to the results of Yarwood(1936). Haustorial penetration from the mycelial weft tends to be confined to the epidermal cells, although infection of the inner tissues does occur from time to time (Klika, 1922). The mycelial weft also gives rise to upright, separate conidiophores, which produce single-celled conidia from about 5 days after the initial infection. Production of conidia follows a diurnal cycle (Yarwood, 1936), being much more active around midday than during darkness, when clover leaves are naturally closed, a process which renders them more difficult to infect. White mycelia and conidia present mainly on the upper leaf surfaces. Inoculation experiments (Yarwood, 1936) showed that infection is more successful during daylight than at night and can take place within 12 hrs of a spore landing on the leaf surface. From this initial infection, a prostrate white/grey mycelium is formed, which is largely superficial, but does produce secondary appressoria and haustoria. The mycelium also gives rise to erect conidiophores, which in turn produce elliptical conidia. A detailed study of their formation was undertaken by Yarwood (1936) who showed that only one conidium was mature at any one time and that this happened at the same time each day on each conidiophores. Production of conidia begins within 5 days of infection. Conidia are carried to other leaves by air currents and are the main method of infection during the growing season. As with other powdery mildews, *E. polygoni* tends to be more severe in warm, dry seasons. Such conditions also exacerbate wilting produced by severe infections, as infected leaves tend to transpire more readily than uninfected leaves (Foex, 1924).

Pathogen

Erysiphe trifolii conidiophores are erect, 45.2 - 84.6 × 6.4 - 10.5 µm in size. Foot cells of conidiophores straight to nearly so, followed by one to three cells. Conidia produced singly, ellipsoid to oblong with slightly obtuse or truncate apices, 25.6 -39.8 × 13.6 -20.4 µm, without fibrosin bodies. These morphological characteristics are fully consistent with anamorph *Pseudoidium* of genus *Erysiphe trifolii* (currently *E. trifoliorum*), (Braun and Cook 2012).

Towards the end of the season, spherical cleistothecia (90-125 μm) may be formed which bear10-30 appendages (Kapoor, 1967). At first they are a light straw colour but become dark and carbonaceous as they mature (Carr, 1971). Each cleistothecium contains five to ten ovoid asci (50-80 X 25-40 μm) which in turn contain two to six (average four) ascospores (20-25 X 10-15 μm). The ascospores, produced toward the end of the season, are forcibly shot out of the cleistothecia and are then carried by air currents to other plants. It is not considered that cleistothecia are particularly important for overwintering in the UK (O'Rourke, 1976), as the fungus can overwinter in mycelial form in host tissues. This stage may, however, be important in the formation of new physiologic races by genetic recombination.

E. polygoni is most severe on red clover but will also attack sweet pea (*Lathyrus odoratus*) and sainfoin as well as white clover. *E. polygoni* exists as specialized physiologic races. Staveley and Hanson (1966a), in the USA, showed 12 distinct races on six clones of red clover. Stavely and Hanson (1966 b, c; 1967) studied the genetics of the fungus and host in great detail and showed that host plant resistance was dominant and that the resistance to most races was monogenically inherited.

Effects on yield and quality

Both yield and quality of clover crops may be reduced by severe attacks of *E. polygoni*, although the effect can be variable (Horsfall, 1930). Carr (1984) stated that the disease could reach quite serious proportions on red clover in the UK, particularly in warm, dry seasons. However, there appears to be little quantitative information available.

Disease control

E. polygoni has been shown to be controlled by fungicides such as tridemorph in small-scale trials (Carr, 1984). However, it is not clear whether this would be economic on a field scale. Seed treatments are not effective. Apply foliar fungicides containing tebuconazole or triadimefon at the first sign of disease usually in early spring. Don't delay as the disease spreads quickly. If disease develops late in the season (within 4 weeks of haying off) then the effect on yield is minimal and spraying is rarely worthwhile.

In the USA, breeding has developed red clover cultivars with resistance to *E. polygoni* (Stavely and Hanson, 1967; Taylor *et al.*, 1990). In Australia, breeding of subterranean clover has also resulted in lines with resistance to *E. polygoni* (Anonymous, 1992). In Poland, Mikolajska and Majchrzak (1988) showed reduced occurrence of *E. polygoni* on red clover when it was planted with a cover crop.

Literature Cited for Powdery Mildew

Agrios GN. 2005. Plant Pathology. 5th ed.New York: Academic press. USA. 952 pp.

Al-hassan, K. K. 1973. Powdery mildew on their hosts in Iraq. FAO Plant Protection Bulletin 21: 88-91.

Allen, D.J. 1983. The pathology of tropical food legumes: disease resistance in crop improvement. John Wiley and Sons, Chichester, 413pp.

Ashri, A. 1971. Evaluation of the World Collection of Safflower, Carthamus tinctorius LI Reaction to Several Diseases and Associations with Morphological Characters in Israel 1. Crop science, 11(2): 253-257.

Ayesu-offei, E.N. 1998. Formae specials of Leveillula taurica infecting peppers and eggplants in Ghana. Trop. Agric. 66: 355-360.

Bayaa, B.O. 1989. New diseases reported from Syria. Arab and Near East Plant Protection Newsletter. 8, 31-32.

Berkeley J. 1847. Sur une nouvelle espe´ce d'oidium, O. tuckery, parasite de la vigne. Gardener's Chronicle 27 de noviembre.

Berkeley, M.J. and Curtis, M.A., 1848. Contributions to the Mycology of North America. S. Converse.

Berlese, A.N. and De-Toni, J.B., 1888. Phycomycetteae. Sylloge fungorum, 7.

Bettiol, W., 1999. Effectiveness of cow's milk against zucchini squash powdery mildew (Sphaerotheca fuliginea) in greenhouse conditions. Crop Protection, 18(8):489-492.

Bose, R.D. 1932. Studies in Indian Pulses, 5 Urd or Black gram (Phaseolus mungo Linn. var. Roxburghii Prain).Indian Journal of Agricultural Science. 2, 625-637.

Bouriquet C. 1946. Les maladies des plantes cultivees B Madagascar. Encylopddie Mycologque. 12: 137-166.

Braun U. 1995. The Powdery Mildews of Europe. Jena: Gustav Fischer Verlag. 337 pp.

Butler, E.J. 1918. Fungi and diseases in plants. Thacker Spink and Co. Calcutta Vol. VI.

Café filho, A.C., Coelho, M.V.S., Souza, V.L. Oídios de Hortaliças. In: Stadnik, M.J., Rivera M.C. Oídios. Jaguariúna: Embrapa Meio Ambiente, 2001. cap.11, p.285-302.

Chorin, M.A.T.H.I.L.D.A., 1961. Powdery mildew on leaves of groundnuts. Bulletin Research Council, Israel D, 10:148-149.

Cohen, Y., Meron, I., Mor, N. and Zuriel, S. 2003. A new pathotype of Pseudoperonospora cubensis causing downy mildew in cucurbits in Israel. Phytoparasitica. 31: 458–466.

Correll, J.C., Klittich, C.J.R. and Leslie, J.F., 1987. Nitrate non utilizing mutants of Fusarium oxysporum and their use in vegetative compatibility tests. Phytopathology, 77(12): 1640-1646.

Darpoux, H., 1946. New or little known diseases of Safflower. In Annales des Epiphyties. Institut National de la Recherche Agronomique.12: 297-315.

Debary, A. 1863. Recherches sur le dcveloppement de quelques champignons parasites. Ann. Sci. Nat. Bot. Biol. Veg. 20: 5-148.

Diaz Moral, D.D. 1993. Enfermedades criptogamicas de la Lenteja (Lens culinaris Medilc) Microflora Asociada a las semillas de Lenteja de Castilla-la Mancha.Servicio de Investigacion 5 Experimentation Agraria. Serie Area de Cultivas, pp. 39.

392   The Plant Mildews

du Toit, L.J., Glawe, D.A., and Pelter, G.Q. 2004. First report of powdery mildew of onion (Allium cepa) caused by Leveillula taurica in the Pacific Northwest. Online. Plant Health Progress.

Fakir, G.A. 1983. Pulse diseases and their control. (In Bengali) Mymensingh, Bangladesh: Bangladesh Agricultural University. 14pp.

Fallik, E., Ziv, O., Grinberg, S., Alkalai, S. and Klein, J.D., 1997. Bicarbonate solutions control powdery mildew (Leveillula taurica) on sweet red pepper and reduce the development of postharvest fruit rotting. Phytoparasitica, 25(1):41-43.

Farlow, W. G. 1876. On a Disease of Olive and Orange Trees. The Monthly Microscopical Journal. 16(3): 111-119.

Franco, D. 1983. Effect of sunflower powdery mildew in the north of Tamaulipas. Revista Mexicana de Fitopatologia. 2: 7-10.

Gattani, M.L., 1962. A Technique for inoculating Wheat with Rusts for Glass-house and Test-tube Culture. Nature. 196(4850):190.

Germar, B., 1934. Some functions of silicic acid in cereals with special reference to resistance to mildew. German.) Z. Pflanzenernaehr. Bodenkd, 35: 102-115.

Golovin, P. N. 1956. Monographic survey of the genus Leveillula Arnaud. Trans. Bot. Inst. USSR, Acad. Sci. Ser. 2(10): 195-308 (Russian).

Gupta, P.K, Roy, J.K., Prasad, M. 2001. Single nucleotide polymorphisms: a new paradigm for molecular marker technology and DNA polymorphism detection with emphasis on their use in plants. Curr Sci. 80: 524–535.

Gupta, S. K., Thind, 2006. Diseases of Cruciferous Vegetables. 170-185.

Heffer, V.K.B. Johnson, M.L. Powelson and N. Shishkoff. 2006. Identification of powdery mildew fungi anno 2006. The Plant Health Instructor. DOI:1094/PH-1-2006-0706.

Hagedorn, D.J. ed., 1984. Compendium of pea diseases (No. 633.37293/H141). St Paul, MN: American Phytopathological Society.

Heluta , V. 1988. Phylogenetic connections among genera of powdery mildew fungi and some questions of systematic of Erysiphales.Biological Journal of Armenia. 41(5): 582-584.

Hirata, K. 1966. Host range and geographical distribution of the powdery mildews, Memeo. Niigata Univ., Japan.

Holliday, P., 1980. Fungus Diseases of Tropical Crops. Cambridge Univ. Press, London, 607 pp.

Iqbal, S.M., Ghafoor, A., Bashir, M. and Baksh, A. 1988. Reaction of faba bean genotypes to various diseases in Pakistan. FABIS Newsletter. 21, 40-42.

Karakaya, A., F.A. Gray, and D.W. Koch. 1993. Powdery mildew of Brassica spp.in Wyoming. Plant Disease. 77: 106

Keshwal, R.L. and Choubay, P.C., 1983. Studies on control of powdery mildew of chillies. Pesticides.

Koch, E., and A. J. Slusarenko. 1990. Fungal pathogens of Arabidopsis thaliana (L.) Heyhn. Botanica Helvetica 100: 257–268.

Koike, S. T. 1997. First report of powdery mildew, caused by Erysiphe cruciferarum, on broccoli raab in California. Plant Disease. 81: 1093.

Kumar, S., and G. S. Saharan. 2002. Sources of multiple disease resistance in Brassica spp. Journal of Mycology and Plant Pathology. 32: 184-188.

Kunkel, B. N. 1996. A useful weed put to work: genetic analysis of disease resistance in Arabidopsis thaliana. Trends in Genetics. 12: 63–69.

Laemmlen, F.F. and Endo, R.M., 1985. Powdery mildew (Oidiopsis taurica) on onion in California. Plant Dis. 69(5): 451.

Lawn, R.J. and Ahn, C.S., 1985. Mungbean, Vigna radiata (L.) Wilczek. Vigna mungo, 584-623.

Lewartowska, E., Jedryczka, M., Frencel,I., Gorski,P and W, Konopka.1994. Occurrence of fungal diseases on winter oilseed rape in cultivation region Zulawy near Gdansk(1987-92). In 33 Research Session of Institute of Plant Protection, Poznan(Poland),1993. Panstwowe Wydawnictwo Rolnicze I Lesne.

Lima ,G., De Curtis, F., Spina, A.M., Cicco,V., and D, Shtienberg. 2002. Survival and activity of biocontrol yeasts against powdery mildew of cucurbits in the field. IOBC-WPRS working Group" Biological Control of Fungal and Bacterial Plant Pathogens"(Elad,Y and J, Kohl eds). Proceedings of the 7th working group meeting on Influence of abiotic and biotic factors on biological agents at Pine Bay, Kusadasi, Turkey, 22-25 May,2002. Bulletin OILB- SROP.25(10): 187-190.

Lima.G., De Curtis, F., Castoria,R, and V, De Cicco. 2003. Integrated control of apple post harvest pathogens and survival of bio-control yeasts in semi-commercial condition. Eur.J.Plant Pathol.109: 341-349.

Lodeman, E. G. 1896. The sparying of plants. Macmillan, London.

Lohnes, D.G and Nickell, C.D., 1994. Effects of powdery mildew alleles Rmd-c, Rmd, and rmd on yield and other characteristics in soybean. Plant Disease. 78: 299–301.

Lohnes, D.G. and Bernard, R.L., 1992. Inheritance of resistance to powdery mildew in soybeans. Plant disease. 76(9): 964-965.

Maffia, L.A.; Mizubuti, E.S.G.; Pedrosa, R.A. 2002. Doenças da cebola. Informe Agropecuário, Belo Horizonte, 23(218): 75-87.

Mains, E. B. 1926. Why resistance to leaf rust, stem rust and powdery mildew. Jour. Ag. Research 32: 201-221.

Mares, H. 1856. Note sur le soufrage des vognes. Bull. Soc. Agr. de l'Herault 43me an.: 86–96.

Mamlouk, O.F., Abu Gharbieh, W.J., Shw, C.G., Al. Mousa, A. and Al-Bana, L. 1984.A checklist of plant diseases in Jordan. Faculty of Agriculture, University of Jordan, Amman, Jordan, pp. 107.

Matheron, M.E. and Porchas, M., 1999a. New fungicides evaluated for control of Sclerotinia leaf drop of lettuce in 1997 and 1998.

Matheron, M.E. and Porchas, M., 2000b. Impact of azoxystrobin, dimethomorph, fluazinam, fosetyl-Al, and metalaxyl on growth, sporulation, and zoospore cyst germination of three Phytophthora spp. Plant Disease. 84(4): 454-458.

Mehta, N, Sangwan, M.S, Saharan, G.S. 2005. Fungal diseases of rapeseed-mustard. In: Saharan GS, Mehta N, MS Sangwan (eds) Diseases of oilseed crops. Indus Publishing Company, FS-5, Tagore Garden, New Delhi-110027, 15-86pp.

Mehta, K.C. 1930. Studies on the annual recurrence of Powdery mildew of wheat and Barley in India. Agric India.25: 283-285.

Mehta, K.C.1931. Annual outbreaks of rusts on wheat and Barley in the plains of India. Indian.J.Agric. Sci.1: 297-301.

Mendes, M.A.S.; Silva, V.L.; Dianese, J.C.; Ferreira, M.A.S.V.; Santos, C.E.N.; Gomes Neto, E.; Urben, A.F.; Castro, C. 1998. Fungos em Plantas no Brasil. Brasília, Embrapa-SPI, Embrapa Cenargen. 555 p.

Milovtzova, M.O. 1937. New species of fungi on the medicinal and essential oil plants of the Ukraine. Trav. Inst. Bot. Univ. Kharkoff. 2:7-13.

Morrall, R. A. A., & mckenzie, D. L. (1977). Susceptibility of five faba bean cultivars to powdery mildew disease in western canada. Canadian Journal of Plant Science, 57(1): 281-283.

Mundkar, B.B.1949. Fungi and Plant disease. Macmillan, London.pp.246.

Narayanaswamy, P. & T. Jaganathan. 1975. A note on powdery mildew disease of pigeonpea (Cajanus cajan (L.) Millsp.). Sci. Cult.41: 133–134.

Nene, Y.L., Sheila, V.K. and Sharma, S.B. 1996. A world list of chickpea and pigeonpea pathogens, 5th ed. ICRISAT, Patancheru, India, on prune trees. Crop Protection 19: 335-341.

394 The Plant Mildews

Ondieki JJ. 1973. The control of powdery mildew of capsicums with certain systemic and non-systemic fungicides. Acta Horticulture (ISHS) 33: 137-142.

Palti , J. 1953. Field observations on the humidity relationships of two powdery mildews in Israel. Palest. J. Bot., Rehovot. Ser. 8: 205-215.

Palti, J.1975. Pseudoperonospora cubensis (Berk & M. A. Curtis) Rost. CMI Descriptions of Pathogenic Fungi and Bacteria, 457: 1-2.

Palti, J. 1959. Oidiopsis diseases of vegetable and legume crops in Israel. Plant Disease Reporter, 43: 221-226.

Paul, N. D., and P. G. Ayres. 1986. The impact of a pathogen on populations of groundsel (Senecio vulgaris) overwintering in the field. Journal of Ecology. 74: 1069-1084.

Peregrine, W.J.H., and Siddiqui, M.A. 1972. A revised and annotated list of plant diseases in Malawi. CAB Phytopathology Paper 16:29.

Photiades, J. and Alexandrou, G. 1979. Food legume research and production in Cyprus. In: Hawtin, G.C. and Chancellor, G.J. (eds) food legume improvement and development. International Center for Agricultural Research in the Dry Areas/ International Development Research Centre, Aleppo, Syria. pp. 75-79.

Prameela, H.A., Viswanath, S. and Anilkumar, T.B., 1989. The predisposing effect of sterility mosaic to powdery mildew infection in pigeonpea. Zentralblatt für Mikrobiologie, 144(2): 111-114.

Price, T.V., 1977. Powdery mildew, a new disease of the winged bean (Psophocarpus tetragonolobus) in Papua New Guinea. Plant disease reporter.

Raj, H., Bhardwaj, M.L., Sharma,I.M ., and N.K.Sharma. 1992. Performance of commercial Okra varieties in relation to disease and insect pests. Indian Journal of Agriculture Sciences. 63: 747-748.

Raju, T.N. 1988. Studies on pigeonpea powdery mildew. Ph.D. thesis, University of agricultural sciences Bangalore. 285pp.

Rawal, P., Sharma, P., Singh, N.D. and Joshi, Arunabh .2013. Evaluation of fungicides, neem bio-formulations and bio-control agent for the management of root rot of safed musli caused by Rhizoctonia solani. J. Mycol. Plant Pathol. 43(30): 297.

Reddy, R.M. 1982. Evaluation of fungicides against major diseases of chilli. M.Sc(Ag) Thesis. TNAU, Coimbatore, India.

Reddy, M.V., S.B.Sharma and Y.L.Nene. 1990b. Pigeonpea.: disease management.303-347.

Reddy, K.S., Pawar, S.E., and C.R. Bhatia. 1994. Inheritance of powdery mildew(Erysiphe polygoni) resistance in mungbean(Vigna radiata). Theor. Appl. Genet. 88: 945-948.

Reddy, M.V., Kannaiyan, J. and Nene, Y.L. 1984. Increased susceptibility of sterility mosaic infected pigeonpea to powdery mildew. Indian Journal of Tropical Plant Diseases. 2: 35-40.

Reed, G. M. 1920. Varietal resistance and susceptibility of oats to powdery mildew, crown rust and smuts. Res. Bull. Univ. Mo. agric. Exp, Sta. no. 37.

Reuveni, M., 2000. Efficacy of trifloxystrobin (Flint), a new strobilurin fungicide, in controlling powdery mildews on apple, mango and nectarine, and rust on prune trees. Crop Protection, 19(5): 335-341.

Reuveni, M., Oppenheim, D. and Reuveni, R., 1998. Integrated control of powdery mildew on apple trees by foliar sprays of mono-potassium phosphate fertilizer and sterol inhibiting fungicides. Crop protection. 17(7): 563-568.

Reuveni, R., Perl, M. and Rotem, J., 1976. Inhibition of shedding of pepper leaves infected with powdery mildew (Leveillula taurica) by application of auxins. Phytoparasitica, 4(3):197.

Riley, R., 1960. The diploidisation of polyploid wheat. Heredity, 15:407-29.

Romero, D., de Vicente, A., Zeriouh, H., Cazorla, F.M., Fernández-Ortuño, D., Torés, J.A. and Pérez-García, A. 2007. Evaluation of biological control agents for managing cucurbit powdery mildew.

Romero, D., Rivera, M.E., Cazorla, F.M., de Vicente, A. and Pérez-García, A. 2003. Effect of mycoparasitic fungi on the development of Sphaerotheca fusca in melon leaves. Mycol. Res. 107, 64–71.

Rothwell, A. 1983. A revised list of plant diseases occurring in Zimbabwe. Kirkia. 12(11): 275.

Schwartz H. F., Mohan S. K. 1995. Compendium of Onion and Garlic Diseases. American Phytopathological Society, St. Paul, MN.70 p.

Saluja, V.K. and V.P. Bhide. 1962. Powdery mildew of Safflower (Carthamus tinctorius L.) caused by Erysiphe cichoracearum DC. in Maharashtra. Ind. Phytopath. 15:291.

Sankhla, H.C., Singh, H.G., Dalela, G.G. and Mathur, R.L. (1967). Occurance of perithecial stage of Erysiphe polygoni on Brassica campestris var sarson and Brassica juncea. Plant Dis. Reptr. 51: 800.

Schnathorst, W.C. 1959. Spread and life cycle of the lettuce powdery mildew fungus. Phytopath. 49:464-468.

Schnathorst, W.C. 1960. Effect of temperature and moisture stress or the lettuce powdery mildew fungus. Phytopath. 50:304-308.

Sepúlveda, R. P. 1987. Investigación y Progreso Agropecuario, La Platina, Santiago (Chile) 39: 29.

Smith, B.J., and L.L.Black. 1986. First report of Colletotrichum acutatum in strawberry in the United States. Plant Dis. 70: 1074.

Sridhar, T. and Sinha, P. 1989. Assessment of loss caused by powdery mildew (Erysiphe cichoracearum) of okra (Hibiscus esculentus) and its control. Indian J. Agr. Sci., 59: 606-607.

Stadnik, M., Bettiol, W., & Saito, M. (2003). Bioprospecting for plant and fungus extracts with systemic effect to control the cucumber powdery mildew / Bioprospektion von pflanzlichen und pilzlichen Extrakten mit systemischem Effekt zur Bekämpfung des Echten Mehltaupilzes an Gurken. Zeitschrift Für Pflanzenkrankheiten Und Pflanzenschutz / Journal of Plant Diseases and Protection, 110(4): 383-393.

Takamatsu, S. 2013. Molecular phylogeny reveals phenotypic evolution of powder mildews (Erysiphales, Ascomycota). J. Gen Plant Pathol. 79: 218-226.

Thind, T., Mohan, C. and Kaur, S., 2002. Promising activity of pencycuron, a phenylurea-based fungicide, for effective management of black scurf of potato. Indian Phytopathology, 55(1): 39-44.

Tur, S.A.J. 1955. Diseases of economic crops in the Sudan.III Fodders, Pulses and Vegetables. In: FAO PlantProtection Bulletin No. 3, pp. 113-116.

Urquhart, E. J., & Punja, Z. K. 2002. Hydrolytic enzymes and antifungal compounds produced by Tilletiopsis species, phyllosphere yeasts that are antagonists of powdery mildew fungi. Canadian Journal of Microbiology, 48(3): 219-229.

Wallace, G.B. 1930.Mycological work, report, Department of Agriculture, Tanganyika, 1928/29, Part 2. Department of Agriculture, Tanganyika, pp. 35-36.

Williamson, C. J., and W. H.S. Macfarlane. 1986. Fungicidal control of powdery mildew and its effect on yield, digestibility and chemical composition of eight forage rape cultivars. Journal of Agricultural Science, Cambridge 107: 385–391.

Wurms, K., Labbe, C., Benhamou, N., & Belanger, R. R. 1999. Effects of Milsana and benzothiadiazole on the ultrastructure of powdery mildew haustoria on cucumber. Phytopathology. 89: 728-736.

Zheng, R. Y. (1984). The genus Brasiliomyces (Erysiphaceae). Mycotaxon 19: 281–289.

Zimmer, D.E., 1961. Powdery mildew on Safflower. Plant Disease Reporter. 45(12).